Operator Theory
Advances and Applications
Vol. 114

Complex Analysis and Related Topics

E. Ramírez de Arellano
M. V. Shapiro
L. M. Tovar
N. L. Vasilevski
Editors

Springer Basel AG

Editors:
Enrique Ramírez de Arellano and
Nikolai L. Vasilevski
Department of Mathematics
CINVESTAV-IPN
Mexico City
Mexico

Michael Shapiro and
Luis Manuel Tovar
Department of Mathematics
ESFM-IPN
Mexico City
Mexico

1991 Mathematics Subject Classification 47-06

A CIP catalogue record for this book is available from the
Library of Congress, Washington D.C., USA

Deutsche Bibliothek Cataloging-in-Publication Data

Complex analysis and related topics / E. Ramírez de Arellano ...,
ed. – Basel ; Boston ; Berlin : Birkhäuser, 2000
(Operator theory ; Vol. 114)
ISBN 978-3-0348-9734-1 ISBN 978-3-0348-8698-7 (eBook)
DOI 10.1007/978-3-0348-8698-7

ISBN 978-3-0348-9734-1

Originally published by Birkhäuser Verlag in 2000
Softcover reprint of the hardcover 1st edition 2000
Printed on acid-free paper produced from chlorine-free pulp. TCF ∞
Cover design: Heinz Hiltbrunner, Basel

ISBN 978-3-0348-9734-1

Contents

Preface

This volume is a collection of up-to-date research and expository papers on different aspects of complex analysis, including relations to operator theory and hypercomplex analysis.

These articles cover many important and essential subjects, including the Schödinger equation, subelliptic operators, Lie algebras and superalgebras, Toeplitz and Hankel operators, reproducing kernels, and Q_p spaces, among others.

Most of the papers were presented at the International Symposium on Complex Analysis and Related Topics held in Cuernavaca (Morelos), Mexico, November 1996, which was attended by about 50 experts in the field.

The conference was held to celebrate the 35th anniversaries of the Center for Research and Advanced Study (CINVESTAV-IPN) and of the School of Physics and Mathematics of the National Polytechnic Institute (ESFM-IPN). This event was made possible due to the financial support of both Institutions and of the Consejo Nacional de Ciencia y Tecnología (CONACyT-MEXICO). Further support was given by IPN to assist in the publishing of this volume.

The Editors

Operator Theory:
Advances and Applications, Vol. 114

Local integrability of systems of m smooth linearly independent complex vector fields on $m+1$ dimensional manifolds

Lilia N. Apostolova

Abstract. Necessary and sufficient conditions for the local integrability of systems of m linearly independent smooth complex vector fields defined on $m+1$ dimensional smooth real manifolds are given. Some similar results for the closely related abstract CR structures of hypersurface type are discussed also.

1. Preliminaries and Some Introductory Notes

In this paper we shall consider a system of m smooth complex valued linearly independent vector fields

$$L_1, \quad L_2, \quad \dots, \quad L_m, \tag{1}$$

defined on an $m+1$-dimensional smooth real manifold M. As we shall consider local properties, we may assume that M is a neighborhood U of the origin 0 in real Cartesian space $\mathbf{R}^{m+1}$.

Such systems are investigated for example in [9], [10], [11], [25], [30] (see also the references given there), but mainly for their solvability. Here we propose some more about their local integrability.

The system (1) is called *formally integrable* if for each point $p \in U$ and pair of vector fields L_j, L_k, j, $k = 1, 2, \dots, m$ there exists a neighborhood $U' \subset U$ of the point p such that the commutator $[L_j, L_k]$ is a linear combination of the vector fields $L_1, L_2, \dots, L_m$ with smooth functions as coefficients over U'.

The system (1) is called *locally integrable (at the origin)* if there exists a neighborhood $U_0 \subset U$ of the origin 0 and a smooth function Z on U_0 with a non-zero differential dZ at 0 satisfying the system of equations

$$L_1 u = 0, L_2 u = 0, \dots, \quad L_m u = 0. \tag{2}$$

1991 *Mathematics Subject Classification.* 35N10, 32F40.

Partially supported by the contract $MM - 525/95$ with the National Fund for Scientific Researches to the Ministry of Education and Sciences of Bulgaria.

The vector fields of the system (1) span a vector subbundle V of the complex(ified) tangent bundle $\mathbf{C}TU = \mathbf{C} \otimes TU$ with the complex dimension of the fiber equal to m.

Let us note that the formal integrability and the local integrability of the system (1) are properties common to each system of vector fields, forming a basis for the module of smooth sections of the vector bundle V over the ring of smooth functions.

To obtain this for the formal integrability it should be remarked that each one of a pair of systems of vector fields forming a basis for the fibers of the bundle V can be obtained from the other by multiplication by an invertible matrix whose ingredients are smooth functions on their defining domain.

To see the simultaneity of the local integrability for each system of vector fields forming a basis for the fibers of the bundle V we shall consider the solutions of the system of equations (2). If a function u is a solution of the system (2), then u is a solution of each equation $Lu = 0$, where L is a linear combination of the vector fields in the system (2) with smooth functions as coefficients. Thus any two systems of the form (1) each of which may be obtained as a linear combination of the vector fields in the other, have the same functions as solutions of (2).

In the terms of the vector bundle V spanned by the vector fields of the system (1) the formal integrability of (1), means that the commutator of any two smooth sections of the bundle V, over an open set where these sections are defined, is a smooth section of V as well; i.e., the sections of V form a Lie algebra with respect to the Lie bracket (commutator) of vector fields. The local integrability of (1) means that the one-dimensional vector subbundle T of the complex cotangent bundle $\mathbf{C}T^*U$ annihilating V admits as local generator an exact one-form.

It can be seen that the bundle V corresponding to the system (1) has a base of the form

$$L_j = \frac{\partial}{\partial t^j} + a^j(t,x)\frac{\partial}{\partial x} \quad \text{for} \quad j = 1, 2, \dots, m, \tag{3}$$

where $a^j(t,x)$ are smooth complex-valued functions on some neighborhood U_0 of the origin 0 in $\mathbf{R}^{m+1}$ with coordinates $t^1, t^2, \dots, t^m, x$ and $a^j(0) = 0$.

To prove this let us consider an arbitrary system of vector fields

$$L_j^0 = \sum_{i=1}^{m+1} a^{ij}(y^1, y^2, \dots, y^{m+1})\frac{\partial}{\partial y^i} \quad \text{for} \quad j = 1, 2, \dots, m,$$

which provides a basis of tangent vectors for the fibre of the vector bundle V at each point in some neighborhood of the origin of $\mathbf{R}^{m+1}$. As the vector fields $L_1^0, L_2^0, \dots, L_m^0$ are linearly independent, the variables $y^1, y^2, \dots, y^{m+1}$ may be relabeled in such a way that the corresponding matrix of functions $A = \|a^{ij}\|_{i,j=1}^m$ is invertible over some neighborhood U_0 of the origin. Then multiplying on the left the column vector with ingredients the vector fields $L_1^0, L_2^0, \dots, L_m^0$ and denoting y^j by t^j for $j = 1, 2, \dots, m$ and y^{m+1} by x, we obtain m new linearly independent smooth complex vector fields $L_1, L_2, \dots, L_m$ of the form (3) with respect to the coordinate chart $(U_0; t^1, t^2, \dots, t^m, x)$. Also these vector fields form a basis for the corresponding fibre of the bundle V at each point of the neighborhood U_0, because the matrix A is invertible there.

A basis of vector fields of the form (3) for the fibres of the bundle V at each point in some coordinate neighborhood U of the origin will be called *a normal basis of vector fields for the bundle V on the local chart* $(U, t^1, t^2, \dots, t^m, x)$.

It can be seen that the system (3) is formed of commuting vector fields if and only if

$$L_j a^k = L_k a^j \quad \text{for} \quad j, k = 1, \dots, m. \tag{4}$$

In this case the system (3) is formally integrable.

In general, the system (1) is formally integrable if and only if there exists a coordinate neighborhood U_0 of the origin and a system of commuting vector fields of the form (3) spanning the same vector bundle V over U_0 as the one spanned by the vector fields (1).

The systems of smooth complex vector fields under consideration are closely related to an important notion in the multivariable complex analysis: abstract CR structures of hypersurface type.

In 1907 H. Poincaré published the very important paper for the theory of functions in several complex variables paper [29]. There he considered for first time the problems for equivalence (local, global or mixed) of the boundaries of domains in $\mathbf{C}^2$ (cf [37]).

The notion of CR structure on a smooth manifold arises as an abstract analogue of the tangent bundle of the boundary of a domain (or of the tangent bundle of a real hypersurface) in the complex Cartesian space $\mathbf{C}^n$ and the subbundle of holomorphic vectors on it. These classical roots determine many of the problems posed for the CR structures and some of the tools and the approach in reserch on CR structures.

In 1968 S. G. Greenfield published a paper [14], where CR structures in an abstract sense are considered. Their basic properties and relations to known phenomena in multivariable complex analysis are given there.

In 1972 A. Andreotti and C. D. Hill published [4] where they illuminate the close relations between the local existence of $N - n$ solutions with linearly independent differentials of a system of n smooth linearly independent complex vector fields on a domain in $\mathbf{R}^N$ and local embedability of the corresponding CR structure in some Cartesian space $\mathbf{C}^{N-n} \times \mathbf{R}^{n-r}$. A theorem for existence of such solutions under the assumption of real-analyticity of the equations with respect to some of variables is proved there.

In the Séminaire Goulaouic-Lions-Schwartz 1974-75, exposé number 9, L. Boutet de Monvel [7] published a proof of the theorem that the compact manifolds M with CR structures of hypersurface type are embeddable when dim $M \geq 5$ and the Levi form has all eigenvalues of the same sign.

In 1973 L. Nirenberg published the expository paper [27] where it is shown how to obtain a CR structure on $\mathbf{R}^3$ which is not locally embeddable using the first example of a smooth linear partial differential equation without (C^1-smooth) solution, given by Hans Lewy [24]; other related questions are discussed. Then next year was published Nirenberg's example of a strongly pseudoconvex CR structure on $\mathbf{R}^3$ without non-constant CR functions [28].

In 1977 J. J. Faran [13] gave examples of non-analytic hypersurfaces in $\mathbf{C}^n$. Let us note here that the Cauchy-Kovalevski theorem solves the problem for embedability of real-analytic CR structures and so these examples show the possibility of non-analytic embeddable CR structures and the need for seeking other tools for proving the embedability results. Such a tool in [2], [27], [28], [5], [32] is the Newlander-Nirenberg theorem for the existence of complex coordinates on an integrable almost complex manifold.

The nearest type of abstract CR structures to those induced on hypersurfaces in $\mathbf{C}^n$ by the complex structure on $\mathbf{C}^n$ are the CR structures of hypersurface type (see definition below in Section 2). In 1982 M. Kuranishi [21] proved that each such CR structure is locally embeddable when dim $M \geq 9$ and the Levi form of the structure is strongly positive. In 1987 M. Akahori [3] extended this result for the case dim $M \geq 7$. As Nirenberg's example [28] solved the problem for dim $M = 3$ there remains only unsolved the case dim $M = 5$. A new short and suggestive proof of the theorems of Kuranishi and Akahori was given by S. M. Webster [35], [36]. His proof illuminates some of the problems arising for the case dim $M = 5$. Some results and information on the problem for global embedability of CR structures of hypersurface type may be found in the paper of R. Dwilewicz [12].

Ample information about research up to 1991 on integrability of systems of smooth complex vector fields, embedability of CR structures and related topics of analysis on strictly pseudoconvex domains in $\mathbf{C}^n$ may be found in the papers of J. J. Kohn [19], [20], A. V. Abrossimov [1], [2], G. M. Henkin [16] and the references given there. See also closely related books of F.Treves [32], [33] and [34], and the books on CR structures of A. Bogges [6], H. Jacobowitz [18], G. Taiani [31] and K. Yano and M. Kon [38] and the references given there.

Recently a collection of papers [8] with editors P. D. Cordaro and H. Jacobowitz appeared. Some expository papers are published and some closely related topics are treated there.

2. Abstract CR Structures of Hypersurface Type

Let $p \in \mathbf{R}^{2m+1}$ with $m \geq 1$. Let $L_1, L_2, \ldots, L_m$ be smooth complex vector fields defined near the point p. The system L_j, $j = 1, 2, \ldots, m$ is called an *abstract CR structure of hypersurface type near p* if

$$L_1, L_2, \ldots, \quad L_m, \bar{L}_1, \bar{L}_2, \ldots, \quad \bar{L}_m \quad \text{are linearly independent at } p \tag{5}$$

and

$$L_1, L_2, \ldots, L_m \quad \text{is formally integrable.} \tag{6}$$

A function Z annihilated by the vector fields of an abstract CR structure of hypersurface type is called a *CR function.*

A Hermitian form on V defined by using $L_p(v', v'') = \frac{1}{2i}[\bar{L}'', L'] \mod (V+\bar{V})$ where $L'(p) = v'$ and $L''(p) = v''$ is called *a Levi form of the CR structure V*. More precisely, if $L_1, L_2, \ldots, L_m$ is an abstract CR structure of hypersurface type near a point $p \in \mathbf{R}^{m+1}$ there exists a real vector field $T = \frac{\partial}{\partial t}$ such that the vector fields $L_1, L_2, \ldots, L_m, \bar{L}_1, \bar{L}_2, \ldots, \bar{L}_m, T$ generate the bundle $\mathbf{C}TU$ over a neighborhood U of the point p. Then the commutators may be decomposed as follows:

$$[\bar{L}_j, L_i] = \sum_{k=1}^{m} \alpha_{ij}^k L_k + \sum_{k=1}^{m} \beta_{ij}^k \bar{L}_k + c_{ij} T.$$

The matrix (c_{ij}) defines the Hermitian form we are interested in. The correctness of this definition and many properties of the Levi form are considered in detail for example in [14], [20].

An abstract CR structure $L_1, L_2, \ldots, L_m$ is called *realizable or embeddable* if it is locally integrable, i.e., if there exist smooth functions $Z^1, Z^2, \ldots, Z^{m+1}$ which are CR functions near p, and $dZ^1, dZ^2, \ldots, dZ^{m+1}$ are linearly independent.

Lemma 2.1. *Let there be given a system of the form* (3)*, satisfying the conditions* (4)*. Then the system of smooth complex-valued vector fields*

$$L'_j = 2\frac{\partial}{\partial \bar{z}^j} + a^j(t,x)\frac{\partial}{\partial x} \quad \textit{for} \quad j = 1, \ldots, m \tag{7}$$

is an abstract CR structure of hypersurface type near the origin in $\mathbf{R}^{2m+1}$.

Here we use the complex variables $z^j = t^j + is^j$, $s = (s^1, \ldots, s^m) \in \mathbf{R}^m$ and the complex derivatives $\frac{\partial}{\partial \bar{z}^j} = \frac{1}{2}(\frac{\partial}{\partial t^j} + i\frac{\partial}{\partial s^j})$. Let us recall also that $\frac{\partial}{\partial z^j} = \frac{1}{2}(\frac{\partial}{\partial t^j} - i\frac{\partial}{\partial s^j})$.

Proof. Indeed, the vector fields $L'_1, L'_2, \ldots, L'_m, \bar{L}'_1, \bar{L}'_2, \ldots, \bar{L}'_m$ are linearly independent near the origin because so are the complex derivatives

$$\frac{\partial}{\partial z^1}, \frac{\partial}{\partial z^2}, \ldots, \frac{\partial}{\partial z^m}, \frac{\partial}{\partial \bar{z}^1}, \frac{\partial}{\partial \bar{z}^2}, \ldots, \frac{\partial}{\partial \bar{z}^m}$$

and $a_j(0) = 0$. The formal integrability of the system follows by the equalities (4). Indeed, from (4) and the definition of the vector fields L'_j, $j = 1, 2, \ldots, m$ it follows that

$$L'_j a^k = L'_k a^j \quad \text{for} \quad j,k = 1,2,\ldots,m \tag{8}$$

which ensures the formal integrability of the system (7). □

The following lemma gives a relation between the integrability of the systems (3) and (7). An idea used in the paper of M. S. Baouendi and L. P. Rothschild [5] to find a solution of the system (3) in a special form applying the Cauchy-Kovalevski theorem is developed in the proof.

Lemma 2.2. *Let the system* (3) *be locally integrable. Then the abstract CR structure* (7) *is realizable. If the abstract CR structure of hypersurface type* (7) *with nondegenerate indefinite Levi form is realizable, then the system* (3) *is locally integrable.*

Proof. Let the system (3) be locally integrable and the function $Z(t,x)$ be a solution of the system of equations (2) in a neighborhood of the origin in $\mathbf{R}^{m+1}$ with non-zero differential in the origin. Then the functions $Z^1 = t^1 + is^1$, $Z^2 = t^2 + is^2, \dots, Z^m = t^m + is^m$, $Z^{m+1} = Z(t,x)$ will annihilate the vector fields (7) on a neighborhood of the origin in $\mathbf{R}^{2m+1}$, as can be checked.

Also it can be verified directly that their differentials are linearly independent in some neighborhood of the origin. This proves the realizability of the system (7) in this case.

Let now the abstract CR structure of hypersurface type with nondegenerate indefinite Levi form be realizable, and $Z^1(t,s,x)$, $Z^2(t,s,x)$, $\dots, Z^{m+1}(t,s,x)$ be CR functions with linearly independent differentials near the origin.

We will find a function $Z(t,x)$ with non-zero differential at the origin which annihilates the vector fields (3), and is of the form

$$Z = F(Z^1, Z^2, \dots, Z^{m+1})$$

where F is a holomorphic function of $m+1$ variables. As we require Z to satisfy (2), we consider the system of equations

$$\frac{\partial F(Z^1, Z^2, \dots, Z^{m+1})}{\partial s^k} = \sum_{j=1}^{m+1} \frac{\partial Z^j}{\partial s^k} \frac{\partial F}{\partial Z^j} = 0 \quad \text{for} \quad k = 1, 2, \dots, m. \tag{9}$$

The coefficients $\frac{\partial Z^j}{\partial s^k}$ are CR functions, because Z^j are and the differential operators $\frac{\partial}{\partial s^k}$ commute with the complex vector fields in (7). As the Levi form of the structure is nondegenerate and indefinite we can apply the Hans Lewy theorem for continuation (see [23] or [17], Theorem 2.6.13) and represent these functions as a composition of holomorphic functions of $m+1$ variables with the given CR functions Z^1, $Z^2, \dots, Z^{m+1}$; i.e., there exist holomorphic functions G_{jk} such that

$$\frac{\partial Z^j}{\partial s^k} = G_{jk}(Z^1, Z^2, \dots, Z^{m+1})$$

for $j = 1, 2, \dots, m+1$ and $k = 1, 2, \dots, m$.

The system (9) becomes

$$\sum_{j=1}^{m+1} G_{jk}(Z^1, Z^2, \dots, Z^{m+1}) \frac{\partial F}{\partial Z^j} = 0 \quad \text{for} \quad k = 1, 2, \dots, m.$$

This is a system of m first order partial differential equations with holomorphic coefficients for a holomorphic function F of $m+1$ variables. Now we can apply Cauchy-Kovalevski theorem. Therefore there exists a holomorphic solution F_0 of the system (9) with non-zero differential.

Then the function $Z(t,x) = F_0(Z^1, Z^2, \dots, Z^{m+1})$ is the required function for the local integrability of the system (3). Indeed, this function does not depend on the variables $s^1, s^2, \dots, s^m$, and it annihilates the vector fields (3). Moreover, its differential is non-zero at the origin because the differential of F_0 at the origin is non-zero. This proves the local integrability of the system (3). □

3. Necessary Condition for Local Integrability

Theorem 3.1. *Let the system of smooth linearly independent complex vector fields*

$$L_1, L_2, \dots, L_m$$

generate a subbundle V of the complex(ified) tangent bundle CTU on some neighborhood U of the origin $0 \in \mathbf{R}^{m+1}$. Let this system be locally integrable (at the origin). Then there exists a coordinate system $(U_0; t_1, \dots, t_m, x)$ near the origin and smooth complex valued functions $A^j(t,x,y)$ for $j = 1, 2, \dots, m$ defined on some neighborhood $U' \times (-\varepsilon, \varepsilon)$ of the origin in $\mathbf{R}^{m+2}$ with the properties

$$\frac{\partial A^j}{\partial \bar{z}} = 0 \text{ on } U' \times (0, \varepsilon) \tag{10}$$

where $z = x + iy$, and

$$A^j(t,x,0) = a^j(t,x) \text{ for } (t,x) \in U', \tag{11}$$

such that the functions $a^j(t,x)$ define a new system of complex vector fields L_1, $L_2, \dots, L_m$ of the form (3) generating the same subbundle V over U'.

In other words for each locally integrable system of smooth linearly independent complex vector fields there exists a system of the form (3) generating the same subbundle V of the complex tangent bundle $\mathbf{C}TU'$ over some neighborhood U' of the origin 0 in $\mathbf{R}^{m+1}$ with coefficients $a^j(t,x)$ which are "boundary values" on the boundary $y = 0$ of smooth functions A^j defined on $U' \times (-\varepsilon, \varepsilon)$ and holomorphic on $U' \times (0, \varepsilon)$ in the variable $z = x + iy$.

To prove this theorem we need the following result, proved by N. Hanges and H. Jacobowitz.

Theorem 3.2. [15] *Let the integer m be bigger or equal to 1. Assume that the system L'_1, $L'_2, \dots, L'_m$ forms an abstract CR structure near the point p of $\mathbf{R}^{2m+1}$. If the structure is realizable, then there exists a real change of coordinates such that the transformed structure is generated by vector fields of the form (7) satisfying (8), (10) and (11) with smooth functions $a^j = a^j(t,s,x)$, $A^j = A^j(t,s,x,y)$ for $j = 1, 2, \dots, m$. Here the variables (t,x,y) are as in Theorem 3.1 and the variable $s = (s^1, s^2, \dots, s^m)$ belongs to some neighborhood of the origin in $\mathbf{R}^m$.*

Proof. [15] We give the proof of this theorem for completeness and because of its constructive character, which illuminate the situation.

From the realizability of the given abstract CR structure of hypersurface type it follows that there exist real coordinates

$$(t^1, t^2, \dots, t^m, s^1, s^2, \dots, s^m, x)$$

near $0 \in \mathbf{R}^{2m+1}$ such that the structure is generated by

$$L'_j = \frac{\partial}{\partial \bar{z}^j} + b^j(t,s,x)\frac{\partial}{\partial x} \quad \text{for} \quad j = 1, 2, \dots, m$$

$z^j = t^j + is^j, j = 1, 2, \dots, m$ with coefficients b_j smooth complex valued functions, defined near 0 with $b_j(0) = 0$ (see for example [1]).

The CR functions with linearly independent differentials of this structure may be taken to be the functions

$$z^j = t^j + is^j \quad \text{for } j = 1, 2, \dots, m,$$
$$v = x + i\phi(t, s, x),$$

with ϕ a real smooth function, $\phi(0) = 0$ and $d\phi(0) = 0$. In fact, the functions z^1, $z^2, \dots, z^m$ annihilate the vector fields L'_1, $L'_2, \dots, L'_m$ and from the realizability of the considered CR structure there follows the existence of the function ϕ with properties given above.

Then this may be extended to an almost complex structure; i.e., there may be constructed an almost complex structure for which the vector fields $L'_1, L'_2, \dots, L'_m$ are tangent vector fields of type $(0, 1)$.

Let (t, s, x, y) be coordinates near $0 \in \mathbf{R}^{2m+2}$, and introduce the vector field

$$L'_{m+1} = \frac{\partial}{\partial x} + (i - \phi_x)\frac{\partial}{\partial y}.$$

Then the vector fields L'_1, $L'_2, \dots, L'_m$, L'_{m+1} give an almost complex structure. The functions z^1, $z^2, \dots, z^m$ and $w = x + i(\phi(t, s, x) + y)$ are holomorphic functions for this structure. This means that the almost complex structure is integrable; moreover it is a complex structure with complex coordinates z^1, $z^2, \dots, z^m$, w.

It may be assumed that this almost complex structure is defined on a small ball B containing $0 \in \mathbf{R}^{2m+2}$. Let us define

$$B(t, s)^+ = \{(x, y) \in \mathbf{R}^2 : (t, s, x, y) \in B \text{ and } y > 0\}$$

for each (t, s). Let $O(t, s)^+ \subset \mathbf{C}$ be the image of $B(t, s)^+$ under the map $w(t, s, \cdot, \cdot)$. Since $O(t, s)^+$ is simply connected there exists a map $z = z(t, s, \cdot)$ mapping $O(t, s)^+$ into the complex plane such that:

1. z is conformal for each (t, s),
2. $\operatorname{Im} z > 0$ on $O(t, s)^+$,
3. $\operatorname{Im} z(t, s, w(t, s, x, 0)) = 0$ for all (t, s, x),
4. z is smooth in all arguments,
5. $\frac{\partial z}{\partial w}(0, 0, 0) \neq 0$.

In particular the map $(t, s, x) \to (t, s, z(t, s, x, 0))$ is a real diffeomorphism. After the change of coordinates

$$(t, s, x, y) \to (t, s, \operatorname{Re} z, \operatorname{Im} z)$$

the almost complex structure becomes

$$L'_j = \frac{\partial}{\partial \bar{z}^j} + L'_j z \frac{\partial}{\partial z} + L'_j \bar{z} \frac{\partial}{\partial \bar{z}} \quad \text{for} \quad j = 1, 2, \dots, m,$$

$$L'_{m+1} = 2\bar{z}_{\bar{w}} \frac{\partial}{\partial \bar{z}}.$$

It follows from the integrability conditions for the almost complex structure that $L'_j z$ is holomorphic in z for $\operatorname{Im} z > 0$ and $j = 1, 2, \dots, m$. The fact that L'_j, $j = 1, 2, \dots, m$ are tangent to $\operatorname{Im} z = 0$ shows that for $j = 1, 2, \dots, m$ we have

$L'_j z = L'_j \mathrm{Re}\, z$ when $\mathrm{Im}\, z = 0$. Hence the CR structure induced on $\mathrm{Im}\, z = 0$ is generated by

$$L'_j = \frac{\partial}{\partial \bar{z}^j} + L'_j \mathrm{Re}\, z \frac{\partial}{\partial \mathrm{Re}\, z} \quad \text{for} \quad j = 1, 2, \ldots, m$$

and (10) and (11) follow with $A_j = L'_j z$. This completes the proof. □

Proof. [Proof of theorem 3.1.] For the subbundle V generated by the vector fields $L_1, L_2, \ldots, L_m$ we construct a basis of vector fields of the form (3)

$$L_j = \frac{\partial}{\partial t^j} + b^j(t,x) \frac{\partial}{\partial x} \quad \text{for} \quad j = 1, 2, \ldots, m,$$

over some coordinate chart $(U_0; t^1, t^2, \ldots, t^m, x)$, $u_0 \subset U$ with center at the origin in $\mathbf{R}^{m+1}$ as was done in Section 1. As was remarked there, this system of vector fields will be locally integrable also. Using the new generators $L_1, L_2, \ldots, L_m$ of V we constuct the corresponding system of vector fields (7):

$$L'_j = 2\frac{\partial}{\partial \bar{z}^j} + b^j(t,x) \frac{\partial}{\partial x} \quad \text{for} \quad j = 1, 2, \ldots, m,$$

where $z^j = t^j + is^j$ for $j = 1, 2, \ldots, m$. According to Lemma 2.1 the system $L'_1, L'_2, \ldots, L'_m$ will be an abstract CR structure of hypersurface type on some neighborhood of the origin in $\mathbf{R}^{2m+1}$.

From Lemma 2.2 it follows that the so obtained CR structure is realizable.

Now we may apply Theorem 3.2 for the realizable CR structure L'_1, $L'_2, \ldots$, L'_m. The functions $A_j(t,s,x,y)$, $j = 1, 2, \ldots, m$, satisfying the condition (10) arise as has been shown in the proof of Theorem 3.2. Their construction is such, that the variables t and s are the same as in the beginning, and the variables x and y are such that the given CR structure has a representation as follows:

$$L'_j = 2\frac{\partial}{\partial \bar{z}^j} + A^j(t,s,x,0) \frac{\partial}{\partial x} \quad \text{for} \quad j = 1, 2, \ldots, m.$$

It remains to restrict the functions $A_j(t,s,x,y), j = 1, 2, \ldots, m$ to the intersection of their domain of definition with the plane in $\mathbf{R}^{2m+2}$ defined by the conditions $s^1 = 0, s^2 = 0, \ldots, s^m = 0$.

The functions $A^j(t,x,y)$ obtained in this way for $j = 1, 2, \ldots, m$ satisfy the condition (10) also. If we define $a^j(t,x) = A^j(t,x,0)$ for $j = 1, 2, \ldots, m$ we obtain a new representation in the coordinate chart $(U_0; t_1, t_2, \ldots, t_m, x)$ for the vector fields $L_1, L_2, \ldots, L_m$ with the functions $b^j(t,x)$ replaced by the functions $a^j(t,x)$, which are "boundary values" of the functions $A_j(t,x,y)$ holomorphic with respect to $z = x + iy$ on the boundary $y = 0$ and replaced by a new vector field $\frac{\partial}{\partial x}$.

From the construction above it follows that the vector fields L_1, $L_2, \ldots, L_m$ generate the same subbundle V as the system we began with and have the properties required in the statement of the theorem. This proves the theorem. □

4. Sufficient Condition for Local Integrability

A. Andreotti and C. D. Hill have proved in [4] a theorem for local integrability of formally integrable systems of l smooth complex vector fields in normal form like (3) in n dimensional real Cartesian space. In the particular case when $l = m$, $n = m + 1$, this theorem is the following.

Theorem 4.1. *Every formally integrable system of the form* (3) *with smooth complex valued functions* $a^j(t,x)$ *for* $j = 1, \dots, m$ *which are real analytic in the variable* x *is locally integrable.*

We will give here a different proof of this theorem under the assumptions that the Levi form of the CR structure (7) corresponding to the system (3) is nondegenerate and indefinite.

The proof of Theorem 4.1 will be based on the following result for realizability of the abstract CR structures of hypersurface type.

Theorem 4.2. *The abstract* CR *structure of hypersurface type of the form* (7) *is realisable if the functions* $a^j(t,x)$ *in* (7) *are real analytic on the variable* x.

Proof. Let us prolong the complex valued real-analytic functions $a^j(t,x)$ as holomorphic functions $A^j(t,z)$ of the complex variable $z = x + iy$ for $(t,x) \in U, -\varepsilon < y < \varepsilon$ for some $\varepsilon > 0$; i.e., we construct holomorphic functions $A^j(t,z)$ in the variable z such that

$$A^j(t,x) = a^j(t,x) \quad \text{for} \quad (t,x) \in U.$$

We shall consider the system of smooth complex vector fields

$$\begin{aligned} L'_j &= 2\frac{\partial}{\partial \bar{z}^j} + A^j(t,z)\frac{\partial}{\partial x}, \quad \text{for} \quad j = 1, \dots, m, \\ L' &= \frac{\partial}{\partial \bar{z}} \end{aligned} \tag{12}$$

on the domain $U' = \{(t,s,x+iy) : (t,x) \in U, s \in \mathbf{R}^m, y \in \mathbf{R}, -\varepsilon < y < \varepsilon\}$ in $\mathbf{R}^{2m+2}$.

This system consists of $m+1$ linearly independent smooth complex vector fields. They commute each other because of the holomorphicity of the functions $A_j(t,z)$ in the variable z and so condition (8) holds also for the holomorphic prolongations $A_j(t,z)$ of the functions $a_j(t,x)$. Such a system of vector fields forms an integrable almost complex structure on the domain U'. The theorem of Newlander-Nirenberg [26] for integrability of integrable almost complex structures ensures that on the set U' there exist complex coordinates $Z^1, Z^2, \dots, Z^{m+1}$. These are smooth complex valued functions which annihilate the vector fields (12) and have linearly independent differentials. Then $Z^1(t,s,x), Z^2(t,s,x), \dots, Z^{m+1}(t,s,x)$ provide the needed functions for the realizability of the abstract CR structure of hypersurface type (7). Indeed, they annihilate the vector fields (7) and their differentials are also linearly independent, as $\frac{\partial Z^j}{\partial x} = -\frac{\partial Z^j}{\partial y}$ holds. This proves the theorem. □

Proof of Theorem 4.1 for systems whose corresponding CR structures has nondegenerate indefinite Levi form. First we construct for the system (3) satisfying the equalities (4), the corresponding abstract CR structure of hypersurface type (7) according to Lemma 2.1. Then we apply Theorem 4.2 to the structure (7) and obtain its realizability. Finally according to Lemma 2.2 we obtain the local integrability of the given system (3) as the condition for the Levi form of the corresponding CR structure (7) to be degenerate indefinite is assumed. This proves the theorem for the case considered. $\diamondsuit$

Remark. Note that Kuranishi's result [21] with Faran's example [13] shows that the condition for real analyticity may be not fulfilled for embeddable strongly pseudoconvex CR structures. On the other hand for the case of nondegenerate and indefinite Levi forms LeBrun [22] has proved that every embeddable five dimensional twistor CR structure is real analytic in some real analytic atlas on the underlying manifold. A partial result in this direction is the theorem of Hanges and Jacobowitz [15] given in Section 3. We give the following

Conjecture. Every embeddable CR structure with nondegenerate indefinite Levi form is real analytic at least in one direction.

In other words, the condition for real analyticity of the functions $a^j(t,x)$ in the variable x in Theorem 4.2 is a necessary and sufficient condition for embeddability of CR structure of hypersurface type of kind (7). The condition for real analyticity in one direction in the conjecture may be expressed as the existence of an atlas for the given embeddable CR structure which has the representation (7) with functions $a^j(t,x)$ real analytic in the variable x in the corresponding coordinates.

Acknowledgements are due to Petar Popivanov who called my attention to the systems under consideration, to Marin S. Marinov for useful discussions and information on the topics and to the referee for useful suggestions on the exposition.

References

[1] A. V. Abrossimov, *Complex differential systems and tangential Cauchy-Riemann equations*, (in Russian), Math. Sbornik, **122** (164), no. 4(12), (1983), 419–434.

[2] A. V. Abrossimov, *On integrability of complex differential systems*, (in Russian), Diff. Uravn. i ih Primen., Moscow, (1984), 131–138.

[3] T. Akahori, *A New Approach to the Local Embedding Theorem of CR-Structures for $n \geq 4$*, Memoirs of the AMS, no. 366, Providence, R.I., 1987.

[4] A. Andreotti and C.D.Hill, *Complex characteristic coordinates and tangential Cauchy-Riemann equations*, Ann. Scuola Norm. Sup. Pisa, 26 Fasc. II (1972), 299–324.

[5] M. S. Baouendi and L. P. Rothschild, *Embeddability of abstract CR structures and integrability of related systems*, Ann. Inst. Fourier, **37**, no. 3, (1987), 131–141.

[6] A. Boggess, *CR Manifolds and Tangential Cauchy-Riemann Complex*, Studies in Advanced Mathematics, CRC Press, (1991).

[7] L. Boutet de Monvel, *Intégration des équations de Cauchy-Riemann induit formelles,* Séminaire Goulaouic-Lions-Schwartz 1974–1975, exposé no. 9.

[8] P. D. Cordaro and H.Jacobowitz (Editors), *Multidimensional Complex Analysis and Partial Differential Equations*, Proceedings of the Brazil-USA Conference on Multidimensional Complex Analysis and Partial Differential Equations, June 12-16, 1995, São Carlos, Brazil, Contemporary Mathemattics, no. 205, AMS, Providence, Rhode Island, 1997.

[9] P. D.Cordaro and J. Hounie, *Local solvability in C_o^∞ of overdetermined systems of vector fields*, J. of Func. Anal., 87, (1989), 231-249.

[10] A. Djuraev, *On the theory of overdetermined systems of first order* (in Russian), Dokl. AN USSR, 321, no. 4, (1991), 664-669.

[11] A. Djuraev, *On some non-elliptic overdetermined systems of first order and their applications*, (in Russian), Dokl. AN USSR, 321, no. 5, (1991), 885–891.

[12] R. Dwilewicz, *Embeddability of smooth Cauchy-Riemann manifolds*, Ann. di Mat., 139, (1985), 15–43.

[13] J. J. Faran, *Non-analytic hypersurfaces in* $\mathbf{C}^n$, Math. Ann., 226, (1977), 121-123.

[14] S. J. Greenfield, *Cauchy-Riemann equations in several variables*, Ann. Sc. Normale Sup., **22** (1968), 275–314.

[15] N. Hanges and H. Jacobowitz, *A remark on almost complex structures with boundary,* Amer. J. of Math., 111, (1989), 53–64.

[16] G. M. Henkin, *The Hans Lewy equation and analysis on pseudoconvex manifolds, I,* (in Russian), Uspehi Math. Nauk, 32, no. 3 (195), (1977), 57–118.

[17] L. Hörmander, *An Introduction to Complex Analysis in Several Variables*, D. Van Nostrand Company, Inc., Princeton, (1966).

[18] H. Jacobowitz, *An Introduction to CR Structures*, Mathematical Surveys and Monographs, no. 32, AMS, (1990).

[19] J. J. Kohn, *Integration of complex vector fields*, Bull. of the AMS, 78, no. 1, (1972), 1–11.

[20] J. J. Kohn, *Methods of partial differential equations in complex analysis*, Several Complex Variables, Proc. of Symposia in Pure Math., AMS, vol. 30, (1977), 215–237.

[21] M. Kuranishi, *Strongly pseudoconvex CR structures over small balls*, Ann. of Math., I, vol. 115, (1982), 451-500, II, vol. 116, (1982), 1-64, III, vol. 116, (1982), 249-330.

[22] C. LeBrun, *Twistor CR manifolds and three-dimensional conformal geometry,* Trans. Amer. Math. Soc., vol. 284, no. 2, (1984), 601–616.

[23] H. Lewy, *On the local character of the solutions of an atypical linear differential equations in three variables and a related theorem for regular functions of two complex variables*, Ann. of Math., 64, no. 3, (1956), 514–522.

[24] H. Lewy, *An example of a smooth linear partial differential equation without solution,* Ann. of Math., 66, no. 1, (1957), 155–158.

[25] G. A. Mendoza and F. Treves, *Local solvability in a class of overdetermined systems of linear pde*, Duke Math. J., 63, no. 2, (1991), 355–377.

[26] A. Newlander and L. Nirenberg, *Complex coordinates in almost complex manifolds,* Annals of Math., (2) 65, (1957), pp.391-404.

[27] L. Nirenberg, *Lectures on Linear Partial Differential Equations*, Reg. Conf. Series in Math., 17, (1973).

[28] L. Nirenberg, *On a question of Hans Lewy*, Russian Math. Survey, 29, (1974), 251-262.

[29] H. Poincaré, *Les fonctions analytiques de deux variables et la représentation conforme*, Rend. Circ. Math. Palermo, 23, (1907), 185-220, (or Oeuvres, IV, 224–289).

[30] J. Tabov, *Formally integrable complex linear systams of PDE's with constant coefficients in* $\mathbf{R}^3$, C. R. Acad. bulg. Sci., 48, no. 9-10, (1995), 11-14; *Formally integrable Mizohata systams in* $\mathbf{R}^3$, Nihonkai Math. J., 6, 2, (1995), 207-214.

[31] G. Taiani, *Cauchy-Riemann (CR) Manifolds*, Math. Department, Pace University, New York 10038, (1989).

[32] F. Treves, *Approximation and Representation of Functions and Distributions Annihilated by a Systems of Complex Vector Fields*, École Polytechnique, Centre de Mathématiques, (Mai 1981).

[33] F. Treves, *Homotopy Formulas in the Tangential Cauchy-Riemann Complex*, Memoirs of the AMS, vol. 87, no. 434 (second of 3 numbers), (1990).

[34] F. Treves, *Hypo Analytic Structures*, Princeton University Press, Princeton, New Jersey, (1992).

[35] S. M. Webster, *On the local solutions of the tangential Cauchy-Riemann equations*, Ann. Inst. Henri Poincaré, 6, no. 3, (1989), 167–182.

[36] S. M. Webster, *On the proof of Kuranishi's embedding theorem*, Ann. Inst. Henri Poincaré, 6, no. 3, (1989), 183–207.

[37] R. O. Wells, *The Cauchy-Riemann equations and differential geometry*, Bull. of AMS, 6, no. 2, (1982), 187-199.

[38] K. Yano and M. Kon, *CR Submanifolds of Kaehlerian and Sasakian Manifolds*, Progress in Math., vol. 30, Birkhäuser, (1983).

Institute of Mathematics and Informatics,
Bulgarian Academy of Sciences,
Acad. G. Bonchev Street, Bl. 8,
1113-Sofia, Bulgaria
E-mail address: `LiliaNA@math.bas.bg` or `LiliaNA@bas.bg`

Operator Theory:
Advances and Applications, Vol. 114

On Asymptotic properties of the one-dimensional Schrödinger equation

Irina V. Astashova

Abstract. For the one-dimensional Schrödinger equation $y''(x) = ip_0|y(x)|^{2\gamma} y(x)$ with $\gamma > 0$, $p_0 \neq 0$, $x \in \mathbf{R}$, asymptotics of all solutions are obtained.

Consider the differential equation

$$y''(x) = ip(x)|y(x)|^{2\gamma}y(x) \tag{1}$$

with $\gamma > 0$, $x \in \mathbf{R}$ and $p(x)$ a continuous real-valued function. This equation is a generalization of the well-known Emden-Fowler type differential equation

$$y''(x) = p(x)|y(x)|^{2\gamma}y(x). \tag{2}$$

Asymptotic properties of its solutions were investigated in detail by F. Atkinson, R. Bellman, I. Kiguradze, A. Knezer, V. Kondratiev, T. Kuzano, A. Myshkis and others. A detailed bibliography can be found in [5]. From another point of view, equation (1) is the one-dimensional Schrödinger equation. Qualitative properties of solutions to the different problems connected with this equation in the n-dimensional case ($n \geq 2$) were described by H. Brezis, N. Hayashi, H. Hirata, T. Kato, C. Kenig, P. Lions, G. Ponce, J. L. Vazquez, L. Vega, and C. Yarur, cf. [1–4,6,7].

The main result of this paper is the following.

Theorem 0.1. *Let $\gamma > 0$, $p(x) = p_0 =$ const $\in \mathbf{R}$. Then all solutions of equation (1) are described as follows:*

(1) *the only solution defined on the whole real axis is the trivial one: $y \equiv 0$;*
(2) *all solutions defined on the semi-axis $(-\infty, x_0)$ or $(x_0, +\infty)$ can be written out explicitly:*

$$|y(x)| = C|x - x_0|^{-1/\gamma},$$
$$\arg\, y(x) = W\, \log|x - x_0| + \phi_0$$

This work was supported by RFFI (Grant No. 96-15-96177).

with arbitrary real x_0, ϕ_0, *and the constants* C *and* W *defined as*

$$C = \left(\gamma\sqrt{\frac{|p_0|}{(\gamma+2)\sqrt{\gamma+1}}}\,|x-x_0|\right)^{-1/\gamma},$$

$$W = \operatorname{sign} p_0 \cdot \frac{\sqrt{\gamma+1}}{\gamma};$$

(3) *for any bounded interval* (x_1, x_2) *there exists an inextensible solution defined only on this interval, and all these solutions are equivalent, near its bounds, to the solutions of the previous kind; i.e.,*

$$|y(x)| = C|x-x_k|^{-1/\gamma}\,(1+o(1)),$$
$$\arg y(x) = W\,\log|x-x_k|\,(1+o(1)), \qquad k=1,2,$$

as $x \to x_1+0$ *or* $x \to x_2-0$ *with the same* C *and* W.

Proof. Note that if $y(x)$ is a solution of (1), then the function $z(x) = Ay(Bx+C)$ with a complex constant A, a real constant C and $B = |A|^\gamma$ is also a solution of this equation.

The pair of functions $(y(x), y'(x))$ generates a curve in $\mathbf{C}^2$. The curves generated by non-trivial solutions lie in $\mathbf{C}^2\backslash\{0\}$. Two solutions $y(x)$ and $y(x+C)$ generate the same curve up to its reparametrization. Consider an equivalence in $\mathbf{C}^2\backslash\{0\}$ such that two solutions $y_1(x)$ and $y_2(x) = Ay_1(|A|^\gamma + C)$ generate the same curve in the quotient space. This equivalence may be given by

$$(z_0, z_1) \sim (Az_0, A|A|^\gamma z_1)$$

for all real $A \neq 0$.

Let the quotient space of $\mathbf{C}^2\backslash\{0\}$ be denoted by Φ. We can endow it with the structure of a C^1-smooth 2-dimensional real manifold by using a two-chart atlas. Each of the charts is a bijection of a subset of Φ onto $\mathbf{C}$. The first one is defined on equivalence classes of pairs (z_0, z_1) with $z_0 \neq 0$. In other words, it is defined for all points of Φ but one, the equivalence class of the pair $(0, 1)$. Define this bijection by means of a complex-valued function

$$u : [(z_0, z_1)] \longmapsto \frac{z_1}{z_0|z_0|^\gamma}.$$

The second chart is defined for classes of pairs (z_0, z_1) with $z_1 \neq 0$ as follows:

$$\tilde{u} : [(z_0, z_1)] \longmapsto \frac{z_0|z_1|^{\frac{\gamma}{\gamma+1}}}{z_1}.$$

Well-definedness and bijectivity of these functions are proved immediately. The changes of coordinate given by the formulae

$$u = \frac{1}{\tilde{u}|\tilde{u}|^\gamma}, \qquad \tilde{u} = \frac{|u|^{\frac{\gamma}{\gamma+1}}}{u}$$

are C^1 maps $\mathbf{R}^2 \setminus \{0\} \longrightarrow \mathbf{R}^2 \setminus \{0\}$. The manifold obtained is homeomorphic to the 2-dimensional sphere.

Let us describe, in the two charts, all curves generated in Φ by solutions of (1). In the first chart we have

$$u = \frac{y'}{y|y|^\gamma} \tag{3}$$

and, by direct calculations,

$$u' = |y|^\gamma \left(ip_0 - \left(\frac{\gamma}{2}+1\right) u^2 - \frac{\gamma}{2}|u|^2 \right).$$

Hence, by parametrizing the curve with t such that $dt = |y|^\gamma dx$ we obtain its description in internal terms:

$$\dot{u} = \frac{du}{dt} = ip_0 - \left(\frac{\gamma}{2}+1\right) u^2 - \frac{\gamma}{2}|u|^2. \tag{4}$$

The same equation in real coordinates (substituting $u = v + iw$)

$$\begin{cases} \dot{v} = w^2 - (\gamma+1)v^2, \\ \dot{w} = p_0 - (\gamma+2)vw. \end{cases} \tag{5}$$

This system has two fixed points: $u_0 = v_0 + iw_0$ and $-u_0$ with

$$v_0 = \sqrt{\frac{|p_0|}{(\gamma+2)\sqrt{\gamma+1}}},$$

$$w_0 = \operatorname{sign} p_0 \sqrt{\frac{|p_0|\sqrt{\gamma+1}}{\gamma+2}}.$$

Similarly, in the second chart curves generated by solutions of (1) are described by the equation

$$\dot{\tilde{u}} = 1 - \frac{ip_0|\tilde{u}|^{2\gamma}}{2(\gamma+1)} \left((\gamma+2)\tilde{u}^2 + \gamma|\tilde{u}|^2\right), \tag{6}$$

Here $\dot{\tilde{u}}$ means $d\tilde{u}/d\tau$ with a new parameter τ such that

$$d\tau = |y'|^{\frac{\gamma}{\gamma+1}} dx.$$

The right sides of equations (4) and (6) are C^1-smooth in the real sense. There is no doubt that, the two parameters t and τ can be "mixed" to produce another one such that all curves in Φ generated by solutions of (1) are trajectories of a C^1-smooth autonomous dynamical system having this parameter as the independent variable.

Since the manifold Φ is compact, any trajectory of the system can be extended to the whole real axis $(-\infty, +\infty)$. Moreover, only complete trajectories (not their parts) are generated by inextensible solutions of (1). Below, the term "trajectory" always means "complete trajectory".

Let us prove that any trajectory of the dynamical system has a unique limit point coinciding with one of the two fixed points. Since the only point of Φ not covered by the first chart is not a fixed point, we may restrict ourselves to equations (4) (or (5)). We can also assume without loss of generality that $p_0 > 0$.

Lemma 0.2. *If a trajectory enters the domain* $K^{++} = \{(v,w) : v > 0, w > 0\}$, *then* $u \to u_0$ *as* $t \to +\infty$.

Proof. Note that K^{++} can be split into the disjoint rectangles

$$R_\theta^+ = \partial\{(v,w) : \theta v_0 \le v \le v_0/\theta,\ \theta w_0 \le w \le w_0/\theta\}, \qquad 0 < \theta \le 1,$$

with R_1^+ consisting of a single point, which is just u_0. Thus, a continuous function $\theta : K \to \mathbf{R}$ is defined such that $u \in R^+_{\theta(u)}$ for all $u \in K$. This function increases along the trajectories of (5) or, in other words, all trajectories are directed toward the interior of the rectangles. Indeed, by using the evident equalities

$$w_0 = v_0\sqrt{\gamma+1} \qquad \text{and} \qquad (\gamma+2)v_0w_0 = p_0,$$

we obtain on R_θ^+ :

$$\begin{aligned}
\dot v\big|_{v=\theta v_0} &= w^2 - (\gamma+1)\theta^2 v_0^2 = w^2 - \theta^2 w_0^2 \ge 0,\\
\dot w\big|_{w=\theta w_0} &= p_0 - (\gamma+2)v\theta w_0 \ge p_0 - (\gamma+2)v_0w_0 = 0,\\
\dot v\big|_{v=v_0/\theta} &= w^2 - (\gamma+1)v_0^2/\theta^2 \le \left(w_0^2 - (\gamma+1)v_0^2\right)/\theta^2 = 0,\\
\dot w\big|_{w=w_0/\theta} &= p_0 - (\gamma+2)vw_0/\theta \le p_0 - (\gamma+2)v_0w_0 = 0.
\end{aligned}$$

All inequalities are strict if not at corners. At corners, at least one inequality is strict. Hence no point of R_θ^+ with $\theta < 1$ can be a limit point of a trajectory. This proves the lemma. □

Note that we have proved that the only trajectory resting in K^{++} for all t consists of the single point u_0; all others have entered the domain from outside.

Lemma 0.3. *Any trajectory having a point in the domain* $K^{--} = \{(v,w) : v < 0, w < 0\}$ *either leaves it or consists just of a single point, which is* $-u_0$.

Proof. The lemma is proved in the same way as the previous one by using the rectangles

$$R_\theta^- = \partial\{(v,w) : -v_0/\theta \le v \le -\theta v_0,\ -w_0/\theta \le w \le -\theta w_0\}, \qquad 0 < \theta \le 1.$$

□

Lemma 0.4. *Any trajectory passing through the domain*

$$K^{+-} = \{(v,w) : v > 0, w < 0\} \qquad \text{or} \qquad K^{-+} = \{(v,w) : v < 0, w > 0\}$$

has entered there from K^{--} *and leaves it to enter* K^{++}.

Proof. Immediate calculations of sign $\dot{v}$ and sign $\dot{w}$ on the boundaries describe all possible domain-to-domain passages. Thus, we have only to prove that no trajectory can stay in the domain considered (neither as $t \to +\infty$, nor as $t \to -\infty$). To see this, consider the function $\arg u$ and calculate its derivative along trajectories:

$$(\arg u)^{\cdot} = \frac{v\dot{w} - \dot{v}w}{v^2 + w^2} = \frac{vp_0 - w^3 - v^2 w}{v^2 + w^2}.$$

Thus the function is strictly increasing along trajectories in K^{+-} and strictly decreasing in K^{-+}, but this is impossible for a limit point of a trajectory staying in the domain considered. □

Remark 0.5. A weak point of the above considerations is just the point $u = 0$ (as well as $\tilde{u} = 0$). Immediate calculations of the 2nd and 3rd derivatives show that trajectories passing through these points come from K^{--} and go to K^{++}.

Now we have a complete description of trajectories on Φ. They are:
- two fixed points $u = u_0$ and $u = -u_0$;
- trajectories joining the two points, i.e., satisfying

$$u = u_0\,(1 + o(1)), \quad t \to +\infty, \qquad \text{and} \qquad u = -u_0\,(1 + o(1)), \quad t \to -\infty.$$

The rest is just a substitution of (3) into the above equalities followed by solution of simple differential equations. □

References

[1] H. Brezis, T. Kato, *Remarks on the Shrödinger operator with singular complex potential,* J. Math. Pures et Appl. **58** (1979) 137–151.

[2] N. Hayashi, H. Hirata, *Global existence and asymptotic behaviour in time of small solutions to the elliptic–hyperbolic Davey–Stewartson system,* Nonlinearity, **9** (1996) 1387–1409, Printed in UK.

[3] T. Kato, *Schrödinger operators with singular potentials,* Israël Math. **13** (1972) 135–148.

[4] C. Kenig, G. Ponce, L. Vega, *Small solutions of non-linear Schrödinger equations,* Ann. Inst. Henri Poincaré **11** (1993) 255–288.

[5] I. T. Kiguradze, T. A. Chanturia, *Asymptotic properties of solutions of nonautonomous ordinary differential equations,* Kluwer Academic Publishier Group, Dordrecht (1993) xiv+331

[6] P. Lions, *Isolated singularities in semilinear problems,* Diff. Equ., **38** (1980) 441–450

[7] J. L. Vazquez, C. Yarur. *Isolated singularities of the solutions of the Schrödinger equation with a radial potential,* Arc. for Rat. Mech. and Anal., **98** (1987) 251–284

Kutuzovski prosp. 30/32, apt. 494,
Moscow, 121165,
Russia
E-mail address: ast@mail.ecfor.rssi.ru

Operator Theory:
Advances and Applications, Vol. 114

On Q_p functions

Rauno Aulaskari

1. Introduction

A new class of functions named Q_p has been recently introduced and studied by several mathematicians. These spaces are situated between the classical Dirichlet space $\mathcal{D}_1$ and the Bloch space $\mathcal{B}$, where $\mathcal{B}$ is in a sense maximal among Möbius invariant function spaces. Further, the spaces Q_p as a function of parameter values p fill the gap between $\mathcal{D}_1$ and $\mathcal{B}$ and join these well-known spaces by certain values of p. Now we will show some features of this research.

Let $\Delta = \{z : |z| < 1\}$ be the unit disk. Let $\varphi_a(z) = \frac{a-z}{1-\bar{a}z}$ be a Möbius transformation of Δ. By $g(z,a)$ we denote the Green's function $\log\left|\frac{1-\bar{a}z}{z-a}\right|$ of Δ with logarithmic singularity at $a \in \Delta$.

We define the following spaces of functions analytic in Δ: For $0 < p < \infty$, we set

$$Q_p = \left\{ f : \sup_{a\in\Delta} \iint_{\Delta} |f'(z)|^2 g^p(z,a)\, dx\, dy < \infty \right\} \tag{1}$$

and

$$Q_{p,0} = \left\{ f : \lim_{|a|\to 1} \iint_{\Delta} |f'(z)|^2 g^p(z,a)\, dx\, dy = 0 \right\}.$$

For $p = 1$, we know that $Q_1 = BMOA$ [Ba] and $Q_{1,0} = VMOA$ [Sa]. If $1 < p < \infty$, then $Q_p = \mathcal{B}$ and $Q_{p,0} = \mathcal{B}_0$ (cf. [AuLa]), where

$$\mathcal{B} = \{ f : B(f) = \sup_{z\in\Delta}(1-|z|^2)|f'(z)| < \infty \}$$

and

$$\mathcal{B}_0 = \{ f : \lim_{|z|\to 1}(1-|z|^2)|f'(z)| = 0 \}.$$

Now $\mathcal{B}$ is called the space of Bloch functions and $\mathcal{B}_0$ the space of little Bloch functions. It is well known that $BMOA \subset \mathcal{B}$ and $VMOA \subset \mathcal{B}_0$.

2. Inclusions

Because, for $1 < p < \infty$, the problem has been solved, we will only concentrate on the case $0 < p \leq 1$.

Theorem 2.1. [AuXiZh, Theorems 2 and 3] *For* $0 < p < q < 1$,

(i) $Q_p \subset Q_q \subset BMOA$,
(ii) $Q_{p,0} \subset Q_{q,0} \subset VMOA$.

Proof. (i) Let $f \in Q_p$. Then

$$\sup_{a\in\Delta} \iint_{\Delta} |f'(z)|^2 g^p(z,a)\, dx\, dy < \infty. \tag{2}$$

For $1 < k < \frac{1-p}{1-q}$, using Hölder's inequality we have

$$\iint_{\Delta} |f'(z)|^2 g^q(z,a)\, dx\, dy = \iint_{\Delta} |f'(z)|^{\frac{2}{k}} g^{\frac{p}{k}}(z,a) |f'(z)|^{2-\frac{2}{k}} g^{q-\frac{p}{k}}(z,a)\, dx\, dy$$
$$\leq \left(\iint_{\Delta} |f'(z)|^2 g^p(z,a)\, dx\, dy\right)^{\frac{1}{k}} \left(\iint_{\Delta} |f'(z)|^2 g^{\frac{qk-p}{k-1}}(z,a)\, dx\, dy\right)^{\frac{k-1}{k}}. \tag{3}$$

By Proposition 1 of [AuLa] we know that for $0 < p < \infty$,

$$(1-|a|^2)^2 |f'(a)|^2 \leq \frac{1}{\pi}\left(\frac{2e}{p}\right)^p \iint_{\Delta} |f'(z)|^2 g^p(z,a)\, dx\, dy. \tag{4}$$

Thus, from $f \in Q_p$ for $0 < p < \infty$, we have $f \in \mathcal{B}$. By Proposition 2 of [AuLa] we know that, for every $t > 1$,

$$\iint_{\Delta} |f'(z)|^2 g^t(z,a)\, dx\, dy \leq J(t) B^2(f),$$

where $J(t) = 2\pi \int_0^1 (\log \frac{1}{r})^t (1-r^2)^{-2} r\, dr$. Note that $\frac{qk-p}{k-1} > 1$ for $1 < k < \frac{1-p}{1-q}$. Then

$$\sup_{a\in\Delta} \iint_{\Delta} |f'(z)|^2 g^{\frac{qk-p}{k-1}}(z,a)\, dx\, dy \leq J\left(\frac{qk-p}{k-1}\right) B^2(f). \tag{5}$$

Hence, by (3) and (5),

$$\iint_{\Delta} |f'(z)|^2 g^q(z,a)\, dx\, dy$$
$$\leq \left(J\left(\frac{qk-p}{k-1}\right) B^2(f)\right)^{\frac{k-1}{k}} \left(\iint_{\Delta} |f'(z)|^2 g^p(z,a)\, dx\, dy\right)^{\frac{1}{k}}. \tag{6}$$

From (2) we have

$$\sup_{a\in\Delta}\iint_{\Delta}|f'(z)|^2 g^q(z,a)\,dx\,dy<\infty.$$

Thus $f\in Q_q$.

(ii) Let $f\in Q_{p,0}$. Then

$$\lim_{|a|\to 1}\iint_{\Delta}|f'(z)|^2 g^p(z,a)\,dx\,dy=0. \tag{7}$$

By (4) $f\in\mathcal{B}$. Using (6) and (7) we obtain

$$\lim_{|a|\to 1}\iint_{\Delta}|f'(z)|^2 g^q(z,a)\,dx\,dy=0.$$

Thus $f\in Q_{q,0}$. □

3. Strict inclusions

Because $Q_p=Q_q=\mathcal{B}$ and $Q_{p,0}=Q_{q,0}=\mathcal{B}_0$ for $1<p<q<\infty$, we can ask if the inclusions in Theorem 2.1 are strict or, in fact, equalities. To study this question we characterize gap series in Q_p and $Q_{p,0}$. We will use the following lemmas in proving the main theorem of this section.

Lemma 3.1. *Let $0<p<\infty$. If (n_k) is an increasing sequence of positive integers satisfying $\frac{n_{k+1}}{n_k}\geq\lambda>1$ for all k, then there is a constant A depending only on p and λ such that*

$$A^{-1}\Big(\sum_{k=1}^{\infty}|a_k|^2\Big)^{\frac12}\leq\Big(\frac{1}{2\pi}\int_0^{2\pi}\Big|\sum_{k=1}^{\infty}a_k e^{in_k\theta}\Big|^p\,d\theta\Big)^{\frac1p}\leq A\Big(\sum_{k=1}^{\infty}|a_k|^2\Big)^{\frac12}$$

for any number a_k, $k=1,2,\ldots$.

The above lemma is due to [Zy].

Lemma 3.2. *Let $\alpha>0$, $p>0$, $n\geq 0$, $a_n\geq 0$, $I_n=\{k:2^n\leq k<2^{n+1},k\in\mathbb{N}\}$, $t_n=\sum_{k\in I_n}a_k$ and $f(x)=\sum_{n=1}^{\infty}a_nx^n$. Then there is a constant K depending only on p and α such that*

$$\frac1K\sum_{n=0}^{\infty}2^{-n\alpha}t_n^p\leq\int_0^1(1-x)^{\alpha-1}f(x)^p\,dx\leq K\sum_{n=0}^{\infty}2^{-n\alpha}t_n^p.$$

The proof of Lemma 3.2 can be found in [MaPa]. By simple calculation we see that the above lemma is still valid for $f(x)=\sum_{n=1}^{\infty}a_nx^{n-1}$, $a_n\geq 0$ (cf. [Mi, p. 108].

The next lemma can be found in [AuXiZh, Theorem 5].

Lemma 3.3. *Let* $I_n = \{k : 2^n \le k < 2^{n+1}, k \in \mathbb{N}\}$ *and let* $f(z) = \sum_{n=0}^{\infty} a_n z^n$ *be analytic in* Δ. *If, for* $0 < p \le 1$,

$$\sum_{n=0}^{\infty} 2^{n(1-p)} (\sum_{k \in I_n} |a_k|)^2 < \infty,$$

then $f \in Q_{p,0}$.

The idea of the proof of the following theorem is found in [Mi, Theorem 2].

Theorem 3.4. [AuXiZh, Theorem 6] *Let* $0 < p \le 1$. *If* $f(z) = \sum_{k=0}^{\infty} a_k z^{n_k}$ *is an analytic function in* Δ *and has Hadamard gaps, that is,*

$$\frac{n_{k+1}}{n_k} \ge \lambda > 1, k = 0, 1, 2, \ldots,$$

then the following statements are equivalent:

(i) $f \in Q_p$,
(ii) $f \in Q_{p,0}$,
(iii) $\sum_{k=0}^{\infty} n_k^{1-p} |a_k|^2 < \infty$.

Proof. (ii)⇒(i). This follows from Theorem 1 (i) in [AuXiZh].

(iii)⇒(ii). Because $n_{k+1}/n_k \ge \lambda > 1$, for all k, the number of Taylor coefficients a_j is at most $[\log_\lambda 2] + 1$ when $n_j \in I_k$, for $k = 0, 1, 2, \ldots$. Then, because $(1/2)n_j < 2^k \le n_j$ whenever $n_j \in I_k$, we have by assumption

$$\sum_{k=0}^{\infty} 2^{k(1-p)} (\sum_{n_j \in I_k} |a_j|)^2 \le ([\log_\lambda 2] + 1) \sum_{k=0}^{\infty} 2^{k(1-p)} \sum_{n_j \in I_k} |a_j|^2 < \infty.$$

Thus, by Lemma 3.3, $f \in Q_{p,0}$.

(i)⇒(iii). Let $I(a) = \iint_\Delta |f'(z)|^2 (1 - |\varphi_a(z)|^2)^p \, dx \, dy$. Then, by Lemmas 3.1 and 3.2,

$$\begin{aligned}
\sup_{a \in \Delta} I(a) &\ge \iint_\Delta |f'(z)|^2 (1 - |z|^2)^p \, dx \, dy \\
&= \iint_\Delta \Big|\sum_{k=1}^{\infty} n_k a_k z^{n_k - 1}\Big|^2 (1 - |z|^2)^p \, dx \, dy \\
&\ge \frac{2\pi}{A^2} \int_0^1 (1 - r^2)^p (\sum_{k=1}^{\infty} n_k^2 |a_k|^2 r^{2(n_k - 1)}) r \, dr \\
&\ge \frac{\pi}{A^2} \int_0^1 (1 - x)^p (\sum_{k=1}^{\infty} n_k^2 |a_k|^2 x^{n_k}) \, dx \\
&\ge \frac{\pi}{A^2 K} \sum_{k=0}^{\infty} 2^{-k(p+1)} t_k,
\end{aligned}$$

where $t_k = \sum_{n_j \in I_k} n_j^2 |a_j|^2 \geq 2^{2k} \sum_{n_j \in I_k} |a_j|^2$. Thus

$$\sup_{a \in \Delta} I(a) \geq \frac{\pi}{A^2 K} \sum_{k=0}^{\infty} 2^{k(1-p)} \sum_{n_j \in I_k} |a_j|^2.$$

By $f \in Q_p$ and $1 - x^2 \leq 2\log(1/x)$, $0 < x < 1$, we have $\sup_{a \in \Delta} I(a) < \infty$. Then

$$\sum_{k=0}^{\infty} 2^{k(1-p)} \sum_{n_j \in I_k} |a_j|^2 < \infty$$

and the theorem is proved. □

For $p = 1$ the theorem was proved by J. Miao [Mi]. Using Theorem 3.4, we can easily prove that the inclusions in Theorem 2.1 are strict.

Remark 3.5. It is a little surprising that in the formulae of (1), for all $1 < p < \infty$, we get the same space $\mathcal{B}$ or $\mathcal{B}_0$ but for all p, $0 < p \leq 1$, separate spaces are obtained. However, this can be understood in the sense that the Bloch space $\mathcal{B}$ is maximal among certain function spaces.

4. Q_p and some well-known function spaces

We take one example showing the relation between Q_p and a well-known function space. For $1 < q < \infty$, we say an analytic function f in Δ is in the (analytic) Besov space B_q provided

$$\iint_{\Delta} |f'(z)|^q (1 - |z|^2)^{q-2}\, dx\, dy < \infty.$$

Note that $B_2 = \mathcal{D}_1$ (the classical Dirichlet space). It is well known that B_q increases with q and $B_q \subset VMOA$ for all $1 < q < \infty$. Using Q_p spaces we can sharpen this result as follows:

Theorem 4.1. [AuCs, Theorem 1 and Corollary 2], *We have*

(i) $B_q \subset \bigcap_{0<p<1} Q_{p,0}$ *for* $1 < q \leq 2$,
(ii) $B_q \subset \bigcap_{1-\frac{2}{q}<p<1} Q_{p,0}$ *for* $2 \leq q < \infty$.

It can be shown that both inclusions in Theorem 4.1 (i) and (ii) are strict and that the lower bound, $1 - 2/q$, of values of p for $2 < q < \infty$, is best possible in the sense that

$$B_q \not\subset Q_{1-\frac{2}{q},0}$$

for $2 < q < \infty$. As an example function we can choose $f(z) = \sum_{n=0}^{\infty} \frac{1}{(n+1)^{\frac{1}{2}} 2^{\frac{n}{q}}} z^{2^n}$.

Other inclusions between Q_p and Lipschitz spaces or mean Lipschitz spaces are also known.

5. Q_p functions with positive Taylor coefficients

In Section 3 we have a criterion for a Hadamard gap series to belong to Q_p for $0 < p \leq 1$. If an analytic function $f(z) = \sum_{n=0}^{\infty} a_n z^n$ is not a Hadamard gap series, is it possible to check by its Taylor coefficients that it belongs to Q_p? The next theorem gives a sufficient condition for arbitrary a_n's and a criterion for positive a_n's.

Theorem 5.1. [AuStXi, Theorem 1.2] *Suppose that* $0 < p \leq 1$ *and that* $f(z) = \sum_{n=0}^{\infty} a_n z^n$ *is an analytic function.*

(i) *The condition*

$$\sup_k \frac{1}{k^p} \sum_{n=0}^{\infty} (n+1)^{1-p} \Big[\sum_{m=0}^{\min(n,k)} \frac{|a_{2n-m+1}|}{(m+1)^{1-p}} \Big]^2 < \infty \tag{8}$$

implies that $f \in Q_p$.

(ii) *If* $a_n \geq 0$ *for all* n *and* $f \in Q_p$ *then condition (8) holds.*

For $BMOA$, the corresponding criterion to (ii) is C. Fefferman's unpublished result, whose published proofs involve some aspect of the duality between H^1 and $BMOA$. In the absence of an analogue to these theories, the proof of Theorem 5.1 only uses the definition of Q_p.

The above theorem is a powerful tool when constructing functions f satisfying $f \in Q_q \setminus Q_p$, $0 < p < q \leq 1$, and some extra condition which excludes the use of lacunary series with Hadamard gaps.

6. Q_p and random power series

Let $\varepsilon_n(\omega)$ be a Bernoulli sequence of random variables on a probability space. This means that each ε_n takes the value $+1$ and -1 with probability $1/2$.

If $f(z) = \sum_{n=0}^{\infty} a_n z^n$ is analytic in Δ, let

$$f_\omega(z) = \sum_{n=0}^{\infty} \varepsilon_n(\omega) a_n z^n.$$

We call f_ω a *random power series* of f.

For $0 \leq p \leq 1$, the weighted Dirichlet space $\mathcal{D}_p$ is the space of analytic functions f in Δ satisfying

$$\|f\|_{\mathcal{D}_p}^2 = \sum_{n=1}^{\infty} n^p |a_n|^2 \approx \iint_\Delta |f'(z)|^2 (1-|z|^2)^{1-p} \, dx \, dy < \infty. \tag{9}$$

It is not difficult to show that $f \in Q_p$ if and only if

$$\sup_{a \in \Delta} \iint_\Delta |f'(z)|^2 (1 - |\varphi_a(z)|^2)^p \, dx \, dy < \infty$$

(cf. [AuXiZh, Proposition 1]). Thus $Q_p \subset \mathcal{D}_{1-p}$ for $0 \le p \le 1$. Note that $\mathcal{D}_0$ is the Hardy space H^2. W. Sledd and D. Stegenga [SlSt] have proved

Theorem 6.1. *There exists $f(z) = \sum_{n=0}^{\infty} a_n z^n \in H^2$ (i.e. $\sum_{n=0}^{\infty} |a_n|^2 < \infty$) but its randomization $f_\omega \notin BMOA$ for any choice of ω.*

By (9) we have the necessary condition: if $f(z) = \sum_{n=0}^{\infty} a_n z^n \in Q_p$ for $0 < p \le 1$, then $\sum_{n=1}^{\infty} n^{1-p} |a_n|^2 < \infty$. Surprisingly, it is also a sufficient condition, almost surely, provided $0 < p < 1$, as the next theorem shows. At the same time this theorem shows a different behaviour of Q_p $(0 < p < 1)$ as compared with $BMOA$.

Theorem 6.2. [AuStZh, Theorem 1] *For any $f(z) = \sum_{n=0}^{\infty} a_n z^n \in Q_p$ (i.e. $\sum_{n=1}^{\infty} n^{1-p} |a_n|^2 < \infty$), $0 < p < 1$, we have $f_\omega \in Q_p$ a.s. (almost surely).*

Proof. Since $\sum_{n=1}^{\infty} n^{1-p} |a_n|^2 < \infty$, $f(z) = \sum_{n=0}^{\infty} a_n z^n \in \mathcal{D}_{1-p}$. By [CoShUl, Theorem 2] $f_\omega \in M(\mathcal{D}_{1-p})$ a.s., where $M(\mathcal{D}_{1-p})$ is the space of pointwise multipliers of $\mathcal{D}_{1-p}$. By Theorem 5.2 in [AuLaXiZh], $M(\mathcal{D}_{1-p}) \subset Q_p$. Thus $f_\omega \in Q_p$ a.s., and the proof is completed. □

7. New trends in Q_p research

For functions in $BMOA$ we know the following boundary criterion: If $f \in H^1$ (Hardy space), then $f \in BMOA$ if and only if

$$\sup_{I \subset \partial\Delta} \frac{1}{|I|^2} \int_I \int_I |f(e^{i\theta}) - f(e^{it})|^2 \, d\theta dt < \infty$$

($|I|$ is the length of an arc $I \subset \partial\Delta$).

M. Essen and J. Xiao have proved the following boundary value criterion for Q_p functions:

Theorem 7.1. [EsXi] *Let $0 < p < 1$ and let $f \in H^1$. Then $f \in Q_p$ if and only if*

$$\sup_{I \subset \partial\Delta} \frac{1}{|I|^p} \int_I \int_I \frac{|f(e^{i\theta}) - f(e^{it})|^2}{|e^{i\theta} - e^{it}|^{2-p}} \, d\theta dt < \infty.$$

This boundary value criterion might have some applications in more general settings. Further, M. Essen and J. Xiao have considered the question when the Blaschke product

$$B(z, z_n) = \prod_n \frac{|z_n|}{z_n} \frac{z_n - z}{1 - \bar{z}_n z}$$

belongs to Q_p for $0 < p < 1$. Note that not all bounded functions belong to Q_p, $0 < p < 1$. O. Resendiz and L. M. Tovar have continued this research by searching for explicit conditions on the zeros $\{z_n\}$ of the Blaschke product which guarantee it to belong to Q_p.

Since the definition (1) of the spaces Q_p is Möbius invariant we can transfer it to arbitrary Riemann surfaces R with Green's functions. The nesting property

$Q_p(R) \subset Q_q(R)$, $0 < p < q < \infty$, and the inclusion $\mathcal{D}_1(R) \subset Q_p(R)$ for all p, $0 < p < \infty$, have been proved. The latter inclusion sharpens T. Metzger's well-known result that the classical Dirichlet space $\mathcal{D}_1(R) \subset BMOA(R)$. Also for any p, $1 < p < \infty$, there exists a Riemann surface R such that $Q_p(R) \subsetneq \mathcal{B}(R)$, where $\mathcal{B}(R)$ is the Bloch space on R. This differs from the situation in the unit disk Δ.

References

[AuCs] R. Aulaskari and G. Csordas, *Besov spaces and the $Q_{q,0}$ classes,* Acta Sci. Math., **60,** (1995) 31–48.

[AuLa] R. Aulaskari and P. Lappan, *Criteria for an analytic function to be Bloch and a harmonic or meromorphic function to be normal,* Complex analysis and its applications, Editors C.-C. Yang et al., Pitman Res. Notes Math. Ser., 305, Longman, Harlow, (1994) 136–146.

[AuLaXiZh] R. Aulaskari, P. Lappan, J. Xiao and R. Zhao, *On α-Bloch spaces and multipliers of Dirichlet spaces,* J. Math. Anal. Appl., 209, (1997) 103–121

[AuStXi] R. Aulaskari, D. A. Stegenga and J. Xiao, *Some subclasses of BMOA and their characterization in terms of Carleson measures,* Rocky Mountain J. Math., **26**, (1996) 485–506.

[AuStZh] R. Aulaskari, D. A. Stegenga and R. Zhao, *Random power series and Q_p,* Proc. 16th Rolf Nevanlinna Colloquium, Editors I. Laine and O. Martio, de Gruyter, Berlin (1996) 247–255.

[AuXiZh] R. Aulaskari, J. Xiao and R. Zhao, *On subspaces and subsets of BMOA and UBC,* Analysis, **15,** (1995) 101–121.

[Ba] A. Baernstein II, *Analytic functions of bounded mean oscillation,* Aspects of contemporary complex analysis, Editors D. A. Brannan and J. G. Clunie, Academic Press, London, (1980) 2–26.

[CoShUl] W. G. Cochran, J. H. Shapiro and D. C. Ullrich, *Random Dirichlet functions: multipliers and smoothness,* Canad. J. Math., **45,** (1993) 255–268.

[EsXi] M. Essen and J. Xiao, *Some results on Q_p spaces,* $0 < p < 1$, J. Reine Angew. Math., 485, (1997) 173–195.

[MaPa] M. Mateljevic and M. Pavlovic, *L^p-behaviour of power series with positive coefficients and Hardy spaces,* Proc. Amer. Math. Soc., **87,** (1983) 309–316.

[Mi] J. Miao, *A property of analytic functions with Hadamard gaps,* Bull. Austral. Math. Soc., **45,** (1992) 105–112.

[Sa] D. Sarason, *Functions of vanishing mean oscillation,* Trans. Amer. Math. Soc., **207,** (1975) 391–405.

[SlSt] W. T. Sledd and D. A. Stegenga, *An H^1 multiplier theorem,* Ark. Mat., **19,** (1981) 265–270.

[Zy] A. Zygmund, Trigonometric series, Cambridge Univ. Press, London (1959).

Department of Mathematics
University of Joensuu
P.O. Box 111
FIN-80101 Joensuu
Finland
E-mail address: Rauno.Aulaskari@joensuu.fi

Operator Theory:
Advances and Applications, Vol. 114

On Green's functions for subelliptic operators

Richard Beals, Bernard Gaveau, and Peter Greiner

Abstract. In this article we give two new derivations of the Green's function, or fundamental solution, for the sublaplacian associated to the hypersurface $\operatorname{Im} z_{n+1} = |z|^{2k}$, $z \in \mathbb{C}^n$. This Green's function, given by an Euler transform, is expressed in variables which smoothly relate the generic step two points to the higher step (step $2k$) points on the $\operatorname{Re} z_{n+1}$ axis.

1. Introduction

A fundamental solution (Green's function) for a class of subelliptic operators that occur in analysis on weakly pseudo-convex hypersurfaces was constructed in [2]. Qualitative results for such operators have been known for some time, but exact formulas are known in only a few cases; see references in [2]. Exact formulas are of interest for a number of reasons. They are closely related to an underlying geometry. Moreover, they can lead to equations of hypergeometric type in several variables. The latter equations are not the most obvious generalizations of the classical hypergeometric equation, but they are natural and have explicit non-trivial solutions. We discuss these points in more detail below.

The operators considered here act on functions in $\mathbb{R}^{2n+1} = \mathbb{R}^{2n} \times \mathbb{R} = \{(x,t)\}$ and have the form

$$\begin{aligned} \Delta_\lambda &= \Delta_{\lambda,k} \\ &= \frac{1}{2}\sum_{j=1}^{n}\left[\left(\frac{\partial}{\partial x_j}\right)^2 + \left(\frac{\partial}{\partial x_{n+j}}\right)^2 + 4k|x|^{2k-2}\left(x_{n+j}\frac{\partial}{\partial x_j} - x_j\frac{\partial}{\partial x_{n+j}}\right)\frac{\partial}{\partial t}\right] \\ &\quad + 2k^2|x|^{4k-2}\left(\frac{\partial}{\partial t}\right)^2 + 2i\lambda k(n+k-1)|x|^{2k-2}\frac{\partial}{\partial t}, \end{aligned} \tag{1.1}$$

where k is some fixed positive integer and λ is a complex parameter. They are connected with the real submanifold

$$M = \left\{\operatorname{Im} z_{n+1} = f\left(\sum_{j=1}^{n} z_j\bar{z}_j\right)\right\} \subset \mathbb{C}^{n+1}, \qquad f(x) = x^k. \tag{1.2}$$

Research partially supported by NSF Grant DMS-9213595, by E.U. Grant "Capital humain et mobilité" and by NSERC Grant OGP0003017.

In fact set $z_j = x_j + ix_{n+j}$, $1 \leq j \leq n$, $t = \text{Re } z_{n+1}$. The vector fields

$$Z_j = \frac{\partial}{\partial z_j} + if'(|z|^2)\bar{z}_j \frac{\partial}{\partial t} = \frac{1}{2}(X_j - iX_{n+j}) \tag{1.3}$$

are the restrictions to M of holomorphic vector fields which are tangent to M and $\{Z_1, \ldots, Z_n, \bar{Z}_1, \ldots, \bar{Z}_n\}$ is a frame for the CR structure on the complexified tangent bundle $\mathbb{C} \otimes T(M)$. Note that

$$\begin{aligned} \Delta_\lambda &= \sum_{j=1}^{n} (Z_j \bar{Z}_j + \bar{Z}_j Z_j + \lambda[\bar{Z}_j, Z_j]) \\ &= (1+\lambda) \sum_{j=1}^{n} \bar{Z}_j Z_j + (1-\lambda) \sum_{j=1}^{n} Z_j \bar{Z}_j. \end{aligned} \tag{1.4}$$

The operator (1.1) is step 2 when $z \neq 0$ and step $2k$ on the t-axis; that is, we need one bracket of the Z_j-s and $\bar{Z}_j$-s to obtain the missing direction $\partial/\partial t$ when $z \neq 0$, but on the t-axis, when $z = 0$, we need $2k - 1$ brackets of the Z_j-s and $\bar{Z}_j$-s to obtain $\partial/\partial t$; such hypoelliptic operators are discussed in [5].

Let (z, t), (w, s) denote arbitrary points of the manifold M, identified to $\mathbb{C}^n \times \mathbb{R}$. The expression for the Green's function of Δ_λ involves five auxiliary functions of the $4n + 2$ variables z, $\bar{z}$, t, w, $\bar{w}$, s. These functions are

$$A_+ = A(z, t; w, s) = \frac{1}{2}[|z|^{2k} + |w|^{2k} - i(t - s)] \qquad A_- = \bar{A}; \tag{1.5}$$

$$p_+ = p(z, t; w, s) = \frac{z \cdot \bar{w}}{A^{1/k}}, \qquad p_- = \bar{p}; \tag{1.6}$$

$$\rho = \rho(z, t; w, s) = \frac{|z|^{2k}|w|^{2k}}{|A|^2}. \tag{1.7}$$

These functions are well defined as long as $|z| + |w| + |t - s| \neq 0$. Note that

$$|p| \leq 1; \qquad p = 1 \text{ if and only if } (z, t) = (w, s) . \tag{1.8}$$

Moreover $\rho = |p|^{2k}$ if $n = 1$.

The following was proved in [2].

Theorem 1.9. *For* $|\,\text{Re}\,\lambda| < 1$ *the function*

$$K_\lambda(z, t; w, s) = c_{n,k} A_+^{-\alpha_+} A_-^{-\alpha_-} F_\lambda \tag{1.10}$$

is a Green's function for Δ_λ*, i.e.*

$$\Delta_\lambda K_\lambda(z, t; w, s) = \delta(z - w)\delta(t - s). \tag{1.11}$$

Here the constant is

$$c_{n,k} = -\frac{\Gamma(n)}{8k\pi^{n+1}} \tag{1.12}$$

and F_λ is an integral transform of an algebraic function of the variables (1.6), (1.7):

$$F_\lambda = \int_0^1 \sigma^{\alpha_+ - 1}(1-\sigma)^{\alpha_- - 1} G_{n,k}\big(\sigma^{1/k}p_+, (1-\sigma)^{1/k}p_-, \sigma(1-\sigma)\rho\big)\, d\sigma. \quad (1.13)$$

The parameters in (1.10) and (1.13) are

$$\alpha_+ = \left(\frac{n+k-1}{k}\right)\left(\frac{1-\lambda}{2}\right), \qquad \alpha_- = \left(\frac{n+k-1}{k}\right)\left(\frac{1+\lambda}{2}\right), \quad (1.14)$$

while the function $G_{n,k}$ is the algebraic function

$$G_{n,k}(p_+, p_-, \rho) = \frac{\varphi_-(\sqrt{1-4\rho})}{\left[\varphi_+(\sqrt{1-4\rho}) - p_+ - p_-\right]^n} \quad (1.15)$$

where

$$\varphi_+(\xi) = \left(\frac{1+\xi}{2}\right)^{1/k} + \left(\frac{1-\xi}{2}\right)^{1/k}, \quad (1.16)$$

$$\varphi_-(\xi) = \frac{1}{\xi}\left[\left(\frac{1+\xi}{2}\right)^{1/k} - \left(\frac{1-\xi}{2}\right)^{1/k}\right].$$

The Green's function K_λ is real-analytic as a function of $(z, \bar{z}, t; w, \bar{w}, s)$ and is holomorphic in λ. It has a meromorphic extension in λ which continues to be a Green's function for Δ_λ except at the poles

$$\lambda_\nu^\pm = \pm\left(1 + \frac{2\nu}{n+k-1}\right), \qquad \nu = 0, 1, 2, \ldots. \quad (1.17)$$

As noted, qualitative results for Δ_λ and similar operators have long been known, at least when λ is purely imaginary. The explicit formula makes possible a complete determination of the singularities (1.17), which are otherwise inaccessible.

In fact, reasons for interest in such formulae go considerably further. For example, the Green's function of the Laplacian is a function of the Euclidean distance. The simplest non-elliptic example is the subelliptic Laplacian on the Heisenberg group, where the Koranyi distance $|z-w|^4 + (t - s - 2\,\mathrm{Im}\, z \cdot \bar{w})^2$ plays the same role. In both cases the exact formula, though not necessary for qualitative conclusions, contains an important geometric invariant. In both cases, also, the simplicity of the formula is intimately tied to the associated transitive group structure. The lack of a transitive group structure associated in this way to operators like Δ_λ means that the analysis is more complicated, but it also means that additional features and structures of interest may – and do – occur.

The operator Δ_λ is invariant with respect to the natural action of the group $U(n) \times \mathbb{R}$ acting on M by $(g, a)(z, t) = (gz, t - a)$, so one looks for a Green's function that is invariant under the associated action on the product manifold $M \times M$,

$$(g, a)(z, t; w, s) = (gz, t - a; gw, s - a), \qquad g \in U(n), \quad a \in \mathbb{R}.$$

A dimension count indicates that it takes $2 \cdot 3 - 2 = 4$ real functions, or two complex functions, to parametrize the orbits if $n = 1$ and $2(2n+1) - (4n-3) = 5$ real functions, or two complex functions and one real function, to parametrize the orbits if $n > 1$. The functions (1.5)-(1.7) provide such a parametrization. Therefore a fundamental solution for Δ_λ that is invariant under the group action has, in principle, a kernel that is a function of these variables. However there is nothing in this argument that distinguishes these particular variables, or that suggests that when the Green's function is expressed in these variables it will have a form that is as simple as the form given in Theorem 1.7. The geometric significance of these particular variables is discussed in [2].

For imaginary λ the hypoellipticity of Δ_λ is a consequence of Hörmander's criterion [5]: brackets of the vector fields X_l of (1.3) span the tangent space to the manifold M at each point. Generic points (z,t), $z \neq 0$ are *step two* points: it is necessary and sufficient to take brackets of two of the X_l to obtain the missing direction $\partial/\partial t$. The points of the manifold $z = 0$ are *step* $2k$ points: one must go to $2k$-fold brackets to obtain the missing direction. Each "step" is a step away from ellipticity and thus in the direction of a stronger singularity of the Green's function. The factorization (1.10) reflects this: the first part $\Phi_\lambda = A_+^{-\alpha_+} A_-^{-\alpha_-}$ accounts precisely for the stronger singularity on the diagonal, at $z = w = 0$. The quotient $K_\lambda/\Phi_\lambda \sim F_\lambda$ must then be regular at $z = w = 0$ but singular otherwise on the diagonal. The variables p, $\bar{p}$, play the role of *uniformizing variables* here; they collapse the diagonal of the step-two manifold

$$(M\backslash\{(0,t)\}) \times (M\backslash\{(0,t)\}) \tag{1.18}$$

to the point $p = \bar{p} = 1$ and thus provide a uniform account of the singularity at the step two points.

As a function of the variables p, $\bar{p}$, and ρ, the numerator F_λ satisfies an equation of hypergeometric type – second order with polynomial coefficients, singular on a submanifold. It is not one of the standard generalized hypergeometric equations, but the fact that it occurs naturally in this problem suggests that there is a rich class of such equations that remain to be understood. The transform (1.13) is a generalization to three variables of an Euler transform. Such transforms intertwine differential operators of hypergeometric type and can permit simple explicit solutions to be found.

Under the assumption that the Green's function has the form (1.10), it can be verified with some pain that the function $G_{n,k}$ satisfies an equation of hypergeomtric type in three variables. Moreover, it must have a singularity of a certain degree at $p = 1$. One derivation of the expression (1.15) for $G_{n,k}$, based on these assumptions, was given in [2]. Because similar equations arise in connection with many other subelliptic operators, it may be useful to have additional ways to derive the form (1.15). Two such ways are given in this paper. In Section 2 we obtain the function ϕ_+ as the solution of a nonlinear Hamiltonian and prove the uniqueness of the form (1.15), subject again to an assumption about its singularity at $u = 1$.

In Section 3 we note that the differential equation satisfied by $G_{n,k}$ separates, and use this fact to solve for $G_{n,k}$.

2. On the uniqueness of $G_{n,k}$.

We set

$$u_\pm = p_\pm^k, \qquad v = \left(\frac{u_+ u_-}{\rho}\right)^{1/k}, \tag{2.1}$$

and

$$D_{u_\pm} = u_\pm \frac{\partial}{\partial u_\pm}, \qquad D_\rho = \rho \frac{\partial}{\partial \rho}. \tag{2.2}$$

When $(z,t) \neq (w,s)$, $\Delta_\lambda(\Phi_\lambda F_\lambda) = 0$ with $\Phi_\lambda = A_+^{-\alpha_+} A_-^{-\alpha_-}$. This yields

$$\begin{aligned}\big[D_{u_+} D_{u_-} &+ v D_\rho (D_\rho + D_{u_+} + D_{u_-} + \beta - 1) \\ &- v\rho (D_\rho + D_{u_+} + \alpha_+)(D_\rho + D_{u_-} + \alpha_-)\big] F_\lambda = 0,\end{aligned} \tag{2.3}$$

where we defined

$$\beta = \frac{n+k-1}{k}, \tag{2.4}$$

see (3.14) of [2]. Next, with a slight abuse of notation, we have

Lemma 2.5. *The function*

$$F(u_+, u_-, \rho) = \int_0^1 \sigma^{\alpha_+} (1-\sigma)^{\alpha_-} G\big(\sigma u_+, (1-\sigma) u_-, \sigma(1-\sigma)\rho\big) \frac{d\sigma}{\sigma(1-\sigma)} \tag{2.6}$$

satisfies equation (2.3), provided that G satisfies the equation

$$\begin{aligned} D_{u_+} D_{u_-} G &+ v D_\rho (D_\rho + D_{u_+} + D_{u_-} + \beta - 1) G \\ &- v\rho (2D_\rho + D_{u_+} + D_{u_-} + \beta)(2D_\rho + D_{u_+} + D_{u_-} + \beta + 1) G = 0. \end{aligned} \tag{2.7}$$

This is Corollary 3.15 of [2]. We look for a solution of (2.7) that is a function of the variables ρ and $y = p_+ + p_-$ only. Acting on such functions the operator that occurs in (2.7) is $v\, T_{n,k}$, where

$$\begin{aligned} T_{n,k} = {} & \frac{\rho^{1/k}}{k^2} \left(\frac{\partial}{\partial y}\right)^2 + D_\rho \left(D_\rho + \frac{1}{k} D_y + \frac{n-1}{k}\right) \\ & - \rho \left(2D_\rho + \frac{1}{k} D_y + \frac{n-1}{k} + 1\right)\left(2D_\rho + \frac{1}{k} D_y + \frac{n-1}{k} + 2\right). \end{aligned} \tag{2.8}$$

We set $x = \rho^{1/k}$ and consider functions of the form

$$G(x,y) = \frac{\Phi(x)}{\big(\Psi(x) - y\big)^n} \tag{2.9}$$

which satisfy

$$T_{n,k} G = 0. \tag{2.10}$$

In the (x, y)-variables we have

$$k^2 T_{n,k} = x\left(\frac{\partial}{\partial y}\right)^2 + D_x(D_x + D_y + n - 1) \qquad (2.11)$$
$$- x^k(2D_x + D_y + n - 1 + k)(2D_x + D_y + n - 1 + 2k).$$

We want to show that $G = G_{n,k}$, that is $\Phi(x) = \varphi_-(\sqrt{1-4x^k})$, $\Psi(x) = \varphi_+(\sqrt{1-4x^k})$.

Theorem 2.12. *If $G(x,y)$ of (2.9) satisfies (2.10) with $\Phi(0) = \Psi(0) = 1$, then*

$$\Psi(x) = \left(\frac{1+\sqrt{1-4x^k}}{2}\right)^{1/k} + \left(\frac{1-\sqrt{1-4x^k}}{2}\right)^{1/k}, \qquad (2.13)$$

$$\Phi(x) = \frac{1}{\sqrt{1-4x^k}}\left[\left(\frac{1+\sqrt{1-4x^k}}{2}\right)^{1/k} - \left(\frac{1-\sqrt{1-4x^k}}{2}\right)^{1/k}\right]. \qquad (2.14)$$

Proof. We rewrite (2.11):

$$\frac{k^2}{x} T_{n,k} = x(1-4x^k)\frac{\partial^2}{\partial x^2} \qquad (2.15)$$
$$+ \left((1-4x^k)D_y - (4n+6k)x^k + n\right)\frac{\partial}{\partial x}$$
$$+ \frac{\partial^2}{\partial y^2} - x^{k-1}(D_y + n - 1 + k)(D_y + n - 1 + 2k).$$

The corresponding derivatives of G may be arranged in increasing powers of $\Psi - y$:

$$\frac{\partial G}{\partial x} = -\frac{n\Phi\Psi_x}{(\Psi-y)^{n+1}} + \frac{\Phi_x}{(\Psi-y)^n},$$

$$\frac{\partial^2 G}{\partial x^2} = \frac{n(n+1)\Phi\Psi_x^2}{(\Psi-y)^{n+2}} - \frac{n(2\Phi_x\Psi_x + \Phi\Psi_{xx})}{(\Psi-y)^{n+1}} + \frac{\Phi_{xx}}{(\Psi-y)^n},$$

$$\frac{\partial G}{\partial y} = \frac{n\Phi}{(\Psi-y)^{n+1}}, \qquad \frac{\partial^2 G}{\partial y^2} = \frac{n(n+1)\Phi}{(\Psi-y)^{n+2}},$$

$$D_y G = \frac{ny\Phi}{(\Psi-y)^{n+1}} = \frac{n\Phi\Psi}{(\Psi-y)^{n+1}} - \frac{n\Phi}{(\Psi-y)^n},$$

$$D_y^2 G = \frac{n(n+1)\Phi\Psi^2}{(\Psi-y)^{n+2}} - \frac{n(2n+1)\Phi\Psi}{(\Psi-y)^{n+1}} + \frac{n^2\Phi}{(\Psi-y)^n},$$

$$D_y\frac{\partial G}{\partial x} = -\frac{n(n+1)\Phi\Psi\Psi_x}{(\Psi-y)^{n+2}} + \frac{n^2\Phi\Psi_x + n(\Phi_x\Psi + \Phi\Psi_x)}{(\Psi-y)^{n+1}} - \frac{n\Phi_x}{(\Psi-y)^n},$$

where the subscript x denotes derivatives with respect to x. Substituting these derivatives into (2.15) yields

$$\begin{aligned}\frac{k^2}{x}T_{n,k}G = &\left[x(1-4x^k)\Psi_x^2 - (1-4x^k)\Psi\Psi_x - x^{k-1}\Psi^2 + 1\right]\frac{n(n+1)\Phi}{(\Psi-y)^{n+2}} \\ &+ \left[(1-4x^k)(\Psi - 2x\Psi_x)\Phi_x + \left(-x(1-4x^k)\Psi_{xx} + (1-4x^k+6kx^k)\Psi_x\right.\right. \\ &\left.\left. + 3(1-k)x^{k-1}\Psi\right)\Phi\right]\frac{n}{(\Psi-y)^{n+1}} \\ &+ \left[x(1-4x^k)\Phi_{xx} - 6kx^k\Phi_x - (1-k)(1-2k)x^{k-1}\Phi\right]\frac{1}{(\Psi-y)^n}.\end{aligned} \tag{2.16}$$

Now (2.16) can vanish if and only if the coefficient of every power of $\Psi - y$ vanishes. This implies three equations for Φ and Ψ:

$$x(1-4x^k)\Psi_x^2 - (1-4x^k)\Psi\Psi_x - x^{k-1}\Psi^2 + 1 = 0, \tag{2.17}$$

$$\begin{aligned}&(1-4x^k)(\Psi - 2x\Psi_x)\Phi_x \\ &+ \left[-x(1-4x^k)\Psi_{xx} + (1-4x^k+6kx^k)\Psi_x + 3(1-k)x^{k-1}\Psi\right]\Phi = 0,\end{aligned} \tag{2.18}$$

$$x(1-4x^k)\Phi_{xx} - 6kx^k\Phi_x - (1-k)(1-2k)x^{k-1}\Phi = 0. \tag{2.19}$$

(i) Solution of (2.17) for Ψ. We set

$$\xi = \sqrt{1-4x^k}, \qquad x = \left(\frac{1-\xi^2}{4}\right)^{1/k}, \qquad \frac{d}{dx} = -\frac{2kx^{k-1}}{\sqrt{1-4x^k}}\frac{d}{d\xi}, \tag{2.20}$$

and rewrite (2.17):

$$4k^2\left(\frac{1-\xi^2}{4}\right)^{2-1/k}\Psi_\xi^2 + 2k\left(\frac{1-\xi^2}{4}\right)^{1-1/k}\Psi_\xi\xi\Psi - \left(\frac{1-\xi^2}{4}\right)^{1-1/k}\Psi^2 + 1 = 0.$$

Completing the square yields

$$\left(2k\frac{1-\xi^2}{4}\frac{\Psi_\xi}{\left(\frac{1-\xi^2}{4}\right)^{1/2k}} + \frac{\xi}{2}\frac{\Psi}{\left(\frac{1-\xi^2}{4}\right)^{1/2k}}\right)^2 = \frac{1}{4}\frac{\Psi^2}{\left(\frac{1-\xi^2}{4}\right)^{1/k}} - 1 = \frac{1}{4}(\tilde{\Psi})^2 - 1,$$

or

$$4k^2\left(\frac{1-\xi^2}{4}\right)^2\left(\frac{d\tilde{\Psi}}{d\xi}\right)^2 = \frac{1}{4}\tilde{\Psi}^2 - 1. \tag{2.21}$$

We note that

$$s = \log\left(\frac{1+\xi}{1-\xi}\right)^{1/k} \Rightarrow \left(\frac{d\tilde{\Psi}}{ds}\right)^2 = \frac{1}{4}\tilde{\Psi}^2 - 1,$$

whose general solution is

$$\tilde{\Psi} = 2\cosh\left(\frac{s}{2} + \tilde{C}\right) = Ce^{s/2} + \frac{1}{C}e^{-s/2}.$$

This yields the general solution of (2.17):

$$\Psi = C\left(\frac{1+\xi}{2}\right)^{1/k} + \frac{1}{C}\left(\frac{1-\xi}{2}\right)^{1/k}, \tag{2.22}$$

where C is an arbitrary constant. Now $\Psi(0) = 1$ implies $C = 1$, and we have derived (2.13).

(ii) Solution of (2.18) for Φ. Again we introduce the variable ξ, and (2.18) takes the form

$$\begin{aligned} &-2k\big[k(1-\xi^2)\Psi_\xi + \xi\Psi\big]\Phi_\xi \\ &+\left[-k^2(1-\xi^2)\Psi_{\xi\xi} + \left(2k^2\xi - 4k\xi - 2k^2\frac{1-\xi^2}{\xi}\right)\Psi_\xi + 3(1-k)\Psi\right]\Phi = 0. \end{aligned} \tag{2.23}$$

Now

$$2k\Psi_\xi = \left(\frac{1+\xi}{2}\right)^{1/k-1} - \left(\frac{1-\xi}{2}\right)^{1/k-1}, \tag{2.24}$$

therefore

$$k(1-\xi^2)\Psi_\xi + \xi\Psi = \left(\frac{1+\xi}{2}\right)^{1/k} - \left(\frac{1-\xi}{2}\right)^{1/k} = \xi\varphi_-(\xi). \tag{2.25}$$

Also

$$\xi\left(\frac{1+\xi}{2}\right)^{1/k-1} - \xi\left(\frac{1-\xi}{2}\right)^{1/k-1} = 2\Psi - 2k\big(\xi\varphi_-(\xi)\big)_\xi \tag{2.26}$$

which yields

$$k^2(1-\xi^2)\Psi_{\xi\xi} = (1-k)\big(2k(\xi\varphi_-(\xi))_\xi - \Psi\big). \tag{2.27}$$

Hence the coefficient of Φ in (2.23) is

$$\begin{aligned} &-2k(1-k)\big(\xi\varphi_-(\xi)\big)_\xi + 4(1-k)\Psi \\ &+(k-2)\xi\left(\left(\frac{1+\xi}{2}\right)^{1/k-1} - \left(\frac{1-\xi}{2}\right)^{1/k-1}\right) - \frac{2k}{\xi}\big(\xi\varphi_-(\xi) - \xi\Psi\big) \\ &= 2k\xi(\varphi_-)_\xi(\xi), \end{aligned}$$

and (2.23) is reduced to

$$-2k\xi\varphi_-(\xi)\frac{d\Phi}{d\xi} + 2k\xi\frac{d\varphi_-}{d\xi}\Phi = 0. \tag{2.28}$$

This yields

$$\Phi = \varphi_-(\xi), \tag{2.29}$$

since $\Phi(0) = 1$. Finally a simple calculation shows that $\Phi = \varphi_-(\xi)$ satisfies (2.19). This completes the proof of Theorem 2.12. □

The normalization $\varphi_-(1) = 1$ may be absorbed in the overall constant $c_{n,k}$. On the other hand the normalization $\varphi_+(1) = 1$ is essential for the correct singularity of $G_{n,k}$ at $p_\pm = 1$, $\rho = 1$. To see this we set

$$\varphi_\ell(\xi) = \ell\left(\frac{1+\xi}{2}\right)^{1/k} + \frac{1}{\ell}\left(\frac{1-\xi}{2}\right)^{1/k}, \tag{2.30}$$

so $\varphi_+ = \varphi_1$. We note that $G_{n,k}$ is singular exactly when its denominator

$$\varphi_+\big(\sqrt{1-4\sigma(1-\sigma)\rho}\big) - \sigma^{1/k}p_+ - (1-\sigma)^{1/k}p_- \tag{2.31}$$

vanishes; (2.31) vanishes identically in $\sigma, \sigma \in [0,1]$, when $z = t$, $w = s$.

Proposition 2.32. *Let* $\varphi_\ell(\sqrt{1-4x^k})$ *denote the general solution of (2.17). Then the function of* σ

$$\varphi_\ell\big(\sqrt{1-4\sigma(1-\sigma)\rho}\big) - \sigma^{1/k}p_+ - (1-\sigma)^{1/k}p_-\,, \qquad \sigma \in [0,1], \tag{2.33}$$

has a zero exactly when $(z,t) = (w,s)$, *if and only if* $\ell = 1$.

Proof. We shall show that for $0 < \ell < 1$ (2.33) has unwanted zeros, and for $\ell > 1$ (2.33) does not vanish. $\varphi_\ell\big(\sqrt{1-4\sigma(1-\sigma)\rho}\big) \geq 0$, so the vanishing of (2.33) implies that p is real, $0 \leq p_\pm \leq 1$, or that $\sigma = \frac{1}{2}$. The following variables turn out to be convenient to work with:

$$\rho_\pm = \frac{u_\pm}{v^{k/2}}, \quad \rho = \rho_+\rho_-, \quad p_\pm = \sqrt{v}\,\rho_\pm^{1/k}. \tag{2.34}$$

(i) $\sigma = 1/2$, $\operatorname{Im} p_+ \neq 0$. Here

$$\frac{1}{2^{1/k}}p_+ + \frac{1}{2^{1/k}}p_- = 2\sqrt{v}\operatorname{Re}\left(\frac{\rho_+}{2}\right)^{1/k} < 2\left(\frac{|\rho_+|}{2}\right)^{1/k}. \tag{2.35}$$

Also

$$\begin{aligned} &2^{1/k}\Big[\varphi_\ell\big(\sqrt{1-4\sigma(1-\sigma)\rho}\big) - \sigma^{1/k}p_+ - (1-\sigma)^{1/k}p_-\Big]_{\sigma=\frac{1}{2}} \qquad (2.36)\\ &> \ell\big(1+\sqrt{1-|\rho_+|^2}\big)^{1/k} + \frac{1}{\ell}\big(1-\sqrt{1-|\rho_+|^2}\big)^{1/k} - 2|\rho_+|^{1/k} \geq 0, \end{aligned}$$

since $a\ell + b\ell^{-1}$ attains its minimum $2\sqrt{ab}$ on $(0,\infty)$ at $\ell = (b/a)^{1/2}$. Therefore (2.33) cannot vanish if p_+ is not real.

(ii) $0 \leq p_\pm \leq 1$, $\ell < 1$. $p_\pm = \ell$ is an unwanted zero of (2.33) at $\sigma = 1$.

(iii) $0 \leq p_\pm \leq 1$, $\ell > 1$. φ_ℓ is a strictly increasing function of ℓ for $\ell \geq 1$. Therefore $\ell > 1 \Rightarrow \varphi_\ell > \varphi_1$, and the nonvanishing of (2.33) follows from

$$\varphi_1\big(\sqrt{1-\sigma(1-\sigma)\rho}\big) - \sigma^{1/k}p_+ - (1-\sigma)^{1/k}p_- \geq 0, \tag{2.37}$$

which is a consequence of (4.3) of [2]. This proves Proposition 2.32.

3. Separating Variables for $G_{n,k}$.

We cannot find variables which separate (1.1). On the other hand $G_{n,k}$ satisfies a separable partial differential equation. We start with (2.3) and introduce the variables $\rho_\pm = u_\pm v^{-k/2}$ and v,

$$\rho_+ = \left(\frac{u_+}{u_-}\rho\right)^{1/2}, \qquad \rho_- = \left(\frac{u_-}{u_+}\rho\right)^{1/2}, \qquad v = \left(\frac{u_+u_-}{\rho}\right)^{1/2k}. \tag{3.1}$$

Note that $\rho = \rho_+\rho_-$. Then

$$D_{u_\pm} = \frac{1}{k}D_v \pm \frac{1}{2}(D_{\rho_+} - D_{\rho_-}), \qquad D_\rho = -\frac{1}{k}D_v + \frac{1}{2}(D_{\rho_+} + D_{\rho_-}),$$

and (2.3) takes the form

$$\begin{aligned}\Big[&\left(\frac{1}{k}D_v + \frac{1}{2}(D_{\rho_+} - D_{\rho_-})\right)\left(\frac{1}{k}D_v - \frac{1}{2}(D_{\rho_+} - D_{\rho_-})\right) \\ &-v\left(\frac{1}{k}D_v - \frac{1}{2}(D_{\rho_+} + D_{\rho_-})\right)\left(\frac{1}{k}D_v + \frac{1}{2}(D_{\rho_+} + D_{\rho_-}) + \frac{n-1}{k}\right) \\ &-v\rho_+\rho_-(D_{\rho_+} + \alpha_+)(D_{\rho_-} + \alpha_-)\Big]F_\lambda = 0.\end{aligned} \tag{3.2}$$

The 3-dimensional Euler transform (1.13) that sends $G_{n,k}$ to F_λ was taken with respect to the variables $u_\pm$, ρ, see (2.25) of [2]. In the $\rho_\pm$, v variables this transform reduces, up to a factor, to a 2-dimensional Euler transform in the $\rho_\pm$ variables, with v as a parameter. Then (3.2) induces an equation for $G_{n,k}$ which contains only the sum and the difference of D_{ρ_+} and D_{ρ_-}. The following is a simple consequence of (2.35) of [2]:

Lemma 3.3. *The function*

$$F_\lambda(\rho_+,\rho_-,v) = \int_0^1 \sigma^{\alpha_+}(1-\sigma)^{\alpha_-}G(\sigma\rho_+,(1-\sigma)\rho_-,v)\frac{d\sigma}{\sigma(1-\sigma)} \tag{3.4}$$

satisfies equation (3.2) provided G satisfies the equation

$$\begin{aligned}\Big[&\left(\frac{1}{k}D_v + \frac{1}{2}(D_{\rho_+} - D_{\rho_-})\right)\left(\frac{1}{k}D_v - \frac{1}{2}(D_{\rho_+} - D_{\rho_-})\right) \\ &-v\left(\frac{1}{k}D_v - \frac{1}{2}(D_{\rho_+} + D_{\rho_-})\right)\left(\frac{1}{k}D_v + \frac{1}{2}(D_{\rho_+} + D_{\rho_-}) + \frac{n-1}{k}\right) \\ &-v\rho_+\rho_-\left(D_{\rho_+} + D_{\rho_-} + \frac{n-1}{k} + 2\right)\left(D_{\rho_+} + D_{\rho_-} + \frac{n-1}{k} + 1\right)\Big]G = 0.\end{aligned} \tag{3.5}$$

It is convenient to introduce the coordinates $\rho = \rho_+\rho_-$, $y = \rho_+/\rho_-$ and v. Then (3.5) takes the following form:

$$\begin{aligned} &- D_y^2 G \\ &+ \left\{ \left[\frac{1}{k^2} D_v^2 - \frac{1}{k^2} v D_v^2 - \frac{n-1}{k^2} v D_v \right] \right. \\ &\left. + v \left[D_\rho^2 + \frac{n-1}{k} D_\rho - \rho \left(2D_\rho + \frac{n-1}{k} + 2 \right) \left(2D_\rho + \frac{n-1}{k} + 1 \right) \right] \right\} G = 0. \end{aligned} \tag{3.6}$$

We write $G(v, y, \rho) = Q(\rho)Y(y)V(v)$ and divide (3.6) by QYV. This separates off $D_y^2 Y/Y = \mu^2$, with eigenvalue μ^2. When divided by v the operator in curly brackets separates, leading to two ordinary differential equations with eigenvalue ν^2. The resulting decoupled system is

$$D_y^2 Y = \mu^2 Y, \tag{3.7}$$

$$(1 - v) D_v^2 V - (n-1) v D_v V + (k^2 \nu^2 v - k^2 \mu^2) V = 0, \tag{3.8}$$

$$\begin{aligned} (1 - 4\rho) D_\rho^2 Q &+ \left[\frac{n-1}{k} - 2 \left(3 + \frac{2(n-1)}{k} \right) \rho \right] D_\rho Q \\ &- \left[\nu^2 + \left(\frac{n-1}{k} + 1 \right) \left(\frac{n-1}{k} + 2 \right) \rho \right] Q = 0. \end{aligned} \tag{3.9}$$

(3.7) yields

$$Y(y) = y^{\pm\mu}. \tag{3.10}$$

To solve (3.8) we set

$$V(v) = v^{k|\mu|} \tilde{V}(v). \tag{3.11}$$

Then $\tilde{V}$ satisfies

$$v(1 - v) \frac{d^2 \tilde{V}}{dv^2} + [c - (a_+ + a_- + 1)v] \frac{d\tilde{V}}{dv} - a_+ a_- \tilde{V} = 0, \tag{3.12}$$

with

$$a_\pm = k|\mu| + \frac{n-1}{2} \pm \sqrt{\left(\frac{n-1}{2} \right)^2 + k^2 \nu^2} = k|\mu| + \frac{n-1}{2} \pm k\gamma, \tag{3.13}$$

$$c = 2k|\mu| + 1 \ \Rightarrow \ a_+ + a_- = c + n - 2. \tag{3.14}$$

Consequently the only solutions of the hypergeometric equation (3.12) which are bounded on $[0, 1]$ are the polynomial solutions; see [6], pp. 44–45. They are

$$\tilde{V}(v) = P_\ell^{(n-2, 2k|\mu|)}(2v - 1), \tag{3.15}$$

where

$$\ell(\ell + n - 2 + 2k|\mu| + 1) = -a_+ a_- \ \Rightarrow \ \ell + k|\mu| + \frac{n-1}{2} = k\gamma, \tag{3.16}$$

$\ell = 0, 1, 2, \dots$ and

$$P_\ell^{(a,b)}(2v-1) = \frac{\Gamma(a+\ell+1)}{\ell!\Gamma(a+b+\ell+1)} \sum_{r=0}^{\ell} \binom{\ell}{r} \frac{\Gamma(a+b+\ell+r+1)}{\Gamma(a+r+1)} (v-1)^r \quad (3.17)$$

denote the Jacobi polynomials, see [6], p. 211. As for (3.9), we set

$$Q(\rho) \;=\; \rho^\alpha \tilde{Q}(\rho), \quad (3.18)$$

where

$$\alpha = -\frac{n-1}{2k} + \gamma \;=\; |\mu| + \frac{\ell}{k} \quad (3.19)$$

is one of the 2 roots of $\alpha^2 + \frac{(n-1)}{k}\alpha - \nu^2 = 0$. With $s = 4\rho$, $\tilde{Q}$ satisfies

$$s(1-s)\,\frac{d^2\tilde{Q}}{ds^2} + \left[2\gamma + 1 - \left(2\gamma + \frac{5}{2}\right)s\right]\frac{d\tilde{Q}}{ds} - \left(\gamma + \frac{1}{2}\right)(\gamma+1)\tilde{Q} = 0.$$

We normalize $\tilde{Q}$ by $\tilde{Q}(0) = 1$. Then

$$\begin{aligned} Q(\rho) = Q_{|\mu|,\ell}(\rho) = \rho^\alpha \tilde{Q}_{|\mu|,\ell}(\rho) &= \rho^\alpha F\left(\gamma + \frac{1}{2}, \gamma+1; 2\gamma+1; 4\rho\right) \quad (3.20)\\ &= \rho^\alpha \frac{1}{\sqrt{1-4\rho}}\left(\frac{2}{1+\sqrt{1-4\rho}}\right)^{2\gamma}, \end{aligned}$$

see [3], p. 101,(6). Summarizing, we look for a G in the form

$$G = \sum_{\mu,\ell} c_{\mu,\ell}\, y^\mu v^{k|\mu|} P_\ell^{(n-2,2k|\mu|)}(2v-1)\rho^{|\mu|+\ell/k}\tilde{Q}_{|\mu|,\ell}(\rho). \quad (3.21)$$

To find the coefficient $c_{\mu,\ell}$ we assume that G becomes $G_{n,1}$ at $\rho = 0$, the step two kernel.

Ansatz 3.22. *$G(p_+, p_-, \rho)$ satisfies*

$$G(p_+, p_-, 0) = \frac{1}{(1 - p_+ - p_-)^n}. \quad (3.23)$$

This is true when $n = 1$, see (1.26) of [2]. It can also be thought of as the quantitative expression of the heuristic notion that once the principal singularity $\Phi_\lambda = A_+^{-\alpha_+} A_-^{-\alpha_-}$, which is needed at the step $2k$ points i.e. on the t-axis, has been factored out, the remaining factor F_λ, the Euler transform of $G_{n,k}$, is needed only for the generic step 2 points. We introduce the variables p_+, p_-, ρ into (3.21) and set $\rho = 0$; (3.17) easily yields

$$G(p_+, p_-, 0) = \sum_{\ell=0}^{\infty} (p_+ p_-)^{\ell} \tag{3.24}$$
$$\cdot \left\{ \sum_{\mu \geq 0} c_{\mu,\ell} \frac{\Gamma(n-2+2k|\mu|+2\ell+1)}{\ell!\Gamma(n-2+2k|\mu|+\ell+1)} p_+^{2k|\mu|} + \sum_{\mu<0} c_{\mu,\ell} \frac{\Gamma(n-2+2k|\mu|+2\ell+1)}{\ell!\Gamma(n-2+2k|\mu|+\ell+1)} p_-^{2k|\mu|} \right\}.$$

Comparing this with a similar expansion of (3.23), we see that we must choose

$$\mu = \frac{m}{2k}, \qquad m = 0, \pm 1, \pm 2, \dots . \tag{3.25}$$

Then (3.23)–(3.25) yield

$$\Gamma(n) c_{m,\ell} = (n-2+|m|+2\ell+1)\,(|m|+\ell+1)_{n-2}, \tag{3.26}$$

where we denoted $c_{\mu,\ell}$ by $c_{m,\ell}$ and $(a)_n = a(a+1)\cdots(a+n-1)$.

Theorem 3.27. *Assuming (3.23), the expansion (3.21) yields* $G = G_{n,k}$*:*

$$G_{n,k} = \sum_{m=-\infty}^{\infty} \sum_{\ell=0}^{\infty} c_{m,\ell} v^{\frac{|m|}{2}} P_{\ell}^{(n-2,|m|)}(2v-1) \rho^{\frac{|m|}{2k}+\frac{\ell}{k}} \tilde{Q}_{|m|,\ell}(\rho) y^{\frac{m}{2k}} \tag{3.28}$$
$$= \frac{\varphi_-(\sqrt{1-4\rho})}{\left[\varphi_+(\sqrt{1-4\rho}) - p_+ - p_-\right]^n},$$

where $c_{m,\ell}$*,* $\tilde{Q}_{|m|,\ell}$ *and* $\varphi_{\pm}$ *are given by (3.26), (3.20) and (1.15), (1.16), respectively.*

Proof. We write (3.21) as follows,

$$\Gamma(n) G = \frac{\rho^{-\frac{n-2}{2k}}}{\sqrt{1-4\rho}} \left(\frac{2}{1+\sqrt{1-4\rho}}\right)^{1/k} \tilde{G}, \tag{3.29}$$

where

$$\tilde{G} = \frac{d}{dW} \sum_{\ell,m} (|m|+\ell+1)_{n-2} v^{\frac{|m|}{2}} y^{\frac{m}{2k}} P_{\ell}^{(n-2,|m|)}(2v-1) W^{n-2+|m|+2\ell+1}, \tag{3.30}$$

$$W = \left(\frac{\sqrt{4\rho}}{1+\sqrt{1-4\rho}}\right)^{1/k} = \left(\frac{1-\sqrt{1-4\rho}}{1+\sqrt{1-4\rho}}\right)^{1/2k}. \tag{3.31}$$

Using [6], p. 212 (3) and p. 41, we write

$$\begin{aligned}
&(|m|+\ell+1)_{n-2}P_\ell^{(n-2,|m|)}(2v-1)\\
&=(-1)^\ell\binom{\ell+|m|}{\ell}(|m|+\ell+1)_{n-2}F(n-2+|m|+\ell+1,-\ell;|m|+1;v)\\
&=(-1)^\ell\binom{\ell+|m|}{\ell}v^{1-(|m|+\ell+1)}\left(\frac{d}{dv}\right)^{n-2}\left[v^{n-2+|m|+\ell}F(|m|+\ell+1,-\ell;|m|+1;v)\right]\\
&=v^{-|m|-\ell}\left(\frac{d}{dv}\right)^{n-2}\left[v^{n-2+|m|+\ell}P_\ell^{(0,|m|)}(2v-1)\right].
\end{aligned}$$

Hence

$$\tilde{G}=\frac{d}{dW}W^{n-1}\left[\sum_{m=-\infty}^{\infty}u^{-\frac{|m|}{2}}y^{\frac{m}{2k}}W^{|m|}\cdot\left(\frac{d}{dv}\right)^{n-2}\left\{v^{|m|+n-2}\sum_{\ell=0}^{\infty}P_\ell^{(0,|m|)}(2v-1)Z^{2\ell}\right\}\right]_{u=v} \tag{3.32}$$

with

$$Z=\left(\frac{v}{u}\right)^{1/2}W. \tag{3.33}$$

Set

$$R=(1+Z^2)\sqrt{1-4vA^2},\qquad A=\frac{Z}{1+Z^2}. \tag{3.34}$$

Then the generating function of the Jacobi polynomials, [6], p. 213, yields

$$\sum_{\ell=0}^{\infty}P_\ell^{(0,|m|)}(2v-1)Z^{2\ell}=\frac{2^{|m|}}{R}(1+Z^2)^{-|m|}(1+\sqrt{1-4vA^2})^{-|m|}.$$

We substitute this into (3.32):

$$\begin{aligned}
\tilde{G}&=\frac{d}{dW}W^{n-1}\left[\left(\frac{d}{dv}\right)^{n-2}\frac{v^{n-2}}{R}\sum_{m=-\infty}^{\infty}\left(\frac{\sqrt{4vA^2}}{1+\sqrt{1-4vA^2}}\right)^{|m|}y^{\frac{m}{2k}}\right]_{u=v}\\
&=\frac{d}{dW}W^{n-1}\left[\left(\frac{d}{dv}\right)^{n-2}v^{n-2}\frac{1}{1+Z^2}\,\frac{1}{1-Av^{1/2}\tilde{y}}\right]_{u=v}\\
&=\frac{d}{dW}W^{n-1}\left[u\left(\frac{d}{d\gamma}\right)^{n-2}\frac{\gamma^{n-2}}{u-\left(\frac{u^{1/2}\tilde{y}}{W}-1\right)\gamma}\right]_{u=v},
\end{aligned}$$

where we set $\gamma=vW^2$ and $\tilde{y}=y^{1/2k}+y^{-1/2k}$. With

$$\omega=\left(\frac{u^{1/2}\tilde{y}}{W}-1\right)\gamma$$

one has

$$\left(\frac{d}{d\gamma}\right)^{n-2} \frac{\gamma^{n-2}}{u - \left(\frac{u^{1/2}\tilde{y}}{W} - 1\right)\gamma} = \left(\frac{d}{d\omega}\right)^{n-2} \frac{\omega^{n-2}}{u-\omega}$$

$$= \left(\frac{d}{d\omega}\right)^{n-2} \left(\frac{\omega^{n-2} - u^{n-2}}{u-\omega} + \frac{u^{n-2}}{u-\omega}\right)$$

$$= \left(\frac{d}{d\omega}\right)^{n-2} \left(\frac{u^{n-2}}{u-\omega} - u^{n-3} - u^{n-4}\omega - \cdots - \omega^{n-3}\right)$$

$$= \frac{\Gamma(n-1)u^{n-2}}{(u-\omega)^{n-1}}.$$

Thus we obtain

$$\tilde{G} = \frac{d}{dW} W^{n-1} v^{n-1} \frac{\Gamma(n-1)}{\left[v - \gamma\left(\frac{v^{1/2}\tilde{y}}{W} - 1\right)\right]^{n-1}} = \frac{\Gamma(n)\, W^{k-2}(1-W^2)}{(1+W^2 - v^{1/2}W\tilde{y})^n}.$$

Finally the identity

$$1 \pm W^2 = \left(\frac{2}{1+\sqrt{1-4\rho}}\right)^{1/k} \left(\left[\frac{1+\sqrt{1-4\rho}}{2}\right]^{1/k} \pm \left[\frac{1-\sqrt{1-4\rho}}{2}\right]^{1/k}\right)$$

and an elementary calculation yield (3.28). This proves Theorem 3.27. □

References

[1] R. Beals, B. Gaveau and P. C. Greiner, *On a geometric formula for the fundamental solution of subelliptic Laplacians,* Math. Nachr. **181** (1996), 81–163.

[2] R. Beals, B. Gaveau and P.C. Greiner, *Uniform hypoelliptic Green's functions,* J. Maths. Pures Appl., **77** (1998), 209–248.

[3] A. Erdélyi et al., *Higher Transcendental Functions, vol. 1,* McGraw-Hill, New York, 1955.

[4] P. C. Greiner, *A fundamental solution for a non-elliptic partial differential operator,* Can. J. Math. **31** (1979), 1107–1120.

[5] L. Hörmander, *Hypoelliptic second order differential equations,* Acta Math. **119** (1967), 147-171.

[6] W. Magnus, F. Oberhettinger and R. P. Soni, *Formulas and Theorems for the Special Functions of Mathematical Physics,* Springer-Verlag New York Inc. 1966.

Richard Beals,
Department of Mathematics,
Yale University,
P.O. Box 208283,
New Haven, Connecticut
06520-8283, USA
E-mail: beals@math.yale.edu

Bernard Gaveau,
Laboratoire Equations aux dérivées partielles et physique mathématique,
Université Pierre et Marie Curie (P.VI),
Boîte courrier 172, Tour 46,
5ème étage, 4 Place Jussieu, 75230 Paris Cedex 05, France
E-mail: gaveau@ccv.jussieu.fr

Peter Greiner,
Department of Mathematics,
University of Toronto,
Toronto, Ontario M5S 3G3,
Canada
E-mail: greiner@math.toronto.edu

Operator Theory:
Advances and Applications, Vol. 114

Clifford analysis on Poincaré space

Jan Cnops

Abstract. In this paper an introduction and an overview will be given of the function theory regarding monogenic functions in Poincaré space.

Research in this area has been started by H. Leutwiler, based on the study of harmonic functions in conformally flat domains (see [1]). Here we take the approach of considering differential forms on the Poincaré manifold, and defining the Hodge-Dirac operator for these. Comparisons are made between the several models of the Poincaré manifold.

1. Introduction

In classical Clifford analysis the solutions of the Dirac operator on Euclidean space are studied. Since this space is flat, no distinction has to be made between the Clifford and the spinor bundles. In the case of general manifolds with a metric, derivation ∂ is replaced by a connection X. Two important connections here are the connection $\nabla\!\!\!/$ acting on spinor bundles, and ∇ acting on vector fields, which can be generalised to the Clifford bundle.

$\nabla\!\!\!/$ leads to the spinor Dirac operator (also called the Atiyah-Singer operator), and ∇ to the Hodge Dirac operator. The behaviour of the first operator regarding conformal transformations is well understood (see e.g. [12]) and its theory on conformally flat manifolds is straightforward. In general the latter operator is more difficult to cope with, but research started by H. Leutwiler leads to a function theory (which is as yet quite incomplete) for the Hodge operator of the Poincaré metric (also called hyperbolic space; it should be pointed out however that the metric is positive definite). One of the intriguing aspects of the theory is that it lives on a homogeneous, boundaryless, manifold, but that there is some kind of 'frontier' (which is infinitely far away from any point of the manifold), and that this frontier plays an important rôle in the theory, making part of it a study in analysis on inhomogeneous manifolds. The results discussed in this paper are spread widely over the literature. As most important sources we refer to [5] and [6] for the Dirac operator on flat space; [11] gives an account on the different models of Poincaré space we use here, including the Laplacian Δ_M for scalar functions. The use of Clifford matrices for conformal mappings on spaces with positive definite metric

1991 *Mathematics Subject Classification.* 53C35, 30G35.
Key words and phrases. hyperbolic space, Poincaré metric, monogenic differential forms.

goes back to Vahlen; for the non-definite case we refer to [2]. The main sources for the function theory are [7]–[10].

2. Classical Clifford analysis

Let V be a finite dimensional real vector space with a non-degenerate inner product $\mathcal{B}$. Over V the Clifford algebra $\mathcal{C}\ell(V)$ is constructed as the free algebra over V modulo the relation

$$\vec{v}^2 = -\mathcal{B}(\vec{v}, \vec{v})$$

for elements $\vec{v}$ of V. There are two special cases of interest for this paper: Minkowski space $\mathbb{R}^{1,n}$ and Euclidean space $\mathbb{R}^{0,n}$.

The space $\mathbb{R}^{1,n}$ is the $n+1$-dimensional vector space over $\mathbb{R}$ carrying the inner product

$$\mathcal{B}(\vec{X}, \vec{Y}) = -x_0 y_0 + \sum_{i=1}^{n} x_i y_i,$$

where $\vec{X} = (x_0; x_1, \ldots, x_n)$ and $\vec{Y} = (y_0; y_1, \ldots, y_n)$, Taking the canonical basis $e_0, e_1, \ldots, e_n$ the defining relation results in the equivalent relations

$$e_0^2 = 1 \qquad e_i^2 = -1 \quad (i = 1, \ldots n) \qquad e_i e_j = -e_j e_i \ (i \neq j).$$

More specifically, let $A = \{k_1, \ldots, k_i\} \subset \{0, \ldots, n\}$ be given so that $k_1 < \ldots < k_i$. We use the short notation e_A for the product $1e_{k_1} \ldots e_{k_i}$ (the factor 1 is added so the definition makes sense when A is empty). Then $\{e_A : A \subset \{0, \ldots, n\}\}$ is a basis of $\mathcal{C}\ell(1,n)$ as a vector space. The space $\mathbb{R}^{0,n}$ is the n-dimensional subspace of $\mathbb{R}^{1,n}$ spanned by $e_1, \ldots, e_n$. For conciseness, the element $\vec{x} = (0; x_1, \ldots, x_n)$ will be identified with $(x_1, \ldots, x_n)$. A basis of $\mathcal{C}\ell(n)$ is given by $\{e_A : A \subset \{1, \ldots, n\}\}$.

If A has k elements, then e_A is called a *k-vector*. Likewise any linear combination of k-vectors is called a k-vector, and the vector space of k-vectors is written as $\mathcal{C}\ell(V)^k$. Each Clifford number can be split into k-vector parts, for k going from 0 to n:

$$a = [a]_0 + [a]_1 + \ldots + [a]_n.$$

Not every element of a Clifford algebra is invertible, so division is not always possible. However, if division is possible we shall sometimes write $\frac{a}{b}$ or a/b for ab^{-1} for aesthetic reasons.

The following (anti)automorphisms will be used (here $\vec{x}$ is an arbitrary vector, and a and b are arbitrary Clifford numbers):

(i) The main antiautomorphism, defined by

$$\bar{\vec{x}} = -\vec{x} \qquad \overline{(ab)} = \bar{b}\bar{a}.$$

(ii) Reversion, defined by

$$\vec{x}^* = \vec{x} \qquad (ab)^* = b^* a^*.$$

(iii) The main automorphism, is defined by

$$\vec{x}' = -\vec{x} \qquad (ab)' = a'b'.$$

These (anti)morphisms will mainly be applied to products of vectors. Explicitly one obtains for a product of k vectors:

$$\begin{aligned} \overline{(\vec{x}_1 \ldots \vec{x}_k)} &= (-\vec{x}_k) \ldots (-\vec{x}_1) \\ (\vec{x}_1 \ldots \vec{x}_k)^* &= \vec{x}_k \ldots \vec{x}_1 \\ (\vec{x}_1 \ldots \vec{x}_k)' &= (-\vec{x}_1) \ldots (-\vec{x}_k). \end{aligned}$$

The wedge product on the Clifford algebra is defined by

$$\vec{x} \wedge \vec{y} = \vec{x}\vec{y} + \mathcal{B}(\vec{x}, \vec{y}) = \frac{1}{2}(\vec{x}\vec{y} - \vec{y}\vec{x}),$$

and can be extended using associativity. In general, for two Clifford numbers a and b, $[a]_k \wedge [b]_\ell = [ab]_{k+\ell}$. In this way, the Clifford algebra over V can be considered as being the exterior algebra over V, to which a product is added containing information on the inner product $\mathcal{B}$. This is reflected in the dot product, given by

$$\vec{x} \cdot \vec{y} = -\mathcal{B}(\vec{x}, \vec{y}) = \frac{1}{2}(\vec{x}\vec{y} + \vec{y}\vec{x}).$$

A (right) module over a Clifford algebra is a vector space U over $\mathbb{R}$, together with a mapping $R : \mathcal{C}\ell(V) \to L(U)$ associating with each Clifford number a a linear transformation $R(a)$ of U, called right multiplication with a, such that $R(a)R(b) = R(ba)$. In a more suggestive notation, if we have $\psi \in U$ we write ψa instead of $R(a)\psi$, and the condition for a right multiplication becomes

$$(\psi a)b = \psi(ab).$$

In Euclidean space the Dirac operator D_E is defined by

$$D_E f(\vec{x}) = \sum_{i=1}^{n} e_i \partial_i f(\vec{x}),$$

where ∂_i is the i-th partial derivative and f is a (sufficiently smooth) function which takes values in the Clifford algebra. For a full overview of the theory of solutions of the equation $D_E f = 0$, so-called monogenic functions, we refer to [5]. Important here is that $D_E^2 = -\Delta$, where Δ is the Laplacian, and so each monogenic function is harmonic and hence analytic. The operator D_E can be separated in a radial and a spherical part:

$$D_E = \xi \left(\partial_r + \frac{1}{r}\Gamma \right),$$

where $r = |\vec{x}|$, and $\vec{x} = r\xi$, ξ a unit vector. The eigenspaces of Γ, as an operator restricted to the sphere, are right modules over $\mathcal{C}\ell(n)$ and the eigenmodules are given by

$$\mathcal{P}_k = \{f \in C^1(S^{n-1}) : \Gamma f = -kf\},$$

the module of *inner spherical monogenic functions* of degree k and

$$\mathcal{Q}_k = \{f \in C^1(S^{n-1}) : \Gamma f = (k+n-1)f\},$$

the module of *outer spherical monogenic functions* of degree k, where k is any nonnegative integer. Clearly $\mathcal{P}_k \subset \mathcal{H}_k$ and $\mathcal{Q}_k \subset \mathcal{H}_{k+1}$, where $\mathcal{H}_k$ is the module of spherical harmonics of degree k. With P_k (Q_k) we shall denote an arbitrary element of $\mathcal{P}_k$ ($\mathcal{Q}_k$). Since Γ is selfadjoint all its eigenmodules are orthogonal between each other. The mapping which maps P_k to $\vec{\xi}P_k$ is an isometry between $\mathcal{P}_k$ and $\mathcal{Q}_k$. Notice that the extensions of a P_k (resp. Q_k), for which we shall use the same symbol and which are defined by

$$\begin{aligned} P_k(r\xi) &= r^k P_k(\xi) \\ Q_k(r\xi) &= r^{-k-n-1} Q_k(\xi) \qquad (r \neq 0) \end{aligned}$$

are monogenic in their domains of definition. For a full review of spherical monogenics see again [5].

3. Conformal mappings in $\mathbb{R}^{1,n}$ and $\mathbb{R}^{0,n}$

For a detailed account on the results in this section we refer to [2]. A conformal mapping of a region in V to some other (or identical) region is a not necessarily sense preserving mapping which preserves angles, taken in the sense defined by the inner product on V. If the dimension of V is bigger than two, any conformal mapping is a Möbius transformation, and can be extended to the whole of V, up to a set of measure zero. We are only interested in Möbius transformations here, so we assume $n > 2$ (the results hold for $n = 2$, apart from some slight modifications). A Möbius transformation g of $\mathbb{R}^{0,n}$ can be extended in two ways to $\mathbb{R}^{1,n}$, one sense preserving and one sense inverting. The relevant results of [2] for $\mathbb{R}^{1,n}$ can be summarised in the following theorem:

Theorem 3.1. *Let G be the set of 2×2 matrices A of the form $\begin{pmatrix} a & b \\ c & d \end{pmatrix}$, with entries a, b, c and d in $\mathcal{C}\ell(1,n)$ satisfying*

(a) *a, b, c, d are products of vectors in $\mathbb{R}^{1,n}$.*
(b) *$b^*d, a^*c, {}^*ab, {}^*cd \in \mathbb{R}^{1,n}$.*
(c) *The pseudo-determinant $\Delta(A) = a^*d - b^*c$ is real and non-zero.*

Then

(i) *G is a group under matrix multiplication.*
(ii) *For any A in G, the mapping $g_A : \mathbb{R}^{1,n} \to \mathcal{C}\ell(1,n)$ given by*

$$g_A(\vec{X}) = \frac{a\vec{X} + b}{c\vec{X} + d}$$

is actually a mapping $\mathbb{R}^{1,n} \to \mathbb{R}^{1,n}$, and is a Möbius transformation. It is clear that if A is multiplied by a non-zero real number λ, then λA still belongs to G, and $g_{\lambda A} = g_A$.

(iii) *Conversely, if g is a Möbius transformation of $\mathbb{R}^{1,n}$, one can find two matrices B and C in G, such that any matrix A for which $g = g_A$ is a real multiple of B or C.*

(iv) *Finally, if g is a Möbius transformation leaving $\mathbb{R}^{0,n}$ invariant, exactly one of these two matrices B and C has entries which are products of vectors in $\mathbb{R}^{0,n}$.*

Since we can take any real multiple of a matrix in G to describe the same Möbius transformation, it will be convenient to normalise the matrices using the pseudo-determinant, and restrict our attention to the group

$$\mathrm{Spin}(2, n+1) = \{A \in G : \Delta(A) = \pm 1\}.$$

Item (iii) of the theorem can then be reformulated as follows: if g is a Möbius transformation of $\mathbb{R}^{1,n}$, one can find two matrices B and C, such that the set of matrices A in $\mathrm{Spin}(2, n+1)$ for which $g = g_A$ is $\{B, C, -B, -C\}$.

4. The Poincaré manifold

There are three important forms in which the Poincaré manifold often appears in the literature: the (half) hyperboloid, the Poincaré ball, and the Poincaré half space.

The first is the most basic one, although it is not widely used. Let H be the hypersurface in $\mathbb{R}^{1,n}$ given by

$$H = \left\{\vec{X} : \vec{X}^2 = 1 \text{ and } \mathcal{B}(\vec{X}, e_0) < 0\right\}.$$

As a hypersurface, it inherits the metric of $\mathbb{R}^{1,n}$ in a natural way. Points on H will be denoted by Greek uppercase letters.

At a point Ξ we have the tangent space given by

$$T_\Xi = \left\{\vec{X} \in \mathbb{R}^{1,n} : \mathcal{B}(\vec{X}, \Xi) = 0\right\}.$$

The inner product of $\mathbb{R}^{1,n}$ restricted to T_Ξ is nondegenerate, and so the Clifford algebra $\mathcal{C}\ell(T_\Xi)$ can be constructed, a Clifford algebra which is a subalgebra of $\mathcal{C}\ell(1, n)$. A function f with domain in H, and values in $\mathcal{C}\ell(1, n)$, is called a section of the Clifford bundle, or a *Clifford section,* if $f(\Xi) \in \mathcal{C}\ell(T_\xi)$ for all Ξ in f's domain. Note that, due to the possible identification of the Clifford algebra with the exterior algebra, Clifford sections can be considered as being differential forms, in the language of differential geometry. The vector space of C^∞ Clifford sections will be denoted by $\Gamma^\infty(H)$.

For Clifford sections, classical derivation in a direction is replaced by a so-called connection, the construction of which we give here.

Let f be a smooth Clifford section and let $(\Xi, \vec{Y})$ be a tangent vector (i.e. $\vec{Y}$ is a vector in T_Ξ). The connection on the Clifford bundle is given by

$$\nabla_{(\Xi,\vec{Y})} f = \Pi_\Xi \, \partial_{(\Xi,\vec{Y})} f.$$

Here $\partial_{(\Xi,\vec{Y})}f$ is ordinary derivation in the $\vec{Y}$ direction, i.e. $\partial_{(\Xi,\vec{Y})}f = \partial_t\Psi(\Xi + t\vec{Y})|_{t=0}$, where Ψ is an arbitrary smooth extension of f in a neighbourhood in $\mathbb{R}^{1,n}$ of Ξ, and Π_Ξ means orthogonal projection onto the Clifford algebra $\mathcal{C}\ell(T_\Xi)$.

One finds that

$$\nabla_{(\Xi,\vec{Y})}f = \partial_{(\Xi,\vec{Y})}f + \frac{\Xi\vec{Y}}{2}f(\Xi) - f(\Xi)\frac{\Xi\vec{Y}}{2}.$$

Let $\vec{Y}_1, \dots, \vec{Y}_n$ be an orthonormal basis of T_Ξ. Then the Dirac operator in Ξ is given by

$$\nabla_H f(\Xi) = \sum_{i=1}^{n} \vec{Y}_i \nabla_{(\Xi,\vec{Y})} f.$$

It is possible to prove that this definition is independent of the orthonormal basis $\vec{Y}_1, \dots, \vec{Y}_n$ chosen; as a matter of fact alternative definitions are possible which do not use such a basis at all.

Notice that $\nabla_H f(\Xi)$ is again in the Clifford algebra $\mathcal{C}\ell(T_\Xi)$: ∇_H maps Clifford sections to Clifford sections (in terms of classical differential geometry, ∇_H is equivalent to the Hodge operator $d + d^*$ acting on differential forms).

A second model for the Poincaré manifold is based on the unit ball B in $\mathbb{R}^{0,n}$. The metric here is conformally equivalent with the Euclidean metric. The latter will be written as ds_E^2. The Poincaré metric in B then is defined as

$$ds_B^2 = \frac{ds_E^2}{(1 - |\vec{x}|^2)^2}.$$

A third model for the Poincaré manifold is Euclidean half space. Let P be given by

$$P = \{\vec{x} \in \mathbb{R}^{0,n} : x_n > 0\}.$$

The Poincaré metric in P then is defined as

$$ds_P^2 = \frac{ds_E^2}{x_n^2}.$$

Let us consider first the relevant mappings on and between these three models, all of which will be thought of as hypersurfaces in $\mathbb{R}^{1,n}$. Let g be a conformal mapping of the hyperboloid H. g can be extended to $\mathbb{R}^{1,n}$ in two ways (one sense preserving, one sense inverting). One of these extensions is an element of $O^+(1,n)$, the group of orthogonal mappings of $\mathbb{R}^{1,n}$ leaving H invariant. On the other hand, any element of $O^+(1,n)$ clearly induces a conformal mapping on H, so we can identify $\mathrm{Conf}(H)$ with $O^+(1,n)$. Obviously then, each conformal mapping on H is an isometry.

One can find conformal mappings in $\mathbb{R}^{1,n}$ mapping any of the models P, H and B to any other; the most straightforward mappings being generalisations of the Cayley transformation. We summarise them in a scheme giving one of the matrices in G (not in $\mathrm{Spin}(2, n+1)$, because we haven't normalised) describing the transformation.

\to from	H	B	P
H	Id	$\begin{pmatrix} 1 & -e_0 \\ e_0 & 1 \end{pmatrix}$	$\begin{pmatrix} 1+e_ne_0 & -e_0+e_n \\ e_0+e_n & 1-e_ne_0 \end{pmatrix}$
B	$\begin{pmatrix} 1 & e_0 \\ -e_0 & 1 \end{pmatrix}$	Id	$\begin{pmatrix} 1 & e_n \\ e_n & 1 \end{pmatrix}$
P	$\begin{pmatrix} 1+e_ne_0 & e_0-e_n \\ e_0-e_n & 1+e_ne_0 \end{pmatrix}$	$\begin{pmatrix} 1 & -e_n \\ -e_n & 1 \end{pmatrix}$	Id

Each of these transformations of course acts as an isometry, and so the conformal groups Conf(P) and Conf(B) actually are groups of isometries acting transitively on the respective manifolds, and both are conjugate to the group $O^+(1,n)$.

Consider now the mapping g,

$$g(\Xi) = \frac{\Xi - e_0}{e_0\Xi + 1},$$

mapping H to B. Fix a point Ξ on H, and let $\vec{x}$ be its image. Since g is an isometry, it induces an isometric map τ_Ξ from T_Ξ, the tangent space of H in Ξ, to the tangent space of B in $\vec{x}$. The latter of course can be readily identified with $\mathbb{R}^{0,n}$. The mapping τ_Ξ can be extended in a canonical way to a mapping

$$\tau_\Xi : \mathcal{C}\ell(T_\Xi) \to \mathcal{C}\ell(T_{\vec{x}}) = \mathcal{C}\ell(n).$$

This way it is possible to map Clifford sections on H to Clifford sections on B, by

$$\begin{aligned} \tau :\ & \Gamma^\infty(H) \to \Gamma^\infty(B) \\ & f \to \tau f \qquad : \tau f(g(\Xi)) = \tau_\Xi(f(\Xi)), \end{aligned}$$

and the Dirac operator on B can be defined by $\nabla_B = \tau\nabla_H\tau^{-1}$. Using the auxiliary function $\alpha(\vec{x}) = 1 - |\vec{x}|^2$, some calculation results in

$$\nabla_B h(\vec{x}) = \alpha(\vec{x}) \sum_{i=1}^n e_i \left(\partial_i h - \frac{1}{2\alpha(\vec{x})} ((e_i \wedge \vec{x})h - h(e_i \wedge \vec{x})) \right).$$

The sum of the derivation terms is the Euclidean Dirac operator D_E, and for a real valued function h we obtain $\nabla_B[h]_0 = \alpha D_E[h]_0$. Also for vector valued h, $h = [h]_1$, the expression can be simplified, and we obtain, after substituting $-D_E\alpha$ for $\vec{x}$,

$$\nabla_B h = \alpha D_E h + \left(\frac{-n}{2}\right)(D_E\alpha)h + \left(\frac{2-n}{2}\right)h(D_E\alpha).$$

Put now $g = h/\alpha$. The following equations are equivalent:

$$\nabla_B h = 0 \quad (1)$$

$$\alpha D_E g - (n-2)(D_E\alpha)\cdot g = 0, \quad (2)$$

still assuming h (and hence g) are vector valued.

A similar treatment can be given to the model P, where the conformal weight function this time is given by $\alpha(\vec{x}) = x_n$. This way we arrive at the equivalence of the following problems:

(i) Find, and study, monogenic vector fields on the Poincaré manifold, that is: vector valued Clifford sections f satisfying

$$\nabla f = 0, \quad (3)$$

where ∇ can be ∇_H, ∇_P or ∇_B, and of course f is a Clifford section in the relevant model.

(ii) Find, and study, functions in P satisfying

$$x_n D_E g - (n-2)e_n \cdot g = 0. \quad (4)$$

(iii) Find, and study, functions in B satisfying

$$(1-|\vec{x}|^2)D_E g + (n-2)\vec{x}\cdot g = 0. \quad (5)$$

The problem under consideration is in several aspects easier to deal with when put in the form (ii) or (iii), and from now on we shall omit the model H from our considerations, the main interest of it being the fact that the connection on the Clifford bundle is defined in the most natural way starting from this model. The models P and B are useful in different ways. When considering local theory of monogenic vector fields, we can consider an arbitrary point of the manifold, which is homogeneous since the isometry group acts transitively. The most straightforward choice is the centre $\vec{o}$ of the ball B. The group of sense preserving isometries leaving $\vec{o}$ invariant then is simply the orthogonal group $SO(n)$, and so it is possible, and relevant, to use the theory of spherical monogenic, and spherical harmonic functions. But also the P model has its advantages, which become obvious when studying function theory near the 'frontier' $\mathbb{R}^{0,n-1}$. Indeed the isometry group of P coincides with the conformal group of $\mathbb{R}^{0,n-1}$. Moreover equation (2) clearly reduces to the equation for monogenic functions in the case $n = 2$ (i.e. holomorphic functions), and for higher dimensions the behaviour of monogenic vector fields near a point of the frontier shares many properties with holomorphic functions, a consideration which was a quite important starting point for the study of such functions in [7].

Remark 4.1. *The operator given by $D_\alpha = \alpha D_E g - (n-2)\cdot(D_E\alpha)\cdot g$ has actually the relation indicated above with the Dirac operator for all conformally flat manifolds, i.e., manifolds which can be considered as domains in $\mathbb{R}^{0,n}$ with metric $ds^2 = ds_E^2/\alpha^2$, where α is a strictly positive smooth function.*

Remark 4.2. *Apart from the Dirac operator given above, there is also the spinor Dirac operator (or Atiyah-Singer operator), given on conformally flat manifolds by*

$$\not{\nabla}_\alpha f = \alpha^{(n-1)/2} D_E(\alpha^{(1-n)/2} f).$$

Obviously study of monogenic functions in this sense brings nothing new above knowledge of classical monogenic functions, since $\not{\nabla}_\alpha f = 0$ *if and only if* $\alpha^{(1-n)/2} f$ *is monogenic in the Euclidean sense.*

According to the Bochner-Weitzenböck formula, the Laplacian of a manifold is, up to a zero order curvature term, minus the square of the Dirac operator. Every scalar valued Clifford section h, however, is in the kernel of this zero order term, and for such a section $\Delta_M h = -\nabla_M^2 h$, where M is the metric manifold considered. Therefore, if $\Delta_M h = 0$, then $\nabla_M h$ is a vector valued monogenic function, and, because of the relation between Euclidean Dirac operator and Hodge Dirac operator on a conformally flat manifold, in this case $D_E h$ satisfies (2). For scalar h, $\Delta_M h$ is given by

$$\Delta_M h = \alpha^2 \Delta_E h - (n-2)\alpha \partial_{D_E \alpha} h.$$

Here Δ_E is the Euclidean Laplacian, α the conformal weight function, and $\partial_{D_E \alpha}$ derivation in the direction of the gradient of α. Explicitly this results in

$$\begin{aligned} \Delta_P &= x_n^2 \Delta_E - (n-2) x_n \partial_n \\ \Delta_B &= (1-|\vec{x}|^2)^2 \Delta_E + 2(n-2)(1-|\vec{x}|^2) r \partial_r. \end{aligned}$$

5. Some results

It would not be feasible to give a detailed account of the results obtained in this area of research, and so we refer the reader to the literature (see [7]-[8], [4], and other papers to be published). Especially we refrain from giving technical results, such as the explicit formulation of certain solutions. What we will do is give an overview of the main similarities and differences with complex analysis.

Two observations play a key rôle in the development in the theory. The first is the central position of the isometry group of the manifold, and certain of its subgroups. This isometry group is $O^+(1,n)$, possibly disguised as a congugate group within the conformal group of $\mathbb{R}^{1,n}$. As a consequence, the theory has applications for representations of this group in a way quite similar to the application of the spinor Dirac operator theory to representations of the group $\mathrm{Spin}^+(1,n)$ (see [6]). On the other hand, the presence of the subgroup $SO(n)$, which can be realised as the (orientation preserving) isotropy group of a point of the manifold, brings into play spherical harmonic functions for the solution of certain problems. The second observation is the absence of a Cauchy integral formula. Stokes' theorem is only valid in a limited form, and so no classical Cauchy integral is possible, and to this date the question of a good alternative for it remains open.

One of the main features is the study of monogenic functions regular at the frontier. In order to define such functions it is necessary to extend the P model across the frontier $\mathbb{R}^{0,n-1}$. If a domain Ω crosses the frontier, a function can be monogenic in Ω (i.e. satisfying (4) there), without being continuous there. We call a function f regular monogenic if, apart from satisfying (4) in Ω it is also analytic (i.e. real-analytic) there. Notice that the analyticity of f in $\Omega \setminus \mathbb{R}^{0,n-1}$ follows from (4). An important theorem shows how f is defined by its values at the frontier, assuming that $\omega = \Omega \cap \mathbb{R}^{0,n-1}$ is not empty:

Theorem 5.1. *A regular monogenic f can be split into two parts*

$$f = f^a + f^b$$

such that f^a is a function independent of x_n, takes values only in $\mathcal{C}\ell(0, n-1)$, and is Euclidean monogenic in its $n-1$ variables, i.e. it satisfies

$$\sum_{i=1}^{n-1} e_i \partial_i f^a = 0.$$

f^b is zero on ω, as well as $\partial_n^k f^b$ for $k < n-2$, and is uniquely defined by the characteristic of f, a function defined on $\mathbb{R}^{0,n-1}$ by

$$\Psi_f(\vec{x}) = \frac{-e_n}{(n-2)!} \partial_n^{n-2} f^b(\vec{x} + e_n x_n).$$

It should be noticed that a monogenic function in Ω which is bounded, but not continuous in ω consists of two functions, each of them defined at one side of the frontier, which can be regularly and monogenically extended across the frontier, although not necessarily to the whole of Ω.

A second result, which stresses the link with complex analysis, is given by the generalisation of the holomorphic functions z^m. Let ξ be a unit vector in $\mathbb{R}^{0,n-1}$. The choice of ξ actually is the choice of a real axis. Define the functions

$$R_m(\xi; \vec{x}) = \xi(-\vec{x}\xi)^m.$$

For the complex case, $n = 2$, there is of course only one choice of a real axis possible, and the function R_m coincides with $\pm z^m$. Actually, if ξ is kept fixed, it is possible to use the so-called paravector formalism, where the variable is replaced by a related variable $\vec{z}$, and $R_m(\xi; \vec{x})$ then is replaced by $\vec{z}^m$. We now have two possibilities: we can vary ξ, and hence the choice of a real axis, or we can vary the degree m. The first leads to a theorem akin to complex analysis:

Theorem 5.2. *Each monogenic function homogeneous of degree m can be written as a finite sum of functions of the form $R_m(\xi; \vec{x})$, where only ξ varies.*

Now (4) is a homogeneous equation, so each analytic solution to it in a neighbourhood of zero can be developed into homogeneous parts. Each of these satisfies (4), and can in turn be decomposed using the functions R_m, so these functions can act as a generalisation of z^m used to develop monogenic functions around a point of the frontier.

Keeping ξ fixed however, we obtain a rather unexpected result. Indeed, the closure T_ξ of the linear span of $\{R_m(\xi;\cdot); m = 0, \dots, \infty\}$ (under a suitable topology) seems a quite obvious object for study, given its similarities with the corresponding space of holomorphic functions. However, it turns out to be a space of regular monogenics, each invariant under the group $SO(\xi)$, the group of rotations in $\mathbb{R}^{0,n}$ leaving the axis ξ invariant, and so conjugate to $SO(n-1)$. The remarkable thing here is that this group does not leave the frontier $\mathbb{R}^{0,n-1}$ invariant, much less equation (4).

The two results above are given as examples of what has been obtained. Most of the other research on the P model concentrated on finding special solutions, most of them in connection with a subgroup of the isometry group, or on the Lie algebra of the isometry group.

The B model is especially suited for the study of functions in the neighbourhood of a point of the manifold. The theory here has not been studied until quite recently, and therefore is not developed very far. Mainly, the results are inspired by the fact that the isotropy group of the point $\vec{o}$ is $SO(n)$. Hence the series development of monogenics in the neighbourhood of $\vec{o}$ (and since the isometry group acts transitively, these results are valid for any point after suitable transformation) is based on the irreducible representations of this group by spaces of monogenics.

References

[1] Ö. Akın and H. Leutwiler, *On the invariance of the solutions of the Weinstein equation under Möbius transformations*, in Gowri Sankaran et al. (eds): Classical and modern potential theory and applications, Nato Adv. Sci. Inst. Ser. C Math. Phys. Sci., Kluwer 1994, 19–29.

[2] J. Cnops, *Vahlen matrices for non-definite metrics*, in R. Ablamowicz et al. (eds.): Clifford algebras with numeric and symbolic computations, Birkhäuser 1996, 155–164.

[3] J. Cnops, *Spherical geometry and Möbius transformations*, in F. Brackx, H. Serras (eds.): Clifford Algebras and their applications in mathematical physics, Kluwer, 1993, 75–84.

[4] J. Cnops, *Representations of* Spin$(1, n)$ *and related groups by Clifford algebra valued functions*, to appear.

[5] R. Delanghe, F. Sommen, V. Souček, Clifford analysis and spinor valued functions, Kluwer Acad. Publ., Dordrecht, 1992.

[6] J. Gilbert and M. Murray, Clifford algebras and Dirac operators in harmonic analysis, Cambridge University Press, 1991.

[7] H. Leutwiler, *Modified Clifford analysis*, Complex Variables **17**(1991), 153–171.

[8] H. Leutwiler, *Modified quaternionic analysis in* $\mathbb{R}^3$, Complex Variables **20**(1992), 19–51.

[9] H. Leutwiler, *More on modified quaternionic analysis in* $\mathbb{R}^3$, Forum Math. **7**(1995), 279–305.

[10] H. Leutwiler: *Remarks on modified Clifford analysis*, Potential theory –ICPT 94 (Kouty, 1994), de Gruyter, 1996, 389–397.

[11] J. Peetre, *Moebius invariant function spaces–the case of hyperbolic space*, Proc. R. Ir. Acad. **92A**(1992), 243–265.

[12] J. Ryan, *Dirac operators, conformal transformations and aspects of classical harmonic analysis*, J. Lie Theory **8**(1998), 67–82.

RUG
Galglaan 2
B-9000 Gent
Belgium
E-mail address: `jc@cage.rug.ac.be`

Operator Theory:
Advances and Applications, Vol. 114

Unitarily invariant trace extensions beyond the trace class

K. Dykema, G. Weiss[1], and M. Wodzicki[2]

Abstract. The existence of unitarily invariant trace extensions of the standard trace on the ideal of finite rank operators in $B(\mathcal{H})$ past the trace class to a strictly larger ideal is proven using matrix forms and a certain commutator trace obstruction.

1. Introduction

This paper proves the existence of unitarily invariant trace extensions of the standard trace from the ideal of finite rank operators to an ideal strictly larger than the ideal of trace class operators. The proof presented here is our early matricial solution of the question about sums of commutators described below and proved in Theorem 2, the main theorem. This question arose in connection with the larger question: which ideals have traces nonvanishing on the ideal of finite rank operators? Theorem 2 also follows as a consequence of the general characterization found in [DFWW].

All ideals herein are two-sided ideals of $B(\mathcal{H})$, the class of all bounded linear operators on a separable, infinite-dimensional, complex Hilbert space $\mathcal{H}$. Let $\mathcal{F}$ denote the ideal of all finite rank operators on $\mathcal{H}$ and let $\mathcal{K}$ denote the ideal of all compact operators on $\mathcal{H}$. It is well-known that for all ideals $\mathcal{I}$, $\{0\} \subset \mathcal{F} \subset \mathcal{I} \subset \mathcal{K} \subset B(\mathcal{H})$ and that the standard trace, Tr, is a unitarily invariant linear functional on the trace class, C_1, hence on all ideals contained in C_1 and, in particular, on $\mathcal{F}$.

Ideals are determined by their characteristic sets $s(\mathcal{I})$ where $\mathcal{I} \leftrightarrow s(\mathcal{I}) := \{s(T) := \langle s_n(T)\rangle : T \in \mathcal{I}\}$ is a 1-1, onto, lattice preserving map from the class of ideals $\mathcal{I} \leftrightarrow \mathcal{S}$, the class of characteristic sets. Here, $\forall T \in \mathcal{K}$, define $s(T) := \langle s_n(T)\rangle$, the sequence of s-numbers of T. *Characteristic sets*, S, are those subsets of $c_o^{++} := \{x = \langle x_n\rangle : x \text{ is a decreasing sequence with } x_n \to 0\}$ that are closed under addition, *ampliation* ($x \in S$ implies $(x_1, x_1, x_2, x_2, x_3, \ldots) \in S$) and *domination* ($x_n \le y_n\ \forall n \ge 1$ and $y = \langle y_n\rangle \in S$ implies $x \in S$) (see [G]).

A *unitarily invariant trace* τ (trace, for short) on an ideal $\mathcal{I}$ of $B(\mathcal{H})$ is a unitarily invariant linear functional on $\mathcal{I}$, i.e., a linear functional where $\tau(T) = \tau(U^*TU)\ \forall T \in \mathcal{I}$ and for all unitary operators $U \in B(\mathcal{H})$.

1991 *Mathematics Subject Classification.* 47B10, 47B47,47D50.

[1]Partially supported by NSF Grant DMS 95-03062 and by a Charles Phelps Taft Memorial Foundation Grant.
[2]Partially suppoorted by NSF Grant DMS 91-57410.

A linear functional, is *unitarily invariant* on $\mathcal{I}$ if and only if it vanishes on the commutator space (also called the commutator ideal), $[\mathcal{I}, B(\mathcal{H})]$, the linear span of all $(\mathcal{I}, B(\mathcal{H}))$-single commutators $AB - BA$ with $A \in \mathcal{I}$ and $B \in B(\mathcal{H})$. This follows easily from the identity $[U^*, UT] = T - UTU^*$ and the fact that every $B(\mathcal{H})$ operator B is representable as a linear combination of 4 unitaries.

Every unitarily invariant trace on $\mathcal{I}$ is a unitarily invariant trace extension of some scalar multiple cTr from $\mathcal{F}$ to $\mathcal{I}$ since: every ideal $\mathcal{I}$ contains the finite rank ideal $\mathcal{F}$; the restriction to $\mathcal{F}$ of any unitarily invariant trace on $\mathcal{I}$ is a unitarily invariant trace on $\mathcal{F}$; and the only unitarily invariant traces on $\mathcal{F}$ are the scalar multiples, $c\mathrm{Tr}$, of the standard trace Tr on $\mathcal{F}$. In other words, any unitarily invariant trace on $\mathcal{I}$ which is nonvanishing on $\mathcal{F}$ is, up to a nonzero scalar multiple, a unitarily invariant trace extension of Tr.

A necessary and sufficient condition for the existence of trace extensions from a smaller ideal $\mathcal{J} \subset \mathcal{I}$ to $\mathcal{I}$ of a trace τ is that:

$$[\mathcal{I}, B(\mathcal{H})] \cap \mathcal{J} \subset \mathcal{J}^o,$$

where $\mathcal{J}^o := \{T \in \mathcal{J} : \tau(T) = 0\}$. To see this, first define τ on $\mathcal{J} + [\mathcal{I}, B(\mathcal{H})]$ via $\tau(J + S) := \tau(J)\ \forall J \in \mathcal{J}$ and $\forall S \in [\mathcal{I}, B(\mathcal{H})]$. Then τ is well-defined if and only if $[\mathcal{I}, B(\mathcal{H})] \cap \mathcal{J} \subset \mathcal{J}^o$. Extending τ from the linear subspace $\mathcal{J} + [\mathcal{I}, B(\mathcal{H})]$ nonuniquely to all of $\mathcal{I}$ is an easy Hamel basis argument.

Hence there is a trace on $\mathcal{I}$ which is nonvanishing on the ideal $\mathcal{F}$ of finite rank operators if and only if $[\mathcal{I}, B(\mathcal{H})] \cap \mathcal{F} \subset \mathcal{F}^o$.

The standard trace Tr on the trace class C_1 is a unitarily invariant trace extension from $\mathcal{F}$ to C_1. A natural ideal properly containing C_1 is the ideal, $\mathcal{I}_{o(1/n)}$, of operators with s-numbers $s_n(T) = o\left(\frac{1}{n}\right)$. The class $s(\mathcal{I}_{o(1/n)})$ satisfies the characteristic set axioms [G] and it properly contains C_1 since every decreasing summable sequence satisfies the $o\left(\frac{1}{n}\right)$ condition while the sequence $\left\langle \frac{1}{n \log n} \right\rangle$ is not summable but also satisfies the condition. This leads to the question of the existence of a unitarily invariant trace extension of Tr from $\mathcal{F}$ (beyond C_1) to $\mathcal{I}_{o(1/n)}$, or equivalently:

Question: If F is a finite rank operator which is a finite sum of commutators $AB - BA$ where $B \in B(H)$ and $A \in K$ with $s_n(A) = o\left(\frac{1}{n}\right)$, must $\mathrm{Tr}\, F = 0$?

Note. There are no positive unitarily invariant trace extensions τ of Tr from $\mathcal{F}$ beyond C_1 to a strictly larger ideal $\mathcal{I}$. Assuming otherwise, for any $T \in \mathcal{I} \backslash C_1$, multiply U^* in its polar decomposition $T = U|T|$ so that we can without loss of generality assume $T \geq 0$. Then $\tau(T) \geq \tau(P_n T) = Tr(P_n T) \to \|T\|_{C_1} = \infty$, where P_n denotes the projection onto the span of the eigenvectors of the largest n-eigenvalues of T. Hence $\tau(T) = \infty$, which is a contradiction.

Let $\left\langle \frac{1}{n} \right\rangle$ denote the sequence $(1, 1/2, 1/3, \dots, 1/n, \dots)$. The main theorem is:

Theorem 2. The ideal $\mathcal{I}_{o(1/n)}$ possesses unitarily invariant trace extensions of the standard trace on the ideal of finite rank operators.

Remark. A necessary and sufficient condition for an ideal $\mathcal{J}$ to possess unitarily invariant trace extensions from $\mathcal{F}$ to $\mathcal{J}$ which are nonvanishing on $\mathcal{F}$ is that diag $\langle \frac{1}{n} \rangle \notin \mathcal{J}$ (i.e., $\langle \frac{1}{n} \rangle \notin S(\mathcal{J})$), but this requires more development (see [DFWW] and a variation on [PT] or a slight variation on the simpler [W1, pp. 34-35]).

2. General block upper-Hessenberg forms

Theorem 1. For every finite sequence of operators $T_1, T_2, \ldots, T_k$ with cyclic vector e (i.e., with $\{p(T_1, T_2, \ldots, T_k)e : p \text{ is a } k\text{-variable polynomial}\}$ dense in $\mathcal{H}$), e extends to a basis on which all $T_1, T_2, \ldots, T_k$ simultaneously have block tri-diagonal matrix forms with the same block sizes:

$$\begin{pmatrix} d_1 & b_1 & * & * & . \\ a_1 & d_2 & b_2 & * & . \\ 0 & a_2 & d_3 & . & . \\ 0 & 0 & . & . & . \\ . & . & . & . & . \end{pmatrix}$$

where a_n, b_n, d_n are finite rectangular matrices with height a_n = width d_{n+1} = height d_{n+1} = width $b_n = O(k^n)$.

Remark. In [WW] two additional related matrix forms are obtained.

(i) In the above matrix, one can replace the $*$'s with 0's as follows. For every finite sequence of operators $T_1, T_2, \ldots, T_k$ with $*$-cyclic vector e (i.e., with $\{p(T_1, T_1*, T_2, T_2*, \ldots, T_k, T_k*)e$: p is a $2k$-variable polynomial $\}$ dense in $\mathcal{H}$), e extends to a basis in which all $T_1, T_2, \ldots, T_k$ simultaneously have blocked tri-diagonal matrix forms with the same block sizes: height a_n = width d_{n+1} = height d_{n+1} = width $b_n = O((2k)^n)$.

(ii) If $AB-BA$ is finite rank, then there exists A', B', and a projection P with $AP = A' \oplus 0$, $BP = B' \oplus 0$ with $A'B' - B'A'$ finite rank and $Tr(A'B' - B'A') = Tr(AB-BA)$ where A' and B' simultaneously have block upper-Hessenberg matrix form with the same block sizes: height a_n = width d_{n+1} = height d_{n+1} = width $b_n = O(n)$.

Proof of theorem 1. The set of nth-order words in $T_1, T_2, \ldots, T_k$ is given by the disjoint union $W_n(T_1, T_2, \ldots, T_k) = \bigcup_{j=1}^{k} W_{n-1}(T_1, T_2, \ldots, T_k)T_j$ which leads to the recursive cardinality equation $|W_n(T_1, T_2, \ldots, T_k)| = k|W_{n-1}(T_1, T_2, \ldots, T_k)|$ with $|W_1(T_1, T_2, \ldots, T_k)| = k$, hence $|W_n(T_1, T_2, \ldots, T_k)| = k^n$.

Now denote the subspace $M_n = \operatorname{span}_{1 \leq j \leq n} W_j(T_1, T_2, \ldots, T_k)e$ having $\dim M_n \leq k^n$, and for all $1 \leq j \leq k$, $T_j M_n \subset M_{n+1}$. Then the subspaces $M_{n+1} \ominus M_n$ yield block upper-Hessenberg forms as above, all the T_j's with the same block sizes and with a_n, b_n, d_n matrix sizes $O(k^n)$. □

3. Matrix forms and commutators

The proof of Theorem 2, the main theorem, is obtained from the interplay between matrix forms and commutators, in particular, a certain diagonal trace obstruction.

Theorem 2. The ideal $\mathcal{J}_{o(1/n)}$ possesses unitarily invariant trace extensions of the standard trace on the ideal of finite rank operators.

Proof. As previously mentioned, Theorem 2 is equivalent to the inclusion $[\mathcal{J}_{o(1/n)}, B(\mathcal{H})] \cap \mathcal{F} \subset \mathcal{F}^0$. So suppose $F = \sum_{k=1}^{m}(A_kB_k = B_kA_k)$ is a finite rank operator with each $A_k \in \mathcal{J}_{o(1/n)}$ and $B_k \in B(\mathcal{H})$ and assume to the contrary that $\operatorname{Tr} F \neq 0$.

First we reduce to the case where $\langle A_k \rangle$, $\langle B_k \rangle$ all share a common cyclic vector. Use the general block upper-Hessenberg forms in Theorem 1 to put each operator in the finite sequences $\langle A_k \rangle$ and $\langle B_k \rangle$ simultaneously into an infinite block upper-triangular matrix with all diagonal blocks in upper-Hessenberg form (or finite block upper-triangular matrix according to the number of cyclic subspaces required). Observe the diagonal after multiplying block upper-triangular matrices. Each compression of $F = \sum_{k=1}^{m}(A_kB_k - B_kA_k)$ to each of these cyclic subspaces, i.e., PFP, is again finite rank and retains its block upper-Hessenberg form.

Since $\sum_{k=1}^{m}(A_kB_k - B_kA_k) \in C_1$, the sequence of the traces of these compressions sums to $Tr\sum_{k=1}^{m}(A_kB_k = B_kA_k) \neq 0$. Therefore at least one compression has non-zero trace. In other words, restricting to a common reducing subspace for all A_k, B_k, the finite sequences $\langle A_k \rangle \subset \mathcal{J}_{o(1/n)}$, $\langle B_k \rangle \subset B(\mathcal{H})$ then have a cyclic vector as defined in Theorem 1, the PFP-compression of $F = \sum_{k=1}^{m}(A_kB_k - B_kA_k)$ also has block upper-Hessenberg form, $\sum_{k=1}^{m}(A_kB_k - B_kA_k)$ is finite rank, and its trace $Tr\sum_{k=1}^{m}(A_kB_k - B_kA_k) \neq 0$. So, without loss of generality, we may assume the set $\langle A_k \rangle$, $\langle B_k \rangle$ has a common cyclic vector and hence, for each $1 \leq k \leq m$,

$$A_k = \begin{pmatrix} d_1(k) & b_1(k) & * & * & . \\ a_1(k) & d_2(k) & b_2(k) & * & . \\ 0 & a_2(k) & d_3(k) & . & . \\ 0 & 0 & . & . & . \\ . & . & . & . & . \end{pmatrix}, \quad B_k = \begin{pmatrix} d'_1(k) & b'_1(k) & * & * & . \\ a'_1(k) & d'_2(k) & b'_2(k) & * & . \\ 0 & a'_2(k) & d'_3(k) & . & . \\ 0 & 0 & . & . & . \\ . & . & . & . & . \end{pmatrix}$$

simultaneously with block upper-Hessenberg forms all with the same block sizes and of order $O((2k)^n)$.

Computing the diagonal of F, the first diagonal block entry $D_1 = \sum_{k=1}^{m}(d_1(k)d_1'(k) - d_1'(k)d_1(k) + b_1(k)a_1'(k) - b_1'(k)a_1(k))$, followed by the nth block diagonal entry, for $n \geq 2$,

$$D_n = \sum_{k=1}^{m}(d_n(k)d_n'(k) - d_n'(k)d_n(k) + b_n(k)a_n'(k) + a_{n-1}(k)b_{n-1}'(k) - a_{n-1}'(k)b_{n-1}(k) - b_n'(k)a_n(k)).$$

Then

$$\begin{aligned}
\mathrm{Tr}\left(\sum_{j=1}^{n} D_j\right) &= \sum_{j=1}^{n} \mathrm{Tr}\, D_j \\
&= \sum_{k=1}^{m}\sum_{j=1}^{n} \mathrm{Tr}(d_j(k)d_j'(k) - d_j'(k)d_j(k)) + \sum_{k=1}^{m}\sum_{j=1}^{n} \mathrm{Tr}(b_j(k)a_j'(k) - a_j'(k)b_j(k)) \\
&+ \sum_{k=1}^{m}\sum_{j=1}^{n} \mathrm{Tr}(a_j(k)b_j'(k) - b_j'(k)a_j(k)) + \mathrm{Tr}(b_n(k)a_n'(k) - b_n'(k)a_n(k)) \\
&= \mathrm{Tr}\sum_{k=1}^{m}(b_n(k)a_n'(k) - b_n'(k)a_n(k)),
\end{aligned}$$

since all matrices $a_n(k), b_n(k), d_n(k), a_n'(k), b_n'(k), d_n'(k)$ are finite rank and all commutators of finite rank operators have trace 0. But also $\mathrm{Tr}\Big(\sum_{j=1}^{n} D_j\Big) \to \mathrm{Tr}\, F \neq 0$. That is, for n sufficiently large,

$$\begin{aligned}
0 &< |\mathrm{Tr}\, F|/2 \\
&\leq \left|\mathrm{Tr}\sum_{k=1}^{m}(b_n(k)a_n'(k) - b_n'(k)a_n(k))\right| \\
&\leq \sum_{k=1}^{m}(\|b_n(k)\|_1\|a_n'(k)\| + \|b_n'(k)\|\|a_n(k)\|_1) \\
&\leq \|B\|\sum_{k=1}^{m}(\|b_n(k)\|_1 + \|a_n(k)\|_1).
\end{aligned}$$

Here $\|\cdot\|$ denotes the uniform operator norm and $\|\cdot\|_1$, denotes the trace norm. So for n sufficiently large, $\sum_{k=1}^{m}(\|b_n(k)\|_1 + \|a_n(k)\|_1)$ is bounded below with $\sum_{k=1}^{m}(\|b_n(k)\|_1 + \|a_n(k)\|_1) \geq \varepsilon := |\mathrm{Tr}\, F|/2\|B\|$.

We next construct an operator $T \in \mathcal{J}_{o(1/n)}$ having matrix block diagonal $\left\langle \sum_{k=1}^{m} (|a_n(k)| + |b_n(k)|) \right\rangle$. It suffices to show how to obtain, for each fixed k, an operator $A' \in \mathcal{J}_{o(1/n)}$ with n^{th} block diagonal entry $|a_n(k)|$. Then constructing $B' \in \mathcal{J}_{o(1/n)}$ with matrix block diagonals $|b_n(k)|$ follows similarly and adding $A' + B'$ and summing over k yields T. Using the polar decompositions $a_n(k) = u_n|a_n(k)|$, $A' = A_k^* \begin{pmatrix} 0 & 0 & 0 & 0 \\ u_1 & 0 & 0 & 0 \\ 0 & u_2 & 0 & . \\ 0 & 0 & . & . \end{pmatrix}$ has block diagonal entry $|a_n(k)|$. Although the blocks of A_k are not all square matrices and hence the polar decompositions are not all square matrices, still this product is well-defined in that all matrix sizes match up properly when performing the multiplication and $A' \in \mathcal{J}_{o(1/n)}$ is achieved.

Let P_n denote the projection onto the domain of $\begin{pmatrix} . & . & . & . & . \\ . & 0 & 0 & 0 & . \\ . & 0 & T & 0 & . \\ . & 0 & 0 & 0 & . \\ . & . & . & . & . \end{pmatrix}$ with block sizes the same as all the A_k's. Then $\dim P_n = O((2k)^n)$ and, for all sufficiently large n, say $n \geq N$, $0 < \varepsilon \leq \sum_{k=1}^{m} (\|b_n(k)\|_1 + \|a_n(k)\|_1) = Tr(P_nTP_n)$. Hence $\dim P_n \leq M(2k)^n$ for some M and, using Fan's Theorem [GK, II.4, Lemma 4.1], for all $n \geq N$,

$$\varepsilon(n-N) \leq \sum_{j=N}^{n} Tr(P_jTP_j) \leq \sum_{j=1}^{n} Tr(P_jTP_j)$$

$$\leq \sum_{j=1}^{\dim P_1 + \dim P_2 + \cdots + \dim P_n} s_j(T) \leq \sum_{j=1}^{c(2k)^n} s_j(T),$$

From $\varepsilon(n-N) \leq \sum_{j=1}^{c(2k)^n} s_j(T)$ for all $n \geq N$, it is routine to show that, for some $\varepsilon' > 0$, $\varepsilon' \log n \leq \sum_{j=1}^{n} s_j(T)$, for all sufficiently large n. But $T \in \mathcal{J}_{o(1/n)}$ implies $s_n(T) = o\left(\frac{1}{n}\right)$, hence $\sum_{j=1}^{n} s_j(T) = o(\log n)$, which is a contradiction. □

References

[DFWW] K. Dykema, T. Figiel, G. Weiss and M. Wodzicki, The commutator structure of operator ideals, Odense preprint.

[G] D. J. H. Garling, On the ideals of operators in Hilbert space, Proc. London Math. Soc. (3) 17(1967), 115-138.

[GK] I. C. Gohberg and M.G. Krein, Introduction to the Theory of Linear Nonselfadjoint Operators. Translation from the Russian by A. Feinstein. Translations of Mathematical Monographs, Vol 18, American Mathematical Society, Providence, Rhode Island, 1969.

[K] N. J. Kalton, Trace-class operators and commutators, J. Funct. Anal. 86(1989), 41-74.

[PT] C. Pearcy and D. Topping, On commutators in ideals of compact operators, Michigan J. Math. 18(1971), 247-252.

[W1] G. Weiss, Commutators and Operators Ideals, dissertation 1975, University of Michigan Microfilm.

[W2] G. Weiss, Commutators of Hilbert-Schmidt operators I, Integral Equations and Operatory Theory, 9(1986), 877-892.

[W3] G. Weiss, Commutators of Hilbert-Schmidt operators II, Integral Equations and Operator Theory, 3/4(1980), 574-600.

[WW] G. Weiss and M. Wodzicki, Commutator trace obstructions and infinite convexity in operator ideals, preprint.

K. Dykema
Department of Mathematics and Computer Science,
Odense Universitet;
DK-5230, Odense M,
Denmark
E-mail address: `dykema@imada.ou.dk`

G. Weiss
Department of Mathematical Sciences,
University of Cincinnati,
Cincinnati OH, 45221-0025
U.S.A.
E-mail address: `gary.weiss@uc.edu`

M. Wodzicki
Department of Mathematics,
University of California Berkeley CA;
94720-3840,
U.S.A.
E-mail address: `wodzicki@math.berkeley.edu`

Operator Theory:
Advances and Applications, Vol. 114

L^2 results for $\overline{\partial}$ in a conic

John Erik Fornæss

1. Introduction

The $\overline{\partial}$ equation is the main quantitative tool in the theory of several complex variables. It has been used extensively in analysis of domains in $\mathbb{C}^n$ and on complex manifolds, see [H]. However, in the theory of several complex variables, one also needs to investigate complex spaces as they occur naturally as soon as one considers zero sets of holomorphic functions.

In this short note we address the question whether it is possible to solve the $\overline{\partial}$ equation with L^2 estimates in complex spaces. The main motivation is that L^2 estimates appear as the most promising tool to solve the Levi problem in Stein spaces.

The general philosophy is that one can study a complex Stein space by reducing to a closed k–dimensional subvariety Z in $\mathbb{C}^n$ and consider Z as a branched cover over some suitable $\mathbb{C}^k$. Next one can remove a hypersurface S from Z such that the projection $\pi : Z \setminus S \to \mathbb{C}^k$ is unbranched. In that case one can solve $\overline{\partial}u = \lambda$ using the classical L^2 theory of Hormander ([H]). Next one can use a detailed geometric analysis of the singular space to make adjustments to this u in such a way that one can obtain L^2 results inside Z when possible as well as identify the obstructions to having L^2 solutions.

The purpose of this note is to analyze conic singularities in $\mathbb{C}^3$. It turns out that L^2 estimates work fine, modulo finitely many obstructions, see Theorem 2.4 and Corollary 2.5.

Note that we work without weights and curvature terms. We plan to continue the investigation in future work.

There is a parallel program to investigate L^∞ estimates, see ([FG]).

2. Two-Dimensions, the general conic case

We will investigate the homogeneous variety $X := \{H(x, y, z) = 0\}$ where H is a homogeneous polynomial of degree $d \geq 2$ in $\mathbb{C}^3$ and where we assume that $\nabla H \neq 0$ on $X^* = X \setminus (0)$. (So we assume that the conic singularity is isolated.)

1991 *Mathematics Subject Classification.* 32B10, 32F20.
Key words and phrases. singularity, Cauchy-Riemann equation, cohomology groups.
The author has been supported by an NSF grant.

The space X has a normal singularity at the origin, [L]. By choosing coordinate axes appropriately, we can assume that H has the form $H = z^d + \sum_{j=0}^{d-1} z^j H_j(x,y)$ where each H_j is homogenous of degree $d-j$. We can consider X as a branched cover over $\mathbb{C}^2(x,y)$. The branching occurs at finitely many lines L_i which we may take to be of the form $L_i = \{y = a_i x\} = \{x = b_i y\}$. We fix i. Then over a sector $S_{i,\epsilon} := \{|y - a_i x| < \epsilon |x|\}$ in $\mathbb{C}^2(x,y)$ we can write $X = \{\Pi_{j=1}^d (z - A_j(x,y)) = 0\}$ for locally defined functions A_j. Over the line L_i at least two of the values A_j must coincide, to create branching. We can reorganize the A_j according to whether their values agree over L_j or not. Then $H = \Pi_{\ell=1}^k \Pi_{j\in I_\ell}(z - A_j(x,y))$.

By perturbing slightly the coordinates, we can assume that $|I_\ell| = 1$ except for $|I_1| = 2$. Hence over each branch line, X has $d-1$ branches, one of which has multiplicity two while all the others are single sheeted. The branching occurs then at lines $z = \alpha_i x,\ y = a_i x$.

We will make one further normalization of the coordinates: For each i, consider the intersection with X and the plane $z = \alpha_i x$ in $\mathbb{C}^3$. In addition to the line $y = a_i x$, this contains also lines over $y = a_i^j x$ for a finite collection of $j's$. After further perturbation, we can assume moreover that the tangent space of X along the lines $z = \alpha_i x, y = a_i^j x$ is neither parallel to the x,y plane nor the x,z plane. Moreover we assume that the points a_i, a_i^j as both i and j vary are all pairwise distinct. We assume that we also can write the lines $y = a_i^j x$ as $x = b_i^j y$. All these conditions can be satisfied after a linear coordinate change.

We will solve the problem $\overline{\partial} u = \lambda$ where λ is a $\overline{\partial}$ closed $(0,1)$ form in a deleted neighborhood of the singular point 0 of X, say the punctured unit ball intersect X. We assume that the induced L^2 norm $\|\lambda\|_{L^2(X)}$ of λ in X^* is finite. We use the notation $\|\cdot\|_{L^2(x,y),U}$ to denote the L^2 norm in the set U using the volume form in the (x,y) coordinates. Similarly, we use the notation $\|\cdot\|_{L^2(X),U}$ if we mean the L^2 norm using the induced volume form on X. Usually the set U is supressed from the notation.

For each branching line L_i there is exactly one preimage B_i in X which is mapped to L_i by projection and on which X is branched. Removing the lines B_i from X, $B = \cup B_i$, we get an unbranched Riemann domain $X' := X \setminus B$ over $\mathbb{C}^2(x,y)$.

We have the Hörmander solution u on X', solving $\overline{\partial} u = \lambda$ in the x,y coordinates. So all norms are calculated using the volume form on the x,y plane. Observe that for $(0,1)$ forms, $\|\lambda\|_{L^2(x,y)} \leq \|\lambda\|_{L^2(X)}$ where $\|\lambda\|_{L^2(x,y)}$ denotes the L^2 norm calculated in the (x,y) coordinates.

Next we analyze the singularities of u at the branch lines B_i. For this it is conventent to compare with the Hörmander solution U_i on Ω_i, the set over $W_i = \{(x,y); |y - a_i x| < b|x|\}$ for a fixed small number b and where Ω_i only refers to the connected component which contains the branch line B_i. In local coordinates we can describe Ω_i as the 0 set of the function

$$(z-\alpha_i x)^2 - x^2 \sum_{k\geq 1} a^i_k \frac{(y-a_i x)^k}{x^k} = 0, a^i_1 \neq 0,$$

In Ω_i we can use x, z as coordinates and we will let U_i be the Hörmander solution in these coordinates. Note that X is close to a flat graph over the (x, z) plane in Ω_i and hence the L^2 norm of U_i in $\mathbb{C}^2(x, z)$ and X agree up to bounded multiples. Notice also that the L^2 norm of U_i in X is larger than the L^2 norm calculated in $L^2(x, y)$.

We have $(y - a_i x)x \sim (z - \alpha_i x)^2$.

Then $v_i := u - U_i$ is holomorphic away from the line $y = a_i x$ and is in L^2 in the sense of (x, y) coordinates. It has at most a local singularity like $1/(z - \alpha_i x)$ since it is integrable in the L^2 sense in the x, y direction.

We can write

$$v_i = \sum_{m\geq -1, n\in\mathbb{Z}} a^i_{m,n} x^n (z-\alpha_i x)^m$$

We estimate the $L^2(x, y)$ norm:

$$\begin{aligned}
\|v_i\|^2_{L^2(x,y)} &\sim \sum |a^i_{m,n}|^2 \int |x|^{2n} |(y-a_i x)x|^m \\
&\sim \sum |a^i_{m,n}|^2 \int |x|^{2n+m} |y-a_i x|^m \\
&\sim \sum |a^i_{m,n}|^2 \int \frac{|x|^{2n+m}|bx|^{m+2}}{m+2} \\
&\sim \int \sum |a^i_{m,n}|^2 |x|^{2n+2m+2} \frac{|b|^{m+2}}{m+2} \\
&\sim \sum |a^i_{m,n}|^2 \frac{|b|^{m+2}}{(n+m+2)(m+2)}
\end{aligned}$$

For the integrals to be finite we get the conditions that $m \geq -1$ and $n+m+1 \geq 0$. In particular, if $m = -1, n \geq 0$.

The next step is to rewrite terms of the form $\frac{x^\ell}{z-\alpha_i x}$ when $\ell \geq d-1$. We can write $x = x - \alpha(y - a_i x) + \alpha(y - a_i x)$. The last term $y - a_i x$ is about $(z - \alpha_i x)^2/x$ and by choosing α we can write the sum of the first two terms as $c_j(x - b^j_i y)$ for any of the $d-1$ other coefficients b^j_i. Hence we can write $x^{d-1} = c^i \Pi^{d-1}_{j=1}(x - b^j_i y) + (y - a_i x)Q_i(x, y)$ where $c = \Pi c_j$ and Q_i is a homogeneous polynomial of degree $d - 2$.

Using this we can rewrite $v_i = v^I_i + v^{II}_i + v^{III}_i + v^{IV}_i$ where

$$v^I_i = \frac{1}{z-\alpha_i x} \sum_{n=0}^{d-2} a^i_{-1,n} x^n$$

$$v^{II}_i = \frac{1}{z-\alpha_i x} [c^i \Pi^{d-1}_{j=1}(x - b^j_i y)] \sum_{n=d-1}^{\infty} a^i_{-1,n} x^{n-(d-1)}$$

$$v_i^{III} = \frac{y-a_ix}{z-\alpha_ix}Q_i(x,y)\sum_{n=d-1}^{\infty} a^i_{-1,n}x^{n-(d-1)}$$

$$v_i^{IV} = \sum_{m\geq 0, n+m+1\geq 0} a^i_{m,n}x^n(z-\alpha_ix)^m$$

We next estimate the L^2 contributions of $v_i^{II}, v_i^{III}, v_i^{IV}$.

Lemma 2.1. *The functions v_i^{II} are in $L^2(x,y)\cap\mathcal{O}(X\setminus B_i)$, $\|v_i^{II}\|_{L^2(x,y)} \lesssim \|\lambda\|_{L^2(X)}$. Moreover $v_i^{II}\in L^2(X\setminus\Omega_i)$, $\|v_i^{II}\|_{L^2(X)} \lesssim \|\lambda\|_{L^2(X)}$.*

Lemma 2.2. *The functions v_i^{III} and v_i^{IV} are in $L^2(\Omega_i)\cap\mathcal{O}(\Omega_i)$ and $\|v_i^{III}\|_{L^2(X)}, \|v_i^{IV}\|_{L^2(X)} \lesssim \|\lambda\|_{L^2(X)}$.*

Proof of Lemma 2.1. We consider first the contribution to the $L^2(x,y)$ norm from Ω_i. There we can write

$$\left|\frac{1}{z-\alpha_ix}[c^i\Pi_{j=1}^{d-1}(x-b_i^jy)]\right| \lesssim \frac{|x|^{d-1}}{\sqrt{|y-a_ix|}\sqrt{|x|}}.$$

We obtain on Ω_i:

$$\begin{aligned}
\|v_i^{II}\|^2_{L^2(x,y),\Omega_i} &\lesssim \sum_{n=d-1}^{\infty}\int |a_{-1,n}|^2|x|^{2n-1}\frac{1}{|y-a_ix|}\\
&\lesssim \sum_{n=d-1}^{\infty}\int |a_{-1,n}|^2|b||x|^{2n}\\
&\lesssim \sum_{n=d-1}^{\infty}\frac{|b|}{n}\\
&\lesssim \|\lambda\|^2_{L^2(X)}.
\end{aligned}$$

Next, we consider small conical neighborhoods $\Omega_i^j := \{|z-\alpha_ix|<\epsilon|x|, |y-a_i^jx|<\epsilon|x|$. On these we have the estimates

$$\left|\frac{1}{z-\alpha_ix}[c^i\Pi_{j=1}^{d-1}(x-b_i^jy)]\right| \lesssim |x|^{d-2}.$$

This allows us to get stronger estimates, this time in $L^2(X)$:

$$\begin{aligned}
\|v_i^{II}\|^2_{L^2(X),\Omega_i^j} &\lesssim \sum_{n=d-1}^{\infty} \int |a_{-1,n}|^2 |x|^{2n-2} dx dz \\
&\lesssim \sum_{n=d-1}^{\infty} \int |a_{-1,n}|^2 \epsilon^2 |x|^{2n} dx \\
&\lesssim \sum_{n=d-1}^{\infty} |a_{-1,n}|^2 \frac{\epsilon^2}{n} \\
&\lesssim \|\lambda\|^2_{L^2(X)}.
\end{aligned}$$

The $L^2(X)$ norm of v_i^{II} in a narrow sector about the other branch lines $z = \alpha_k x, y = a_k x$ is similar using the estimate

$$|\frac{1}{z - \alpha_i x} [c^i \Pi_{j=1}^{d-1} (x - b_i^j y)]| \lesssim |x|^{d-2}.$$

in the same way one can estimate the other components over $\{|y - a_i x| < b|x|\}$ as well as the other components over $\{|y - a_i^j x| < \epsilon |x|\}$. Over the remaining parts of X, the conic has a bounded slope as a graph over the (x, y) plane and hence the $L^2(x, y)$ and $L^2(X)$ norms are comparable there.

Since in this region there are no longer any singularities we still get that the $L^2(X)$ norm of v_i^{II} in this set is bounded by a bounded multiple of the L^2 norm of λ. □

Proof of Lemma 2.2. We first consider the function v_i^{III}. Since the part $|\frac{y - a_i x}{z - \alpha_i x} Q_i(x, y)| \lesssim |x|^{d-2}$ this goes like the piece of v_i^{II} over Ω_i^j. So we only have to consider the function v_i^{IV}.

$$\begin{aligned}
\|v_i^{IV}\|^2_{L^2(X),\Omega_i} &= \sum_{m \geq 0, n+m+1 \geq 0} |a^i_{m,n}|^2 \int |x|^{2n} |z - \alpha_i x|^{2m} \\
&\sim \sum_{m \geq 0, n+m+1 \geq 0} |a^i_{m,n}|^2 \int |x|^{2n} \frac{|b|^{m+1} |x|^{2m+2}}{2m+2} \\
&\lesssim \sum_{m \geq 0, n+m+1 \geq 0} |a^i_{m,n}|^2 \frac{|b|^{m+1}}{(2m+2)(2n+2m+4)} \\
&\lesssim \|\lambda\|_{L^2(X)}
\end{aligned}$$

□

Note that with these estimates, we can assume that the terms v_i^{III} and v_i^{IV} all vanish simply because we can absorb them into the solutions U_i. Similarly, we can relace u by $u - \sum_i v_i^{II}$ to get another solution where the only possible singularities preventing the solution from belonging to $L^2(X)$ are the terms v_i^I. In conclusion we get the following description of an improved Hörmander solution:

Theorem 2.3. *Let $\lambda \in L^2(X)$ be a $\overline{\partial}$ closed $(0,1)$ form on a deleted ball around 0 in X. Then there exists a solution u to the equation $\overline{\partial} u = \lambda$ which is in $L^2(X)$ except for singular terms of the form $v_i = \frac{1}{z-\alpha_i x}(d_0^i + d_1^i x + \cdots + d_{d-2}^i(x^{d-2}))$ in a sector around each of the branch lines. These singular contributions belong to $L^2(x,y)$ but not to $L^2(X)$.*

Note that there are $d(d-1)$ of the points $\{a_i\}$.

The next result shows that each of those give rise to obstructions to solving $\overline{\partial}$ in L^2.

Theorem 2.4. *Let $\chi(t)$ be a cut-off function in a small disc around the origin in $\mathbb{C}$. Suppose that $\chi(t) \equiv 1$ in a neighborhood of the origin. Then for any positive integer $1 \leq \ell \leq d-2$, the form $\lambda = \frac{x^\ell}{z-\alpha_i}\overline{\partial}\chi(\frac{z-\alpha_i x}{x})$ is a $\overline{\partial}$ closed form in $L^2(X_1^*)$ (in the unit ball say) for which there is no L^2 solution u.*

Proof. It is clear that λ is L^2 and $\overline{\partial}$ closed. Suppose that there is an L^2 solution u. Then $v := u - \chi(\frac{z-\alpha_i x}{x})\frac{x^\ell}{z-\alpha^i x}$ is holomorphic except on the line where $y = a_i x$ and where it has a local singularity like $\frac{1}{z-\alpha_i x}$. Hence $(z-\alpha_i x)v$ is holomorphic on the punctured neighborhood of 0 and hence, using normality of the singularity, has a power series expansion:

$$(z-\alpha_i x)v(x,y,z) = \mathcal{O}(z-\alpha_i x) + \sum_{j,k\geq 0} A_{j,k}x^j y^k.$$

Next calculate the values of this function on the lines $z = \alpha_i x, y = a_i x$ or $y = a_i^j x$. For the line $y = a_i x$, we get the equality: $\sum_{j+k=\ell} A_{j,k}a_i^k = 1$ while from the $d-1$ other lines we get the equations $\sum_{j+k=\ell} A_{j,k}(a_i^j)^k = 0$. The last $d-1$ eqations in $d-1$ unknowns force all the coefficients $A_{j,k}$ to be zero, contradicting the first equation. □

Corollary 2.5. *The dimension of the quotient space of $\overline{\partial}$ closed L^2 forms modulo those with L^2 solutions, (i.e. a cohomology group) grows like the cube of the degree of the conic singularity.*

References

[FG] Fornæss, J. E., Gavosto, E.; *The Cauchy Riemann Equations on Complex Spaces,* Duke Journal, 93, (1998) 453-477.

[H] Hörmander, L.; *L^2 estimates and existence theorems for the $\overline{\partial}$ operator,* Acta Math. 113 (1965), 89-152.

[L] Laufer, H.; *Normal Two-Dimensional Singularities,* Ann. of Math. Studies No. 71, Princeton University Press, Princeton 1971.

Mathematics Department
The University of Michigan
Ann Arbor, MI 48109, USA
E-mail address: `fornaess@umich.edu`

Operator Theory:
Advances and Applications, Vol. 114

Lie superalgebras of supermatrices of complex size. Their generalizations and related integrable systems

P. Grozman and D. Leites

Abstract. We distinguish a class of simple filtered Lie algebras $LU_{\mathfrak{g}}(\lambda)$ of polynomial growth whose associated graded Lie algebras are not simple. We describe presentations of such algebras. A contraction sends $LU_{\mathfrak{g}}(\lambda)$ into algebras of the same class studied by Donin, Gurevich and Shnider; $LU_{\mathfrak{g}}(\lambda)$ are quantizations of the DGS algebras.

The Lie algebra $\mathfrak{GL}(\lambda)$ of matrices of complex size is the simplest example; it is $LU_{\mathfrak{sl}(2)}(\lambda)$. The dynamical systems associated with it in the space of pseudodifferential operators in the same way as the KdV hierarchy is associated with $\mathfrak{sl}(n)$ are those studied by Gelfand–Dickey and Khesin–Malikov. For $\mathfrak{g} \neq \mathfrak{sl}(2)$ we get generalizations of $\mathfrak{GL}(\lambda)$ and the corresponding dynamical systems, in particular, their superized versions.

Our presentation of Lie superalgebras $LU_{\mathfrak{s}}(\lambda)$ is related to the presentation of simple finite dimensional Lie superalgebras $\mathfrak{s}$ in terms of a certain pair of generators. There are remarkably few such relations, e.g., just 9 relations for $\mathfrak{sl}(n)$.

This is an expanded transcript of the talk given at the International Symposium on Complex Analysis and Related Topics, Cuernavaca, Mexico, November 18 – 22, 1996. We thank A. Turbiner and N. Vasilevski for hospitality.

0.0. History

About 1966, V. Kac and B. Weisfeiler began the study of simple *filtered* Lie algebras of *polynomial growth.* Kac first considered the $\mathbb{Z}$*-graded* Lie algebras associated with the filtered ones and classified *simple graded* Lie algebras of *polynomial growth*

1991 *Mathematics Subject Classification.* 17B01, 17A70; 17B35, 17B66.
Key words and phrases. Defining relations, principal embeddings, Lie superalgebra, Schrödinger operator, matrices of complex size, Gelfand–Dickey bracket, KdV hierarchy, W-algebras, quantized Lie algebras.
We are thankful to V. Kornyak (JINR, Dubna) who checked the generators and relations with an independent program and compared convenience of the Serre relations with that of ours. Financial support of the Swedish Institute and NFR is gratefully acknowledged. We are thankful to B. Feigin and Shi Kangjie Laoshi for their shrewd questions and to S. Shnider, G. Post and M. Vasiliev for the timely information.

under a technical assumption; he conjectured the inessential nature of the assumption. It took more than 20 years to get rid of the assumption: see very technical papers by O. Mathieu, cf. [K] and references therein. For a similar list of simple $\mathbb{Z}$-graded Lie *super*algebras of polynomial growth see [KS], [LSc].

The Lie algebras Kac distinguished (or rather the algebras of derivations of their nontrivial central extensions, the *Kac–Moody* algebras) have proved very interesting in applications. These algebras aroused such interest that the study of filtered algebras was arrested for two decades. Little by little, however, the simplest representative of the new class of simple filtered Lie superalgebras (of polynomial growth), namely, the Lie algebra $\mathfrak{GL}(\lambda)$ of matrices of complex size, and its projectivization, i.e., the quotient modulo the constants, $\mathfrak{pGL}(\lambda)$, drew its share of attention [F], [KM], [KR].

While we typed this paper, Shoikhet [Sh] published a description of representations of $\mathfrak{GL}(\lambda)$; we are thankful to M. Vasiliev who informed us of still other applications of generalizations of $\mathfrak{GL}(\lambda)$, see [BWV], [KV].

This paper begins a systematic study of a new class of Lie algebras: simple filtered Lie algebras of polynomial growth (SFLAPG) for which the graded Lie algebras associated with the filtration considered are not simple; $\mathfrak{pGL}(\lambda)$ is our first example. Actually, another example of a Lie algebra of class SFLAPG was known even before the notion of Lie algebras was introduced. Indeed, the only deformation (physicists call it *quantization*) Q of the Poisson Lie algebra $\mathfrak{po}(2n)$ sends $\mathfrak{po}(2n)$ into $\mathfrak{diff}(n)$, the Lie algebra of differential operators with polynomial coefficients; the restriction of Q to $\mathfrak{h}(2n) = \mathfrak{po}(2n)/center$, the Lie algebra of Hamiltonian vector fields, sends $\mathfrak{h}(2n)$ to the projectivization $\mathfrak{pdiff}(n) = \mathfrak{diff}(n)/\mathbb{C} \cdot 1$ of $\mathfrak{diff}(n)$. The Lie algebra $\mathfrak{pdiff}(n)$ escaped Kac's classification, though it is the deform of an algebra from his list, because its intrinsically natural filtration given by $\deg q_i = -\deg \partial_{q_i} = 1$ is not of polynomial growth while the graded Lie algebra associated with the filtration of polynomial growth (given by $\deg q_i = \deg \partial_{q_i} = 1$) is not simple.

Observe that from the point of view of dynamical systems the Lie algebra $\mathfrak{diff}(n)$ is not very interesting: it does not possesses a nondegenerate bilinear symmetric form; we will consider its subalgebras that do.

In what follows we will usually denote the associative (super)algebras by Latin letters; the Lie (super)algebras associated with them by Gothic letters; e.g., $\mathfrak{gl}(n) = L(\mathrm{Mat}(n))$, $\mathfrak{diff}(n) = L(\mathrm{Diff}(n))$, where the functor L replaces the dot product by the bracket.

0.1. The construction. Problems related

Each of our Lie algebras (and Lie superalgebras) $LU_{\mathfrak{g}}(\lambda)$ is realized as a quotient of the Lie algebra of global sections of the sheaf of twisted D-modules on the flag variety, cf. [Ka], [Di]. The general construction consists of the preparatory step 0), the main steps 1) and 2) and two extra steps 3) and 4).

We distinguish two cases: A) $\dim \mathfrak{g} < \infty$ and $\mathfrak{g}$ possesses a Cartan matrix and B) $\mathfrak{g}$ is a simple vectorial Lie (super)algebra.

Let $\mathfrak{g} = (\underset{\alpha<0}{\oplus} \mathfrak{g}_\alpha) \oplus \mathfrak{h} \oplus (\underset{\alpha>0}{\oplus} \mathfrak{g}_\alpha) = \mathfrak{g}_- \oplus \mathfrak{h} \oplus \mathfrak{g}_+$, where $\mathfrak{g}_+ = \underset{\alpha>0}{\oplus} \mathfrak{g}_\alpha$ and $\mathfrak{g}_- = \underset{\alpha<0}{\oplus} \mathfrak{g}_\alpha$, be one of the simple $\mathbb{Z}$-graded Lie algebras of polynomial growth, either finite dimensional or of vector fields, represented as the sum of its maximal torus (usually identical with the Cartan subalgebra) $\mathfrak{h}$ and the root subspaces $\mathfrak{g}_\alpha$ corresponding to an order in the set R of roots.

Observe that each order of R is in one-to-one correspondence with a system of simple roots. For the finite dimensional Lie algebras $\mathfrak{g}$ all systems of simple roots are equivalent, the equivalence is established by the Weyl group. For Lie superalgebras and infinite dimensional Lie algebras of vector fields there are inequivalent systems of simple roots; nevertheless, there is an analog of the Weyl group and the passage from system to system is described in [PS].

For vectorial Lie algebras and Lie superalgebras, even the dimension of the superspaces $X = (\mathfrak{g}_-)^*$ associated with systems of simple roots can vary. Fortunately, only *essential* (see [PS]) systems of simple roots are essential in the construction of Verma modules (roughly speaking, the space of functions on X) in which we will realize $LU_\mathfrak{g}(\lambda)$. For example, for each simple vectorial Lie algebra there is basically only ONE essential system. For elucidation of this paragraph and the list of essential systems of simple roots see [LSc].

0) From representation theory it is clear that there exists a realization of the elements of $\mathfrak{g}$ by differential operators of degree ≤ 1 on the space $X = (\mathfrak{g}_-)^*$. The realization has rank $\mathfrak{g}$ parameters (coordinates of the highest weight of the $\mathfrak{g}$-module). For the algorithms of construction and its execution in some cases see [BMP], [B], [BGLS].

Let $\tilde{\mathfrak{g}}$ be the image of $\mathfrak{g}$ with respect to this realization. Set $U_\mathfrak{g}(\lambda) \subset \mathfrak{diff}(\mathfrak{g}_-^*)$, where $\lambda = (\lambda_1, \dots, \lambda_n) \in \mathfrak{h}^*$, be the associative subalgebra generated by $\tilde{\mathfrak{g}}$ for the generic λ (i.e., when the Verma module M^λ with highest weight λ is irreducible) and the quotient of the latter algebra modulo the maximal ideal $J(\lambda)$ otherwise.

We will write

$$U_\mathfrak{g}(\lambda) = \begin{cases} \tilde{S}^\cdot(\tilde{\mathfrak{g}}) \subset \mathfrak{diff}(\mathfrak{g}_-) & \text{for generic } \lambda \\ \tilde{S}^\cdot(\tilde{\mathfrak{g}})/J(\lambda) & \text{otherwise.} \end{cases}$$

Roughly speaking, $U_\mathfrak{g}(\lambda)$ is "$\mathfrak{gl}$"(L^λ), where L^λ is the quotient of M^λ modulo the maximal submodule $I(\lambda)$ that can be determined and described with the help of the Shapovalov form, see [K], and $\tilde{S}^\cdot(\tilde{\mathfrak{g}})$ is the symmetric algebra generated by $\tilde{\mathfrak{g}}$. Clearly, $\tilde{S}^\cdot(\tilde{\mathfrak{g}})$ is smaller than $S^\cdot(\mathfrak{g})$ due to the relations between the differential operators. To explicitly describe the generators of $J(\lambda)$ is the main technical problem. We solve it in this paper for $\operatorname{rk}\mathfrak{g} = 1$. The general case will be considered elsewhere.

1) Let $LU_\mathfrak{g}(\lambda)$ be the Lie algebra whose space is the same as that of $U_\mathfrak{g}(\lambda)$ and the bracket is the commutator.

2) S. Montgomery suggested [M] a construction of simple Lie superalgebras:

$$Mo\text{: a central simple } \mathbb{Z}\text{-graded algebra} \mapsto \text{a simple Lie superalgebra.} \qquad (Mo)$$

Observe that the associative algebras $U_\mathfrak{g}(\lambda)$ constructed from simple Lie algebras $\mathfrak{g}$ are central simple. In [LM] we will consider *Montgomery superalgebras* $Mo(U_\mathfrak{g}(\lambda))$

and compare them with the Lie superalgebras $LU_{\mathfrak{s}}(\lambda)$ constructed from Lie superalgebras $\mathfrak{s}$. Montgomery's functor often produces new Lie superalgebras, e.g., if $\mathfrak{g}$ is equal to $\mathfrak{f}_4$ or $\mathfrak{e}_i$, though not always: $Mo(U_{\mathfrak{sl}(2)}(\lambda)) \cong LU_{\mathfrak{osp}(1|2)}(\lambda)$.

3) An outer automorphism a of $\mathfrak{G} = LU_{\mathfrak{g}}(\lambda)$ or $Mo(U_{\mathfrak{g}}(\lambda))$ might single out a new simple Lie subsuperalgebra $a_0(\mathfrak{G})$, the set of fixed points of $\mathfrak{G}$ under a.

For example, the intersection of $LU_{\mathfrak{sl}(2)}(\lambda)$ with the set of skew-adjoint differential operators is a new Lie algebra $\mathfrak{o}/\mathfrak{sp}(\lambda)$ while the intersection of $Mo(U_{\mathfrak{sl}(2)}(\lambda)) = LU_{\mathfrak{osp}(1|2)}(\lambda)$ with the set of superskew-adjoint operators is the Lie superalgebra $\mathfrak{osp}(\lambda+1|\lambda)$. For the description of the outer automorphisms of $LU_{\mathfrak{g}}(\lambda)$ and $Mo(U_{\mathfrak{g}}(\lambda))$ and the subalgebras of their fixed points see [LAS].

4) The deformations of Lie superalgebras obtained via steps 1) – 3) may lead to new algebras of class SFLAPG, cf. [Go]. A. Sergeev observed that it is interesting to find out what Lie algebras and Lie superalgebras we can get by applying the above constructions 1) – 3) to the quantum deformation $U_q(\mathfrak{g})$ of $U(\mathfrak{g})$. Does the dependence on q disappear or provide us with a deformation of $LU_{\mathfrak{g}}(\lambda)$?

Remark . The above procedure can be also applied to (twisted) loop algebras $\mathfrak{g} = \mathfrak{h}^{(k)}$ and the stringy algebras; the result will be realized with differential operators of infinitely many indeterminates; they remind vertex operators. The algebra $LU_{\mathfrak{h}^{(k)}}(\lambda)$ is a polynomial one but not of polynomial growth in the sense of Kac [K].

0.2. Another description of $U_{\mathfrak{g}}(\lambda)$

For the finite dimensional simple $\mathfrak{g}$ there is an alternative description of $U_{\mathfrak{g}}(\lambda)$ as the quotient of $U(\mathfrak{g})$ modulo the central character, i.e., the ideal J_λ generated by rank $\mathfrak{g}$ elements $C_i - k_i$, where the C_i is the i-th Casimir element and the k_i is the (computed by Harish-Chandra and Berezin) value of C_i on M^λ. This description of $U_{\mathfrak{g}}(\lambda)$ goes back, perhaps, to Kostant, cf. [Ka]. From this description it is clear that, after the shift by ρ, the half sum of positive roots, we get

$$LU_{\mathfrak{g}}(\sigma(\lambda)) \cong LU_{\mathfrak{g}}(\lambda) \quad \text{for any } \sigma \in W(\mathfrak{g})$$

and similarly for $Mo(U_{\mathfrak{g}}(\lambda))$. In particular, over $\mathbb{R}$, it suffices to consider the λ that belong to one Weyl chamber only.

For vectorial Lie algebras the latter description is inapplicable. For example, let $\mathfrak{g} = \mathfrak{vect}(n)$. The highest weight Verma modules are (for the standard filtration of $\mathfrak{g}$) identical with Verma modules over $\mathfrak{sl}(n+1)$, but the center of $U(\mathfrak{vect}(n))$ consists of constants only. It is a research problem to describe the generators of J_λ in such cases.

Though the center of $U(\mathfrak{g})$ is completely described by A. Sergeev for all simple finite dimensional Lie superalgebras [S], the problem of describing the *generators of the ideal* J_λ is open for Lie superalgebras $\mathfrak{g}$ even if $\mathfrak{g}$ is of the form $\mathfrak{g}(A)$ (i.e., if $\mathfrak{g}$ has Cartan matrix A) different from $\mathfrak{osp}(1|2n)$: for them the center of $U(\mathfrak{g})$ is not noetherian and it is *a priori* unclear if J_λ has infinitely many generators. (As we will show elsewhere, J_λ is generated for Lie superalgebras $\mathfrak{g}$ of the form $\mathfrak{g}(A)$ by the first rk$\mathfrak{g}$ Casimir operators and finitely many extra elements.)

0.3. Our result

The main result is the statement of the fact that the above constructions 1) – 4) yield a new class of simple Lie (super) algebras (some of which have nice properties).

Observe that our Lie algebras $LU_{\mathfrak{g}}(\lambda)$ are quantizations of the Lie algebras considered in [DGS] which are also of class SFLAPG and are contractions of our algebras. Indeed, Donin, Gurevich and Shnider consider the Lie algebras of functions on the orbits of the coajoint representation of $\mathfrak{g}$ with respect to the Poisson bracket. These DGS Lie algebras are naturally realized as the quotients of the polynomial algebra in an even number of indeterminates modulo a inhomogeneous ideal that singles out the orbit; we realize the result of quantization of DGS algebras by differential operators.

In this paper we consider the simplest case of the superization of this construction: replace $\mathfrak{sl}(2)$ with $\mathfrak{osp}(1|2)$. The cases of higher ranks will be considered elsewhere. The Khesin–Malikov construction [KM] can be applied almost literally to the Lie (super)algebras $LU_{\mathfrak{g}}(\lambda)$ such that $\mathfrak{g}$ admits a (super)principal embedding, see [GL2].

Our main theorems: 2.6 and 4.3. The structure of the algebras $LU_{\mathfrak{g}}(\lambda)$ (real forms, automorphisms, root systems) will be described elsewhere, see e.g., [LAS].

Observe that while the polynomial Poisson Lie algebra has only one class of nontrivial deformations and all the deformed algebras are isomorphic, the dimension of the space of parameters of deformations of Lie algebras of Donin, Gurevich and Shnider is equal to rank of $\mathfrak{g}$ and all of the deforms are pairwise nonisomorphic, generally.

0.4. The defining relations

The notion of defining relations is clear for a nilpotent Lie algebra. This is one of the reasons why the most conventional way to present a simple Lie algebra $\mathfrak{g}$ is to split it into the direct sum of a (commutative) Cartan subalgebra and 2 maximal nilpotent subalgebras $\mathfrak{g}_{\pm}$ (positive and negative). There are about $(2 \cdot \operatorname{rk}\mathfrak{g})^2$ relations between the $2 \cdot \operatorname{rk}\mathfrak{g}$ generators of $\mathfrak{g}_{\pm}$. The generators of $\mathfrak{g}_{+}$ together with the generators of $\mathfrak{g}_{-}$ generate $\mathfrak{g}$ as well. In $\mathfrak{g}$, there are about $(3 \cdot \operatorname{rk}\mathfrak{g})^2$ relations between these generators; the relations additional to those in $\mathfrak{g}_{+}$ or $\mathfrak{g}_{-}$, i.e., between the positive and the negative generators, are easy to grasp. Though numerous, all these relations — called *Serre relations* — are neat and this is another reason for their popularity. These relations are good to deal with not only for humans but for computers as well, cf. Sec. 7.3.

Nevertheless, it so happens that the Chevalley-type generators and, therefore, the Serre relations are not always available. Besides, as we will see, there are problems in which other generators and relations naturally appear, cf. [GL2].

Though not so transparent as for nilpotent algebras, the notion of generators and relations makes sense in the general case. For instance, with the principal embeddings of $\mathfrak{sl}(2)$ into $\mathfrak{g}$ one can associate only **two** elements that generate $\mathfrak{g}$; we call them *Jacobson's generators*, see [GL1]. We explicitly describe the associated with the principal embeddings of $\mathfrak{sl}(2)$ presentations of simple Lie algebras, finite dimensional and certain infinite dimensional; namely, the Lie algebra "of matrices of complex size" realized as a subalgebra of the Lie algebra $\mathfrak{diff}(1)$ of differential operators in 1 indeterminate or of $\mathfrak{gl}_+(\infty)$, see §2.

The relations obtained are rather simple, especially for nonexceptional algebras. In contradistinction with the conventional presentation there are just 9 relations between Jacobson's generators for $\mathfrak{p}\mathfrak{O}\mathfrak{L}(\lambda)$ series (actually, 8 if $\lambda \in \mathbb{C}\setminus\mathbb{Z}$) and not many more for the other algebras.

It is convenient to present $\mathfrak{p}\mathfrak{O}\mathfrak{L}(\lambda)$ as the Lie algebra generated by two differential operators: $X^+ = u^2d/du - (\lambda - 1)u$ and $Z_{\mathfrak{sl}} = d^2/du^2$; its Lie subalgebra $\mathfrak{o}/\mathfrak{sp}(\lambda)$ of skew-adjoint operators — a hybrid of Lie algebras of series $\mathfrak{o}$ and $\mathfrak{sp}$ (do not confuse with the Lie superalgebra of $\mathfrak{osp}$ type!) — is generated by the same X^+ and $Z_{\mathfrak{o}/\mathfrak{sp}} = d^3/du^3$; to make relations simpler, we always add the third generator $X^- = -d/du$. For integer $\lambda = n$ each of these algebras has an ideal of finite codimension and the quotient modulo the ideal is the conventional $\mathfrak{sl}(n)$ and $\mathfrak{o}(2n+1)$ or $\mathfrak{sp}(2n)$, respectively.

In this paper we superize [GL1]: replace $\mathfrak{sl}(2)$ with its closest relative, $\mathfrak{osp}(1|2)$. We denote by $\mathfrak{p}\mathfrak{O}\mathfrak{L}(\lambda|\lambda+1)$ the Lie superalgebra generated by $\nabla^+ = x\partial_\theta + x\theta\partial_x - \lambda\theta$, $Z = \partial_x\partial_\theta - \theta\partial_x{}^2$ and $U = \partial_\theta - \theta\partial_x$, where x is an even indeterminate and θ is an odd one. We define $\mathfrak{osp}(\lambda+1|\lambda)$ as the Lie subsuperalgebra of $\mathfrak{p}\mathfrak{O}\mathfrak{L}(\lambda|\lambda+1)$ generated by ∇^+ and Z. The presentations of $\mathfrak{p}\mathfrak{O}\mathfrak{L}(\lambda|\lambda+1)$ and $\mathfrak{osp}(\lambda+1|\lambda)$ are associated with the *superprincipal* embeddings of $\mathfrak{osp}(1|2)$. For $\lambda \in \mathbb{C}\setminus\mathbb{Z}$ these algebras are simple. For integer $\lambda = n$ each of these algebras has an ideal of finite codimension and the quotient modulo the ideal is the conventional $\mathfrak{sl}(n|n+1)$ and $\mathfrak{osp}(2n+1|2n)$, respectively.

0.5. Some applications

(1) Integrable systems like continuous Toda lattice or a generalization of the Drinfeld–Sokolov construction are based on the superprincipal embeddings in the same way as the Khesin–Malikov construction [KM] is based on the principal embedding, cf. [GL2]. (2) To q-quantize the Lie algebras of type $\mathfrak{p}\mathfrak{O}\mathfrak{L}(\lambda)$ à la Drinfeld, using only Chevalley generators, is impossible; our generators indicate a way to do it.

0.6. Related topics

We would like to draw attention of the reader to several other classes of Lie algebras. One of the reasons is that, though some of these classes have empty intersections with the class of Lie algebras we consider here, they naturally spring to mind and are, perhaps, deformations of our algebras in some, yet unknown, sense.

• *Krichever-Novikov algebras*, see [SH] and references therein. The KN-algebras are neither graded, nor filtered (at least, with respect to the degree considered usually). Observe that so are our algebras $LU_{\mathfrak{g}}(\lambda)$ with respect to the degree induced from $U(\mathfrak{g})$, so a search for a better grading is a tempting problem.

• *Odessky* or *Sklyanin algebras*, see [FO] and references therein.

• *Continuum algebras*, see [SV] and references therein. In particular cases these algebras coincide with Kac–Moody or loop algebras, i.e., have a continuum analogue of the Cartan matrix. But to suspect that $\mathfrak{GL}(\lambda)$ has a Cartan matrix is wrong, see Sec. 2.2

§1. Recapitulation: finite dimensional simple Lie algebras

This section is a continuation of [LP], where the case of the simplest base (system of simple roots) is considered and where non-Serre relations for simple Lie algebras first appear, though in a different setting. This paper is also the direct superization of [GL1]; we recall its results. For presentations of Lie superalgebras with Cartan matrix via Chevalley generators, see [LS], [GL3].

What are "natural" generators and relations for a *simple finite dimensional Lie algebra*? The answer is important in questions when it is needed to identify an algebra $\mathfrak{g}$ given its generators and relations. (Examples of such are Estabrook–Vahlquist prolongations, Drinfeld's quantum algebras, symmetries of differential equations, integrable systems, etc.).

1.0. Defining relations

If $\mathfrak{g}$ is nilpotent, the problem of its presentation has a natural and unambiguous solution: representatives of the homology $H_1(\mathfrak{g}) \cong \mathfrak{g}/[\mathfrak{g}, \mathfrak{g}]$ are the generators of $\mathfrak{g}$ and the elements from $H_2(\mathfrak{g})$ correspond to relations.

On the other hand, if $\mathfrak{g}$ is simple, then $\mathfrak{g} = [\mathfrak{g}, \mathfrak{g}]$ and there is no "most natural" way to select generators of $\mathfrak{g}$. The choice of generators is not unique.

Still, among algebras with the property $\mathfrak{g} = [\mathfrak{g}, \mathfrak{g}]$ the simple ones are distinguished by the fact that their structure is very well known. By trial and error people discovered that for finite dimensional simple Lie algebras, there are certain "first among equal" sets of generators:

1) *Chevalley generators* corresponding to positive and negative simple roots;

2) a pair of generators that generate any finite dimensional simple Lie algebra associated with the *principal* $\mathfrak{sl}(2)$*-subalgebra* (considered below).

The relations associated with Chevalley generators are well-known, see e.g., [OV], [K]. These relations are called *Serre relations.*

The possibility to generate any simple finite dimensional Lie algebra by two elements was first claimed by N. Jacobson; for the first (as far as we know) proof see [BO]. We do not know what generators Jacobson had in mind; [BO] take for them linear combinations of positive and negative root vectors with generic coefficients; nothing like a "natural" choice that we suggest to refer to as *Jacobson's generators* was ever proposed.

To generate a simple algebra with only two elements is tempting but nobody yet had explicitly described relations between such generators, perhaps, because to check whether the relations between these elements are nice-looking is impossible without a modern computer (cf. an implicit description in [F]). As far as we could test, the relations for any other pair of generators chosen in a way distinct from ours are too complicated. There seem to be, however, one exception cf. [GL2].

1.1. The principal embeddings

There exists only one (up to equivalence) embedding $r : \mathfrak{sl}(2) \longrightarrow \mathfrak{g}$ such that $\mathfrak{g}$, considered as $\mathfrak{sl}(2)$-module, splits into $\operatorname{rk}\mathfrak{g}$ irreducible modules, cf. [D] or [OV]. This embedding is called *principal* and, sometimes, *minimal* because for the other embeddings (there are plenty of them) the number of irreducible $\mathfrak{sl}(2)$-modules is $> \operatorname{rk}\mathfrak{g}$. Example: for $\mathfrak{g} = \mathfrak{sl}(n)$, $\mathfrak{sp}(2n)$ or $\mathfrak{o}(2n+1)$ the principal embedding is the one corresponding to the irreducible representation of $\mathfrak{sl}(2)$ of dimension n, $2n$, $2n+1$, respectively.

For completeness, let us recall what the irreducible $\mathfrak{sl}(2)$-modules with highest weight look like. (They are all of the form $M^\mu = L^\mu$, $\mu \notin \mathbb{Z}_+$, and L^n, $n \in \mathbb{Z}_+$, described below.) Select the following basis in $\mathfrak{sl}(2)$:

$$X^- = \begin{pmatrix} 0 & 0 \\ -1 & 0 \end{pmatrix}, \quad H = \begin{pmatrix} 1 & 0 \\ 0 & -1 \end{pmatrix}, \quad X^+ = \begin{pmatrix} 0 & 1 \\ 0 & 0 \end{pmatrix}.$$

The $\mathfrak{sl}(2)$-module M^μ is illustrated with a graph whose nodes correspond to the eigenvectors $l_{\mu-2i}$ of H with the weight indicated;

$$\dots \overset{\mu-2i-2}{\circ} - \overset{\mu-2i}{\circ} - \dots - \overset{\mu-2}{\circ} - \overset{\mu}{\circ}$$

the edges depict the action of $X^\pm$ (the action of X^+ is directed to the right, that of X^- to the left: $X^- l_{\mu-2i} = l_{\mu-2i-2}$ and

$$X^+ l_{\mu-2i} = X^+((X^-)^i l_\mu) = i(\mu - i + 1) l_{\mu-2i+2}; \quad X^+(l_\mu) = 0. \tag{1.1}$$

As follows from (1.1), the module M^n for $n \in \mathbb{Z}_+$ has an irreducible submodule isomorphic to M^{-n-2}; the quotient, obviously irreducible, as follows from the same (1.1), will be denoted by L^n.

There are principal $\mathfrak{sl}(2)$-subalgebras in every finite dimensional simple Lie algebra, though, generally, not in infinite dimensional ones, e.g., not in affine Kac-Moody algebras. The construction is as follows. Let $X_1^\pm, \dots, X_{\operatorname{rk}\mathfrak{g}}^\pm$ be Chevalley generators of $\mathfrak{g}$, i.e., the generators corresponding to simple roots. Let the images of $X^\pm \in \mathfrak{sl}(2)$ in $\mathfrak{g}$ be

$$X^- \mapsto \sum X_i^-; \quad X^+ \mapsto \sum a_i X_i^+$$

and select the a_i from the relations $[[X^+, X^-], X^\pm] = \pm 2X^\pm$ true in $\mathfrak{sl}(2)$. For $\mathfrak{g}$ constructed from a Cartan matrix A there is a solution if and only if A is invertible.

In Table 1.1 a simple finite dimensional Lie algebra $\mathfrak{g}$ is described as the $\mathfrak{sl}(2)$-module corresponding to the principal embedding (cf. [OV], Table 4). The table introduces the number $2k_2$ used in relations.

Table 1.1. $\mathfrak{g}$ **as the** $\mathfrak{sl}(2)$**-module**

$\mathfrak{g}$	the $\mathfrak{sl}(2)$-spectrum of $\mathfrak{g} = L^2 \oplus L^{2k_2} \oplus L^{2k_3} \dots$	$2k_2$
$\mathfrak{sl}(n)$	$L^2 \oplus L^4 \oplus L^6 \dots \oplus L^{2n-2}$	4
$\mathfrak{o}(2n+1)$, $\mathfrak{sp}(2n)$	$L^2 \oplus L^6 \oplus L^{10} \dots \oplus L^{4n-2}$	6
$\mathfrak{o}(2n)$	$L^2 \oplus L^6 \oplus L^{10} \dots \oplus L^{4n-2} \quad \oplus L^{2n-2}$	6
$\mathfrak{g}_2$	$L^2 \oplus L^{10}$	10
$\mathfrak{f}_4$	$L^2 \oplus L^{10} \oplus L^{14} \oplus L^{22}$	10
$\mathfrak{e}_6$	$L^2 \oplus L^8 \oplus L^{10} \oplus L^{14} \oplus L^{16} \oplus L^{22}$	8
$\mathfrak{e}_7$	$L^2 \oplus L^{10} \oplus L^{14} \oplus L^{18} \oplus L^{22} \oplus L^{26} \oplus L^{34}$	10
$\mathfrak{e}_8$	$L^2 \oplus L^{14} \oplus L^{22} \oplus L^{26} \oplus L^{34} \oplus L^{38} \oplus L^{46} \oplus L^{58}$	14

One can show that $\mathfrak{g}$ can be generated by two elements: $x := X^+ \in L^2 = \mathfrak{sl}(2)$ and a lowest weight vector $z := l_{-r}$ from an appropriate module L^r other than L^2 from Table 1.1. For the role of this L^r we take either L^{2k_2} if $\mathfrak{g} \neq \mathfrak{o}(2n)$ or the last module L^{2n-2} in the above table if $\mathfrak{g} = \mathfrak{o}(2n)$. (Clearly, z is defined up to proportionality; we will assume that a basis of L^r is fixed and denote $z = t \cdot l_{-r}$ for some $t \in \mathbb{C}$ that can be fixed at will, cf. §3.)

The exceptional choice for $\mathfrak{o}(2n)$ is occasioned by the fact that by choosing $z \in L^r$ for $r \neq 2n-2$ instead, we generate $\mathfrak{o}(2n-1)$.

We call the above x and z, together with $y := X^- \in L^2$ for good measure, *Jacobson's generators.* The presence of y considerably simplifies the form of the relations, though slightly increases their number. (One might think that taking the symmetric to z element l_r will improve the relations even more but in reality just the opposite happens.)

Concerning $\mathfrak{g} = \mathfrak{o}(2n)$ see sec. 7.2.

1.2. Relations between Jacobson's generators

First, observe that if an ideal of a free Lie algebra is homogeneous (with respect to the degrees of the generators of the algebra), then the number and the degrees of the defining relations (i.e., the generators of the ideal) is uniquely defined provided the relations are homogeneous. This is obvious.

A simple Lie algebra $\mathfrak{g}$, however, is the quotient of a free Lie algebra $\mathfrak{F}$ modulo a inhomogeneous ideal, $\mathfrak{I}$, the ideal without homogeneous generators. Therefore, we can speak about the number and the degrees of relations only conditionally. Our condition is the possibility to express any element $x \in \mathfrak{I}$ via the generators $g_1, \dots$ of $\mathfrak{I}$ by a formula of the form

$$x = \sum [c_i, g_i], \text{ where } c_i \in \mathfrak{F} \text{ and } \deg c_i + \deg g_i \leq \deg x \text{ for all } i. \qquad (*)$$

Under condition $(*)$ the number of relations and their degrees are uniquely determined. Now we can explain why we need an extra generator y: without y the weight relations would have been of very high degree.

We divide the relations between the Jacobson generators into the types corresponding to the number of occurrences of z in them: **0**. Relations in $L^2 = \mathfrak{sl}(2)$; **1**. Relations coming from the $\mathfrak{sl}(2)$-action on L^{2k_2}; **2**. Relations coming from $L^{2k_2} \wedge L^{2k_2}$; $\geq$ **3**. Relations coming from $L^{2k_2} \wedge L^{2k_2} \wedge L^{2k_2} \wedge \dots$ with ≥ 3 factors; among the latter relations we distinguish one — of type "∞" — the relation that shears the dimension. (For small rank $\mathfrak{g}$ the relation of type ∞ can be of the above types.)

Observe that apart form relations of type ∞ the relations of type ≥ 3 are those of type 3 except for $\mathfrak{e}_7$ which satisfies stray relations of type 4 and 5, cf. [GL1].

The relations of type 0 are the well-known relations in $\mathfrak{sl}(2)$

$$\textbf{0.1.}\ \ [[x,y],x] = 2x, \qquad \textbf{0.2}\ \ [[x,y],y] = -2y. \qquad \text{(Rel 0)}$$

The relations of type 1 mirror the fact that the space L^{2k_2} is the $(2k_2+1)$-dimensional $\mathfrak{sl}(2)$-module. To simplify notation we denote: $z_i = (\operatorname{ad} x)^i z$. Then the type 1 relations are:

1.1. $[y,z]=0$, **1.2.** $[[x,y],z] = -2k_2 z$, **1.3.** $z_{2k_2+1} = 0$ with $2k_2$ from Table 1.1.. (Rel1)

1.3. Theorem. *For the simple finite dimensional Lie algebras all the relations between the Jacobson generators are the above relations* (Rel 0), (Rel 1) *and the relations from* [GL1].

In §3 these relations are reproduced for the classical Lie algebras.

§2. The Lie algebra $\mathfrak{p}\mathfrak{GL}(\lambda)$ as a quotient algebra of $\mathfrak{diff}(1)$ and a subalgebra of $\mathfrak{sl}_+(\infty)$

2.1. $\mathfrak{GL}(\lambda)$ is endowed with a trace

The Poincaré-Birkhoff-Witt theorem states that, as spaces, $U(\mathfrak{sl}(2)) \cong \mathbb{C}[X^-, H, X^+]$. We also know that to study representations of $\mathfrak{g}$ is the same as to study representations of $U(\mathfrak{g})$. Still, if we are interested in irreducible representations, we do not need the whole of $U(\mathfrak{g})$ and can do with a smaller algebra, easier to study.

This observation is used now and again; Feigin applied it in [F] writing, actually, (as deciphered in [PH], [GL1], [Sh]) that setting

$$X^- = -\frac{d}{du}, \quad H = 2u\frac{d}{du} - (\lambda - 1), \quad X^+ = u^2\frac{d}{du} - (\lambda-1)u \qquad (2.1)$$

we obtain a morphism of $\mathfrak{sl}(2)$-modules and moreover, of associative algebras: $U(\mathfrak{sl}(2)) \longrightarrow \mathbb{C}[u, d/du]$. The kernel of this morphism is the ideal generated by $\Delta - \lambda^2 + 1$, where $\Delta = 2(X^+X^- + X^-X^+) + H^2$. Observe, that this morphism is not an epimorphism, either. The image of this morphism is our Lie algebra of matrices of "complex size".

Remark. In their proof of certain statements from [F] that we will recall, [PH] made use of the well-known fact that the Casimir operator Δ acts on the irreducible $\mathfrak{sl}(2)$-module L^μ (see sec 1.1) as the scalar operator of multiplication by $\mu^2 + 2\mu$. The passage from [PH]'s λ to [F]'s μ is done with the help of a shift by the weight ρ, a half sum of positive roots, which for $\mathfrak{sl}(2)$ can be identified with 1, i.e., $(\lambda-1)^2 + 2(\lambda-1) = \lambda^2 - 1$ for $\lambda = \mu - 1$.

Consider the Lie algebra $LU(\mathfrak{sl}(2))$ associated with the associative algebra $U(\mathfrak{sl}(2))$. Set

$$U_\lambda = U(\mathfrak{sl}(2))/(\Delta - \lambda^2 + 1). \tag{2.2}$$

It is easy to see that, as an $\mathfrak{sl}(2)$-module,

$$LU_\lambda = L^0 \oplus L^2 \oplus L^4 \oplus \cdots \oplus L^{2n} \oplus \ldots \tag{2.3}$$

It is not difficult to show (see [PH] for details) that the Lie algebra LU_n for $n \in \mathbb{N} \setminus \{0,1\}$ contains an ideal I_n and the quotient LU_n/I_n is the conventional $\mathfrak{gl}(n)$. In [PH] it is proved that for $\lambda \neq \mathbb{Z}$ the Lie algebra LU_λ has only one ideal — the space L^0 of constants. Therefore, set

$$\mathfrak{p}\mathfrak{G}\mathfrak{L}(\lambda) = \mathfrak{gl}(\lambda)/L^0, \text{ where } \mathfrak{gl}(\lambda) = \begin{cases} LU_\lambda & \text{for } \lambda \notin \mathbb{N} \setminus \{0,1\} \\ LU_n/I_n & \text{for } n \in \mathbb{N} \setminus \{0,1\}. \end{cases} \tag{2.4}$$

The definition directly implies that $\mathfrak{G}\mathfrak{L}(-\lambda) \cong \mathfrak{G}\mathfrak{L}(\lambda)$, so speaking about real values of λ we can confine ourselves to the nonnegative values, cf. Sec. 0.2.

Observe, that $\mathfrak{G}\mathfrak{L}(\lambda)$ is endowed with a trace. This follows directly from (2.3) and the fact that

$$\mathfrak{G}\mathfrak{L}(\lambda) \cong L^0 \oplus [\mathfrak{G}\mathfrak{L}(\lambda), \mathfrak{G}\mathfrak{L}(\lambda)].$$

Therefore, $\mathfrak{p}\mathfrak{G}\mathfrak{L}(\lambda)$ can be identified with $\mathfrak{S}\mathfrak{L}(\lambda)$, the subalgebra of the traceless matrices in $\mathfrak{G}\mathfrak{L}(\lambda)$. We can normalize the trace at will, for example, if we set $\mathrm{tr}(id) = \lambda$, then the trace that our trace induces on the quotient of $LU_{\mathfrak{sl}(2)}(n)$ modulo $J(n)$ coincides with the usual trace on $\mathfrak{gl}(n)$ for $n \in \mathbb{N}$.

Another way to introduce the trace was suggested by J. Bernstein. We decipher its description in [KM] as follows. Look at the image of $H \in \mathfrak{sl}(2)$ in $\mathfrak{gl}(M^\lambda)$. Bernstein observed that though the trace of the image is an infinite sum, the sum of the first $D+1$ summands is a polynomial in D, call it $\mathrm{tr}(H)$. It is easy to see that $\mathrm{tr}(H)$ vanishes if $D = \lambda$.

Similarly, for *any* $x \in LU_{\mathfrak{g}}(\lambda)$ considered as an element of $\mathfrak{gl}(M^\lambda)$ set

$$\mathrm{tr}(x; D) = \sum_{i=1}^{D} x_{ii}.$$

Let $D = D(\lambda)$ be the value of the dimension of the irreducible finite dimensional $\mathfrak{g}$-module with highest weight λ, for an exact formula see [D], [OV]. Set $\mathrm{tr}(x) = \mathrm{tr}(x; D(\lambda))$; as is easy to see, this formula determines the trace on $LU_{\mathfrak{g}}(\lambda)$ for arbitrary values of λ.

2.2. There is no analog of Cartan matrix for $\mathfrak{SL}(\lambda)$

Are there *Chevalley generators*, i.e., elements $X_i^{\pm}$ of degree ± 2 and H_i of degree 0 (the *degree* is the weight with respect to the $\mathfrak{sl}(2) = L^2 \subset \mathfrak{SL}(\lambda)$) such that

$$[X_i^+, X_j^-] = \delta_{ij} H_i, \quad [H_i, H_j] = 0 \text{ and } [H_i, X_j^{\pm}] = \pm A_{ij} X_j^{\pm}? \qquad (2.5)$$

The answer is **NO**: $\mathfrak{SL}(\lambda)$ is too small. To see what the problem is, consider the following elements of degree ± 2 from L^4 and L^6 of $\mathfrak{gl}(\lambda)$:

$$\deg = -2 : -4uD^2 - 2(\lambda - 2)D$$
$$\deg = 2 : -4u^3D^2 + 6(\lambda - 2)u^2D - 2(\lambda - 1)(\lambda - 2)u$$

$$\deg = -2 : 15u^2D^3 - 15(\lambda - 3)uD^2 + 3(\lambda - 2)(\lambda - 3)D$$
$$\deg = 2 : 15u^4D^3 - 30(\lambda - 3)u^3D^2 + 18(\lambda - 2)(\lambda - 3)u^2D$$
$$-3(\lambda - 1)(\lambda - 2)(\lambda - 3)u$$

To satisfy (2.5), we can complete $\mathfrak{gl}(\lambda)$ by considering infinite sums of its elements, but the completion erases the difference between different λ's:

Proposition . *For $\lambda \neq \rho$ the completion of $\overline{\mathfrak{SL}(\lambda)}$ generated by Jacobson's generators* (see Tables) *is isomorphic to $\mathfrak{pdiff}(1)$, the quotient of the Lie algebra of differential operators with formal coefficients modulo constants.*

Thoug $\mathfrak{SL}(\lambda)$ has no Cartan matrix, it has its analog, a nonlinear *Cartan operator*, which makes it one of the first examples of Saveliev-Vershik continuum algebras.

2.3. The outer automorphism of $LU_{\mathfrak{g}}(\lambda)$

The invariants of the mapping $X \mapsto -SX^tS$ for;

$$X \in \mathfrak{gl}(n), \text{ where } S = \begin{cases} antidiag(1, \dots, 1) & \text{for } n \in 2\mathbb{N} + 1 \\ antidiag(1, \dots, 1, -1, \dots, -1) & \text{for } n \in 2\mathbb{N}. \end{cases} \qquad (2.6)$$

constitute $\mathfrak{o}(n)$ if $n \in 2\mathbb{N} + 1$ and $\mathfrak{sp}(n)$ if $n \in 2\mathbb{N}$. By analogy, Feigin defined $\mathfrak{o}(\lambda)$ and $\mathfrak{sp}(\lambda)$ as subalgebras of $\mathfrak{gl}(\lambda) = \underset{k \geq 0}{\oplus} L^{2k}$ invariant with respect to the involution

$$X \mapsto \begin{cases} -X & \text{if } X \in L^{4k} \\ X & \text{if } X \in L^{4k+2}, \end{cases} \qquad (2.7)$$

the analogue of (2.6). Since $\mathfrak{o}(\lambda)$ and $\mathfrak{sp}(\lambda)$ — the subalgebras of $\mathfrak{gl}(\lambda)$ singled out by the involution (2.7) — differ by a shift of the parameter λ, it is natural to denote them uniformly (but so as not to confuse with the Lie superalgebras of series $\mathfrak{osp}$), namely, by $\mathfrak{o}/\mathfrak{sp}(\lambda)$. For integer values of the parameter it is clear that

$$\mathfrak{o}/\mathfrak{sp}(\lambda) = \begin{cases} \mathfrak{o}(\lambda) \Subset I_\lambda & \text{if } \lambda \in 2\mathbb{N} + 1, \\ \mathfrak{sp}(\lambda) \Subset I_\lambda & \text{if } \lambda \in 2\mathbb{N}, \end{cases} \quad \text{where } I_\lambda \text{ is an ideal.}$$

In the realization of $\mathfrak{GL}(\lambda)$ by differential operators the transposition is the passage to the adjoint operator; hence, $\mathfrak{o}/\mathfrak{sp}(\lambda)$ is a subalgebra of $\mathfrak{GL}(\lambda)$ consisting of self-skew-adjoint operators with respect to the involution

$$a(u)\frac{d^k}{du^k} \mapsto (-1)^k\frac{d^k}{du^k}a(u)^*. \tag{2.8}$$

The superization of this formula is straightforward: via the Sign Rule.

2.4. The Lie algebra $\mathfrak{gl}(\lambda)$ as a subalgebra of $\mathfrak{gl}_+(\infty)$

Recall that $\mathfrak{gl}_+(\infty)$ often denotes the Lie algebra of infinite (in one direction; the index + indicates that) matrices with nonzero elements inside a strip (depending on the matrix) along the main diagonal and containing it. The subalgebras $\mathfrak{o}(\infty)$ and $\mathfrak{sp}(\infty)$ are naturally defined, while $\mathfrak{sl}(\infty)$ is, by abuse of language, sometimes used to denote $\mathfrak{p}\mathfrak{GL}(\infty)$.

The realization (2.1) provides an embedding $\mathfrak{GL}(\lambda) \subset \mathfrak{sl}_+(\infty) =$ "$\mathfrak{sl}(M^\lambda)$", so for $\lambda \neq \mathbb{N}$ the Verma module M^λ with highest weight μ is an irreducible $\mathfrak{GL}(\lambda)$-module.

Proposition . *The completion of $\mathfrak{gl}(\lambda)$ (generated by the elements of degree ± 2 with respect to $H \in \mathfrak{sl}(2) \subset \mathfrak{gl}(\lambda)$) is isomorphic for any noninteger λ to $\mathfrak{gl}_+(\infty) =$ "$\mathfrak{gl}(M^\lambda)$".*

2.5. The Lie algebras $\mathfrak{GL}(*)$ and $\mathfrak{o}/\mathfrak{sp}(*)$ for $* \in \mathbb{C}P^1 = \mathbb{C} \cup \{*\}$

The "dequantization" of the relations for $\mathfrak{GL}(\lambda)$ and $\mathfrak{o}/\mathfrak{sp}(\lambda)$ (see §3) is performed by passage to the limit as $\lambda \longrightarrow \infty$ under the change:

$$t \mapsto \begin{cases} t/\lambda & \text{for } \mathfrak{GL}(\lambda) \\ t/\lambda^2 & \text{for } \mathfrak{o}/\mathfrak{sp}(\lambda). \end{cases}$$

So the parameter λ above can actually run over $\mathbb{C}P^1 = \mathbb{C} \cup \{*\}$, not just $\mathbb{C}$. In the realization with the help of deformation, cf. 2.7 below, this is obvious. Denote the limit algebras by $\mathfrak{GL}(*)$ and $\mathfrak{o}/\mathfrak{sp}(*)$ in order to distinguish them from $\mathfrak{sl}(\infty)$ and $\mathfrak{o}(\infty)$ or $\mathfrak{sp}(\infty)$ from sec. 2.4.

It is clear that it is impossible to embed $\mathfrak{GL}(*)$ and $\mathfrak{o}/\mathfrak{sp}(*)$ into the "quadrant" algebra $\mathfrak{sl}_+(\infty)$: indeed, $\mathfrak{GL}(*)$ and $\mathfrak{o}/\mathfrak{sp}(*)$ are subalgebras of the whole "plane" algebras $\mathfrak{sl}(\infty)$ and $\mathfrak{o}(\infty)$ or $\mathfrak{sp}(\infty)$.

2.6. Theorem. *For Lie algebras $\mathfrak{GL}(\lambda)$ and $\mathfrak{o}/\mathfrak{sp}(\lambda)$, $\lambda \in \mathbb{C}P^1$, all the relations between the Jacobson generators are the relations of types* 0, 1 *with* $2k_2$ *found from Table* 1.1 *and the borrowed from* [GL1] *relations from Tables in* §3.

§3. Jacobson's generators and relations between them

In what follows the E_{ij} are the matrix units; $X_i^\pm$ stand for the conventional Chevalley generators of $\mathfrak{g}$. For $\mathfrak{GL}(\lambda)$ and $\mathfrak{o}/\mathfrak{sp}(\lambda)$ the generators $x = u^2d/du - (\lambda - 1)u$ and $y = -d/du$ are the same; $z_{\mathfrak{sl}} = td^2/du^2$ while $z_{\mathfrak{o}/\mathfrak{sp}} = td^3/du^3$. For $n \in \mathbb{C} \setminus \mathbb{Z}$ there is no shearing relation of type ∞; for $n = * \in \mathbb{C}P^1$ the relations are obtained with the substitution 2.5. The parameter t can be taken equal to 1 we kept it explicit to clarify "dequantization" of relations as $\lambda \longrightarrow \infty$.

$\mathfrak{SL}(*)$.

$$\begin{aligned} &2.1. \quad 3[z_1, z_2] - 2[z, z_3] = 24t^2 y, \\ &3.1. \quad [z, [z, z_1] = 0, \\ &3.2. \quad 4[[z, z_1], z_3]]] + 3[z_2, [z, z_2]] = -576t^2 z. \end{aligned}$$

$\mathfrak{o}/\mathfrak{sp}(*)$.

$$\begin{aligned} &2.1. \quad 2[z_1, z_2] - [z, z_3] = 72tz, \\ &2.2. \quad 9[z_2, z_3] - 5[z_1, z_4] = 216tz_2 - 432t^2 y, \\ &3.1. \quad [z, [z, z_1] = 0, \\ &3.2. \quad 7[[z, z_1], z_3] + 6[z_2, [z, z_2]] = -720t[z, z_1]. \end{aligned}$$

$\mathfrak{sl}(n)$ for $n \geq 3$. Generators:

$$x = \sum_{1 \leq i \leq n-1} i(n-i)E_{i,i+1}, \qquad y = \sum_{1 \leq i \leq n-1} E_{i+1,i}, \qquad z = t \sum_{1 \leq i \leq n-2} E_{i+2,i}.$$

Relations:

$$\begin{aligned} &2.1. \quad 3[z_1,\ z_2] - 2[z,\ z_3] = 24t^2(n^2-4)y, \\ &3.1. \quad [z,\ [z, z_1]] = 0, \\ &3.2. \quad 4[z_3,\ [z, z_1]] - 3[z_2,\ [z, z_2]] = 576t^2(n^2-9)z. \\ \infty = n-1. &\quad (\operatorname{ad} z_1)^{n-2} z = 0. \end{aligned}$$

For $n = 3, 4$ the degree of the last relation is lower than the degree of some other relations, this yields simplifications.

$\mathfrak{o}(2n+1)$ for $n \geq 3$. Generators:

$$x = n(n+1)(E_{n+1,2n+1} - E_{n,n+1}) + \sum_{1 \leq i \leq n-1} i(2n+1-i)(E_{i,i+1} - E_{n+i+2,n+i+1}),$$

$$y = (E_{2n+1,n+1} - E_{n+1,n}) + \sum_{1 \leq i \leq n-1} (E_{i+1,i} - E_{n+i+1,n+i+2}),$$

$$z = t\big((E_{2n-1,n+1} - E_{n+1,n-2}) - (E_{2n+1,n-1} - E_{2n,n}) + \sum_{1 \leq i \leq n-3} (E_{i+3,i} - E_{n+i+1,n+i+4})\big).$$

Relations:

$$\begin{aligned}
\mathbf{2}.1.\quad & 2[z_1,\, z_2] - [z,\, z_3] = 144t(2n^2 + 2n - 9)z, \\
\mathbf{2}.2.\quad & 9[z_2,\, z_3] - 5[z_1,\, z_4] = 432t(2n^2 + 2n - 9)z_2 \\
& + 1728t^2(n-1)(n+2)(2n-1)(2n+3)y, \\
\mathbf{3}.1.\quad & [z,\, [z, z_1]] = 0, \\
\mathbf{3}.2.\quad & 7[z_3,\, [z, z_1]] - 6[z_2,\, [z, z_2]] = 2880t(n-3)(n+4)[z, z_1], \\
\infty = \boldsymbol{n}.\quad & (\operatorname{ad} z_1)^{n-1} z = 0.
\end{aligned}$$

$\underline{\mathfrak{sp}(2n)}$ for $n \geq 3$. Generators:

$$x = n^2 E_{n,2n} + \sum_{1 \leq i \leq n-1} i(2n-i)(E_{i,i+1} - E_{n+i+1,n+i}),$$

$$y = E_{2n,n} + \sum_{1 \leq i \leq n-1} (E_{i+1,i} - E_{n+i,n+i+1}),$$

$$z = t\Big((E_{2n,n-2} + E_{2n-2,n}) - E_{2n-1,n-1} + \sum_{1 \leq i \leq n-3} (E_{i+3,i} - E_{n+i,n+i+3})\Big).$$

Relations:

$$\begin{aligned}
\mathbf{2}.1.\quad & 2[z_1,\, z_2] - [z,\, z_3] = 72t(4n^2 - 19)z, \\
\mathbf{2}.2.\quad & 9[z_2,\, z_3] - 5[z_1,\, z_4] = 216t(4n^2 - 19)z_2 + 1728t^2(n^2-1)(4n^2-9)y, \\
\mathbf{3}.1.\quad & [z,\, [z, z_1]] = 0, \\
\mathbf{3}.2.\quad & 7[z_3,\, [z, z_1]] - 6[z_2,\, [z, z_2]] = 720t(4n^2 - 49)[z, z_1], \\
\infty = \boldsymbol{n}.\quad & (\operatorname{ad} z_1)^{n-1} z = 0.
\end{aligned}$$

For Jacobson generators and corresponding defining relations for the exceptional Lie algebras see [GL1].

§4. Lie superalgebras

4.0. Linear algebra in superspaces

Superization has certain subtleties, often disregarded or expressed too briefly. We will dwell on them a bit, see [L2].

A *superspace* is a $\mathbb{Z}/2$-graded space; for a superspace $V = V_{\bar{0}} \oplus V_{\bar{1}}$ denote by $\Pi(V)$ another copy of the same superspace: with the shifted parity, i.e., $(\Pi(V))_{\bar{i}} = V_{\bar{i}+\bar{1}}$.

A superspace structure in V induces that in the space $\operatorname{End}(V)$. A *basis of a superspace* is always a basis consisting of *homogeneous* vectors; let $Par = (p_1, \dots, p_{\dim V})$ be an ordered collection of their parities, called the *format* of (the basis of) V. A square *supermatrix* of format (size) Par is a $\dim V \times \dim V$ matrix whose ith row and ith column are said to be of parity p_i. The matrix unit E_{ij} is supposed to be of parity $p_i + p_j$ and the bracket of supermatrices (of the same format) is defined via Sign Rule: *if something of parity p moves past something of parity q the* *sign* $(-1)^{pq}$ *accrues; the formulas defined on homogeneous elements are extended to arbitrary ones via linearity.* For example: $[X,Y] = XY - (-1)^{p(X)p(Y)}YX$; the sign $\wedge$ in what follows is also understood in supersence, etc.

Usually, *Par* is considered to be of the form $(\bar{0},\dots,\bar{0},\bar{1},\dots,\bar{1})$. Such a format is called *standard.* The Lie superalgebra of supermatrices of size *Par* is denoted by $\mathfrak{gl}(Par)$, usually $\mathfrak{gl}(\bar{0},\dots,\bar{0},\bar{1},\dots,\bar{1})$ is abbreviated to $\mathfrak{gl}(\dim V_{\bar{0}}|\dim V_{\bar{1}})$.

For $\dim V_{\bar{0}} = \dim V_{\bar{1}} \pm 1$ we will often use another format, the *alternating* one, $Par_{alt} = (\bar{0},\bar{1},\bar{0},\bar{1},\dots)$.

The *supertrace* is the map $\mathfrak{gl}(Par) \longrightarrow \mathbb{C}$, $(A_{ij}) \mapsto \sum(-1)^{p_i}A_{ii}$. The supertraceless matrices constitute a Lie subsuperalgebra, $\mathfrak{sl}(Par)$.

To the linear map F of superspaces there corresponds the dual map F^* between the dual superspaces; if A is the supermatrix corresponding to F in a format *Par*, then to F^* the *supertransposed* matrix A^{st} corresponds:

$$(A^{st})_{ij} = (-1)^{(p_i+p_j)(p_i+p(A))}A_{ji}.$$

The supermatrices $X \in \mathfrak{gl}(Par)$ such that

$$X^{st}B + (-1)^{p(X)p(B)}BX = 0 \quad \text{for a homogeneous matrix } B \in \mathfrak{gl}(Par)$$

constitute the Lie superalgebra $\mathfrak{aut}(B)$ that preserves the bilinear form on V with matrix B.

The superspace of bilinear forms is denoted by $Bil_C(M,N)$ or $Bil_C(M)$ if $M{=}N$. The *upsetting of forms* $uf : Bil_C(M,N) \to Bil_C(N,M)$, is defined by the formula

$$B^{uf}(n,m) = (-1)^{p(n)p(m)}B(m,n).$$

A form $B \in Bil_C(M)$ is called *supersymmetric* if $B^{uf} = B$ and *superskew-symmetric* if $B^{uf} = -B$.

Given bases $\{m_i\}$ and $\{n_j\}$ of C-modules M and N and a bilinear form $B : M \otimes N \to C$, we assign to B the matrix

$$({}^{mf}B)_{ij} = (-1)^{p(m_i)p(B)}B(m_i,n_j).$$

For a nondegenerate supersymmetric form whose matrix in the standard format is

$$B_{m,2n} = \begin{pmatrix} 1_m & 0 \\ 0 & J_{2n} \end{pmatrix}, \quad \text{where } J_{2n} = \begin{pmatrix} 0 & 1_n \\ -1_n & 0 \end{pmatrix}.$$

The usual notation for $\mathfrak{aut}(B_{m|2n})$ is $\mathfrak{osp}^{sy}(m|2n)$ or just $\mathfrak{osp}(m|2n)$. (Observe that the passage from V to $\Pi(V)$ sends the supersymmetric forms to superskew-symmetric ones, preserved by $\mathfrak{osp}^{sk}(m|2n)$ which is isomorphic to $\mathfrak{osp}(m|2n)$ but has a different matrix realization.)

We will need the orthosymplectic supermatrices in the alternating format; in this format we take the matrix $B_{m,2n}(\text{alt}) = \text{antidiag}(1,\dots,1,-1,\dots,-1)$ with the only nonzero entries on the side diagonal, the last n being -1's. The Lie superalgebra of such supermatrices will be denoted by $\mathfrak{osp}(\text{alt}_{m|2n})$, where, as is easy to see, either $m = 2n \pm 1$ or $m = 2n$.

There is a 1-parameter family of deformations $\mathfrak{osp}_\alpha(4|2)$ of the Lie superalgebra $\mathfrak{osp}(4|2)$; its only explicit description we know is in terms of Cartan matrix apart from [BGLS] of course.

4.1. The superprincipal embeddings

Not every simple Lie superalgebra, even a finite dimensional one, hosts a superprincipal $\mathfrak{osp}(1|2)$-subsuperalgebra. Let us describe those that do. (Aside: an interesting problem is to describe *semiprincipal* embeddings into $\mathfrak{g}$, defined as the ones with the least possible number of irreducible components.)

We select the following basis in $\mathfrak{osp}(1|2) \subset \mathfrak{sl}(\bar{0}|\bar{1}|\bar{0})$:

$$X^- = \begin{pmatrix} 0 & 0 & 0 \\ 0 & 0 & 0 \\ -1 & 0 & 0 \end{pmatrix}, \quad H = \begin{pmatrix} 1 & 0 & 0 \\ 0 & 0 & 0 \\ 0 & 0 & -1 \end{pmatrix}, \quad X^+ = \begin{pmatrix} 0 & 0 & 1 \\ 0 & 0 & 0 \\ 0 & 0 & 0 \end{pmatrix}.$$

$$\nabla^- = \begin{pmatrix} 0 & 0 & 0 \\ 1 & 0 & 0 \\ 0 & 1 & 0 \end{pmatrix}, \quad \nabla^+ = \begin{pmatrix} 0 & 1 & 0 \\ 0 & 0 & -1 \\ 0 & 0 & 0 \end{pmatrix}.$$

The highest weight $\mathfrak{osp}(1|2)$-module $\mathcal{M}^\mu$ is illustrated with a graph whose nodes correspond to the eigenvectors l_i of H with the weight indicated; the horisontal edges depict the $X^\pm$-action (the X^+-action is directed to the right, that of X^- to the left; each horizontal string is an irreducible $\mathfrak{sl}(2)$-submodule; two such submodules are glued together into an $\mathfrak{osp}(1|2)$-module by the action of $\nabla^\pm$ (we set $\nabla^+(l_n) = 0$ and $\nabla^-(l_i) = l_{i-1}$; the corresponding edges are not depicted below); we additionally assume that $p(l_\mu) = \bar{0}$:

$$\begin{array}{l} \dots \overset{\mu-2i}{\circ} \longleftrightarrow \overset{\mu-2i+2}{\circ} \longleftrightarrow \dots \quad \dots \quad \dots \longleftrightarrow \overset{\mu-2}{\circ} \longleftrightarrow \overset{\mu}{\circ} \\ \dots \overset{\mu-2i+1}{\circ} \longleftrightarrow \overset{\mu-2i+3}{\circ} \longleftrightarrow \dots \longleftrightarrow \overset{\mu-3}{\circ} \longleftrightarrow \overset{\mu-1}{\circ} \end{array}$$

As follows from the relations of type 0 below in sec 4.2, the module $\mathcal{M}^n$ for $n \in \mathbb{Z}_+$ has an irreducible submodule isomorphic to $\Pi(\mathcal{M}^{-n-1})$; the quotient, obviously irreducible as follows from the same formulas, will be denoted by $\mathcal{L}^n$.

Serganova completely described superprincipal embeddings of $\mathfrak{osp}(1|2)$ into a simple finite dimensional Lie superalgebra [LSS] (the main part of her result was independently obtained in [vJ]).

As the $\mathfrak{osp}(1|2)$-module corresponding to the superprincipal embedding, a simple finite dimensional Lie superalgebra $\mathfrak{g}$ is presented in Table 4.1 (the missing simple algebras $\mathfrak{g}$ do not contain a superprincipal $\mathfrak{osp}(1|2)$):

The Lie superalgebra $\mathfrak{g}$ of type $\mathfrak{osp}$ that contains a superprincipal subalgebra $\mathfrak{osp}(1|2)$ can be generated by two elements. For such elements we can take $X := \nabla^+ \in \mathcal{L}^2 = \mathfrak{osp}(1|2)$ and a lowest weight vector $Z := l_{-r}$ from the module $M = \mathcal{L}^r$ or $\Pi(\mathcal{L}^r)$, where for M we take $\Pi(\mathcal{L}^3)$ if $\mathfrak{g} \neq \mathfrak{osp}(2n|2m)$ or the last module with the even highest weight vector in the above table (i.e., $\mathcal{L}^{2n-2}$ if $\mathfrak{g} = \mathfrak{osp}(2n|2n)$ and $\mathcal{L}^{2n}$ if $\mathfrak{g} = \mathfrak{osp}(2n+2|2n)$).

Table 4.1. $\mathfrak{g}$ that admits a superprincipal subalgebra: as the $\mathfrak{osp}(1|2)$-module

$\mathfrak{g}$	$\mathfrak{g} = \mathcal{L}^2 \oplus (\underset{i>1}{\oplus} \mathcal{L}^{2k_i})$	$\oplus$	$(\underset{j}{\oplus} \Pi(\mathcal{L}^{m_j}))$
$\mathfrak{sl}(n\vert n+1)$	$\mathcal{L}^2 \oplus \mathcal{L}^4 \oplus \mathcal{L}^6 \dots \oplus \mathcal{L}^{2n-2}$		$\oplus\Pi(\mathcal{L}^1) \oplus \Pi(\mathcal{L}^3) \oplus \dots \oplus \Pi(\mathcal{L}^{2n-1})$
$\mathfrak{osp}(2n-1\vert 2n)$ $(n>1)$	$\mathcal{L}^2 \oplus \mathcal{L}^6 \oplus \mathcal{L}^{10} \dots \oplus \mathcal{L}^{4n-6}$		$\oplus\Pi(\mathcal{L}^3) \oplus \Pi(\mathcal{L}^7) \oplus \dots \oplus \Pi(\mathcal{L}^{4n-1})$
$\mathfrak{osp}(2n+1\vert 2n)$	$\mathcal{L}^2 \oplus \mathcal{L}^6 \oplus \mathcal{L}^{10} \dots \oplus \mathcal{L}^{4n-2}$		$\oplus\Pi(\mathcal{L}^3) \oplus \Pi(\mathcal{L}^7) \oplus \dots \oplus \Pi(\mathcal{L}^{4n-1})$
$\mathfrak{osp}(2\vert 2) \cong \mathfrak{sl}(1\vert 2)$	$\mathcal{L}^2$		$\oplus\Pi(\mathcal{L}^1)$
$\mathfrak{osp}(4\vert 4)$	$\mathcal{L}^2 \oplus \mathcal{L}^6$		$\oplus\Pi(\mathcal{L}^3) \oplus \Pi(\mathcal{L}^3)$
$\mathfrak{osp}(2n\vert 2n)$	$\mathcal{L}^2 \oplus \mathcal{L}^6 \oplus \mathcal{L}^{10} \dots \oplus \mathcal{L}^{4n-2}$	$\oplus\mathcal{L}^{2n-2}$	$\oplus\Pi(\mathcal{L}^3) \oplus \Pi(\mathcal{L}^7) \oplus \dots \oplus \Pi(\mathcal{L}^{4n-1})$
$\mathfrak{osp}(2n+2\vert 2n)$	$\mathcal{L}^2 \oplus \mathcal{L}^6 \oplus \mathcal{L}^{10} \dots \oplus \mathcal{L}^{4n+2}$	$\oplus\mathcal{L}^{2n}$	$\oplus\Pi(\mathcal{L}^3) \oplus \Pi(\mathcal{L}^7) \oplus \dots \oplus \Pi(\mathcal{L}^{4n-1})$
$\mathfrak{osp}_\alpha(4\vert 2)$	$\mathcal{L}^2$	$\oplus\mathcal{L}^2$	$\oplus\Pi(\mathcal{L}^3)$

To generate $\mathfrak{sl}(n|n+1)$ we have to add to the above X and Z a lowest weight vector U from $\Pi(\mathcal{L}^1)$. (Clearly, Z and U are defined up to factors that we can select at our convenience; we will assume that a basis of L^r is fixed and denote $Z = t \cdot l_{-r}$ and $U = s \cdot l_{-1}$ for $t, s \in \mathbb{C}$.)

We call the above X and Z, together with U, and fortified by $Y := X^- \in L^2$ the *Jacobson's generators.* The presence of Y considerably simplifies the form of the relations, though slightly increases the number of them.

4.2. Relations between Jacobson's generators

We repeat the arguments from Sec. 1.2. Since we obtain the relations recurrently, it could happen that a relation of higher degree implies a relation of a lower degree. This did not happen when we studied $\mathfrak{S}\mathfrak{L}(\lambda)$, but does happen in what follows, namely, relation 1.2 implies 1.1.

We divide the relations between Jacobson's generators into the types corresponding the number of occurence of z in them: **0**. Relations in $\mathfrak{sl}(1|2)$ or $\mathfrak{osp}(1|2)$; **1**. Relations coming from the $\mathfrak{osp}(1|2)$-action on $\mathcal{L}^{2k_1}$; **2**. Relations coming from $\mathcal{L}^{2k_1} \wedge \mathcal{L}^{2k_1}$; **3**. Relations coming from $\mathcal{L}^{2k_1} \wedge \mathcal{L}^{2k_2}$; ∞. Relation that shear the dimension.

The relations of type 0 are the well-known relations in $\mathfrak{sl}(1|2)$, those of them that do not involve U (marked with an $*$) are the relations for $\mathfrak{osp}(1|2)$. The relations of type 1 that do not involve U express that the space $\mathcal{L}^{2k_2}$ is the $\mathfrak{osp}(1|2)$-module with highest weight $2k_2$. To simplify notation we denote: $Z_i = \operatorname{ad} X^i Z$ and $Y_i = \operatorname{ad} X^i Y$.

$$\begin{array}{llllll} \mathbf{0.1^*}. & [Y, Y_1] = 0, & \mathbf{0.2^*}. & [Y_2, Y] = 2Y, & \mathbf{0.3^*}. & [Y_2, X] = -X, \\ \mathbf{0.4}. & [Y, U] = 0, & \mathbf{0.5}. & [U, U] = -2Y; & \mathbf{0.6}. & [U, Y_1] = 0, \\ \mathbf{0.7}. & [[X, X], [X, U]] = 0, & \mathbf{0.8}. & [Y_2, U] = U. & & \end{array}$$

$$\mathbf{1.1}.\ [Y, Z] = 0 \Longleftarrow \mathbf{1.2}.\ [[X, Y], Z] = 0, \quad \mathbf{1.3}.\ Z_{4k_2} = 0, \quad \mathbf{1.4}.\ [Y_2, Z] = 3Z.$$

4.3. Theorem. *For the Lie superalgebras indicated, all the relations between Jacobson's generators are the above relations of types* 0, 1 *and the relations from* §6.

§5. The Lie superalgebra $\mathfrak{sl}(\lambda+1|\lambda)$ as the quotient of $\mathfrak{diff}(1|1)$ and a subalgebra of $\mathfrak{sl}_+(\infty|\infty)$

There are several ways to superize $\mathfrak{sl}_+(\infty|\infty)$. For a description of "the best" one from a certain point of view see [E]. For our purposes any version of $\mathfrak{sl}_+(\infty|\infty)$ will do.

5.1

The Poincaré-Birkhoff-Witt theorem states that $U(\mathfrak{osp}(1|2)) \cong \mathbb{C}[X^-, \nabla^-, H, \nabla^+, X^+]$, as superspaces. Set $U_\lambda = U(\mathfrak{osp}(1|2))/(\Delta - (\lambda^2 - 9/4))$. Denote: $\partial_x = \partial/\partial x$, $\partial_\theta = \partial/\partial\theta$ and set

$$\begin{aligned} X^- &= -\partial_x, & \nabla^- &= \partial_\theta - \theta\partial_x,\\ H &= 2x\partial_x + \theta\partial_\theta(\lambda-1), & \nabla^+ &= x\partial_\theta + x\theta\partial_x - \lambda\theta,\\ X^+ &= x^2\partial_x - (\lambda-1)x. \end{aligned}$$

These formulas establish a morphism of $\mathfrak{osp}(1|2)$-modules and, moreover, of associative superalgebras: $U_\lambda \longrightarrow \mathbb{C}[x, \theta, \partial_x, \partial_\theta]$.

In what follows we will need a well-known fact: the Casimir operator

$$\Delta = 2(X^+X^- + X^-X^+) + \nabla^+\nabla^- - \nabla^-\nabla^+ + H^2$$

acts on the irreducible $\mathfrak{osp}(1|2)$-module $\mathcal{L}^\mu$ as the scalar operator of multiplication by $\mu^2 + 3\mu$. (The passage from μ to λ is done with the help of a shift by the weight ρ, which for $\mathfrak{osp}(1|2)$ can be identified with $\frac{3}{2}$.)

Consider the Lie superalgebra $LU(\mathfrak{osp}(1|2))$ associated with the associative superalgebra U_λ. It is easy to see that, as $\mathfrak{osp}(1|2)$-module,

$$LU_\lambda = \mathcal{L}^0 \oplus \mathcal{L}^2 \oplus \cdots \oplus \mathcal{L}^{2n} \oplus \cdots \oplus \Pi(\mathcal{L}^1 \oplus \mathcal{L}^3 \oplus \ldots) \tag{5.1}$$

In the same way as for Lie algebras we show that LU_n contains an ideal I_n for $n \in \mathbb{N}\setminus\{0\}$ and the quotient LU_n/I_n is the conventional $\mathfrak{sl}(n|n+1)$. It is clear that for $\lambda \neq \mathbb{Z}$ the Lie algebra LU_λ has only one ideal — the space $\mathcal{L}^0$ of constants — and $LU_\lambda = \mathcal{L}^0 \oplus [LU_\lambda, LU_\lambda]$; hence, there is a supertrace on LU_λ. This justifies the following notation:

$$\mathfrak{SL}(\lambda|\lambda+1) = \mathfrak{GL}(\lambda|\lambda+1)/\mathcal{L}^0, \text{ where } \mathfrak{GL}(\lambda|\lambda+1) = \begin{cases} U_\lambda & \text{for } \lambda \neq \mathbb{N}\setminus\{0\}\\ LU_n/I_n & \text{otherwise.}\end{cases} \tag{5.2}$$

The definition directly implies that $\mathfrak{SL}(-\lambda-2|-\lambda-1) \cong \mathfrak{SL}(-\lambda|\lambda+1)$, so speaking about real values of λ we can confine ourselves to the nonnegative values.

Define $\mathfrak{osp}(\lambda+1|\lambda)$ as the Lie subsuperalgebra of $\mathfrak{sl}(\lambda+1|\lambda)$ invariant with respect to the involution

$$X \to \begin{cases} -X & \text{if } X \in \mathcal{L}^{4k} \text{ or } X \in \Pi(\mathcal{L}^{4k\pm1})\\ X & \text{if } X \in \mathcal{L}^{4k\pm2} \text{ or } X \in \Pi(\mathcal{L}^{4k\pm3}),\end{cases} \tag{5.3}$$

which is the analogue of the map

$$X \to -X^{st} \quad \text{for} \quad X \in \mathfrak{gl}(m|n). \tag{5.4}$$

5.2. The Lie superalgebras $\mathfrak{SL}(*+1|*)$ and $\mathfrak{osp}(2*+1|*)$, for $* \in \mathbb{C}P^1 = \mathbb{C} \cup \{*\}$ The "dequantization" of the relations for $\mathfrak{SL}(\lambda+1|\lambda)$ and $\mathfrak{osp}(\lambda+1|\lambda)$ is performed by passage to the limit as $\lambda \longrightarrow \infty$ under the change $t \mapsto t/\lambda$. We denote the limit algebras by $\mathfrak{SL}(*+1|*)$ and $\mathfrak{osp}(*+1|*)$ in order not to confuse them with $\mathfrak{sl}(\infty+1|\infty)$ and $\mathfrak{osp}(\infty+1|\infty)$, respectively.

§6. Tables. The Jacobson generators and relations between them

Table 6.1. Infinite dimensional case

- $\mathfrak{osp}(\lambda|\lambda+1)$. Generators:

$$X = x\partial_\theta + x\theta\partial_x - \lambda\theta, \ \ Y = \partial_x, \ \ Z = t(\partial_x\partial_\theta - \theta{\partial_x}^2).$$

Relations:

$$\begin{aligned} &\mathbf{2.1.} \quad 3[Z,Z_3]+2[Z_1,Z_2] = 6t(2\lambda+1)Z, \\ &\mathbf{2.2.} \quad [Z_1,Z_3] = 2t^2(\lambda-1)(\lambda+2)Y + 2t(2\lambda+1)Z_1, \\ &\mathbf{3.1.} \quad [Z_1,[Z,Z]] = 0. \end{aligned}$$

- $\mathfrak{osp}(*|*+1)$. Relations: the same as 3.1 plus the following relations:

$$\begin{aligned} &\mathbf{2.1.} \quad 3[Z,Z_3]+2[Z_1,Z_2] = 12tZ, \\ &\mathbf{2.2.} \quad [Z_1,Z_3] = 2t^2Y + 4tZ_1. \end{aligned}$$

- $\mathfrak{sl}(\lambda|\lambda+1)$ for $\lambda \in \mathbb{C}P^1$. Generators (for $\lambda \in \mathbb{C}$): the same as for $\mathfrak{osp}(\lambda|\lambda+1)$ and $U = \partial_\theta - \theta\partial_x$.

Relations: the same as for $\mathfrak{osp}(\lambda|\lambda+1)$ plus the following:

$$\begin{aligned} &\mathbf{1.5.} \quad 3[Z,[X,U]] - [U,Z_1] = 0, &&\mathbf{2.3.} \quad [Z,[U,Z]] = 0, \\ &\mathbf{1.6.} \quad [[X,U],Z_1] = 0, &&\mathbf{2.4.} \quad [Z_1,[U,Z]] = 0. \end{aligned}$$

Table 6.2. Finite dimensional algebras

In this table E_{ij} are the matrix units; $X_i^{\pm}$ stand for the conventional Chevalley generators of $\mathfrak{g}$.

- $\mathfrak{sl}(n+1|n)$ for $n \geq 3$. Generators:

$$X = \sum_{1\leq i\leq n} \left((n-i+1)E_{2i-1,2i} - iE_{2i,2i+1}\right), \quad Y = \sum_{1\leq i\leq 2n-1} E_{i+2,i},$$

$$U = \sum_{1\leq i\leq 2n} (-1)^{i+1}E_{i+1,i}, \quad Z = \sum_{1\leq i\leq 2n-2} (-1)^{i+1}E_{i+3,i}.$$

Relations: those for $\mathfrak{SL}(\lambda|\lambda+1)$ with $\lambda = n$ and an extra relation to shear the dimension:

$$(\operatorname{ad} Z)^n([X,X]) = 0.$$

For $n = 1$ the relations degenerate in relations of type 0.

• $\mathfrak{osp}(2n+1|2n)$. Generators:

$$X = \sum_{1\le i\le n} ((2n-i+1)(E_{2i-1,2i}+E_{4n+2-2i,4n+3-2i}) \\ - i(E_{2i,2i+1} - E_{4n+1-2i,4n+2-2i})),$$

$$Y = E_{2n+2,2n} + \sum_{1\le i\le 2n-1} (E_{i+2,i} - E_{4n+2-i,4n-i}),$$

$$Z = -E_{2n+2,2n-1} - E_{2n+3,2n} + \sum_{1\le i\le 2n-2} ((-1)^i E_{i+3,i} + E_{4n+2-i,4n-1-i}).$$

Relations: those for $\mathfrak{osp}(2\lambda+1|2\lambda)$ with $\lambda = n$ and an extra relation to shear the dimension (the form of the relation is identical to that for $\mathfrak{sl}(n+1|n)$).

• $\mathfrak{osp}_\alpha(4|2)$. Generators: As $\mathfrak{osp}(1|2)$-module, the algebra $\mathfrak{osp}_\alpha(4|2)$ has 2 isomorphic submodules. The generators X and Y belong to one of them. It so happens that we can select Z from either of the remaining submodules and still generate the whole Lie superalgebra. The choice (a) is from $\Pi(\mathcal{L}^3)$; it is unique (up to a factor). The choices (b) and (c) are from $\mathcal{L}^2$; none of them seem to give simpler relations.

$$X \quad -\tfrac{\alpha+1}{\alpha}X_1^+ + \tfrac{\alpha}{\alpha+1}X_2^+ + \tfrac{1}{\alpha(\alpha+1)}X_3^+, \quad Y = [X_1^-,X_2^-]+[X_1^-,X_3^-]+[X_2^-,X_3^-],$$

$$Z \begin{cases} \text{a)} & -[[X_1^-,X_2^-],X_3^-]]; \\ \text{b)} & -(1+2\alpha)[X_1^-,X_2^-]+\alpha^2(2+\alpha)[X_1^-,X_3^-]+(\alpha-1)(1+\alpha)^2[X_2^-,X_3^-]; \\ \text{c)} & -[X_2^-,X_3^-]-(\alpha+1)[X_1^-,X_2^-]. \end{cases}$$

Relations of type 0 are common for cases a) – c):

0.1 $[Y,[Y,[X,X]]] = 4Y;$ **0.2** $[Y_1[X,X]]] = -2X;$

The other relations are as follows.

Relations a):

1.1 $[Y_1, Z_1] = 3Z,$ **1.2** $(\operatorname{ad}[X,X])^3 Z_1 = 0;$

2.1 $[Z,Z] = 0;$ **2.2** $[Z_1,[[X,X],Z]] = -4\frac{\alpha^2+\alpha+1}{\alpha(\alpha+1)}Z,$

3.1 $[\operatorname{ad}[X,X](Z_1),[Z_1,\operatorname{ad}[X,X](Z_1)]] =$
$-\frac{16}{\alpha(\alpha+1)}Y + 8\frac{\alpha^2+\alpha+1}{\alpha(\alpha+1)}[Z_1,\operatorname{ad}[X,X](Z_1)] + 16\frac{(\alpha^2+\alpha+1)^2}{\alpha^2(\alpha+1)^2}Z_1.$

Relations b):

1.1 $[Y_1,Z_1] = 2Z;$ **1.2** $(\operatorname{ad}[X,X])^2 Z_1 = 0;$

2.1* $[Z_1,Z_1] = 2[Z,[Z,[X,X]]] - 18\alpha^2(1+\alpha)^2 Y + 4(1-\alpha)(2+\alpha)(1+2\alpha)Z;$

3.1 $(\operatorname{ad} Z)^3 X = 0,$

3.2* $[[Z,Z_1],(\operatorname{ad}[X,X])^2 Z_1] =$
$(-1+\alpha)(2+\alpha)(1+2\alpha)[Z,[Z,[X,X]]] + 12(1-\alpha)\times$
$(2+\alpha)(1+2\alpha)\alpha^2(1+\alpha)^2 Y + 8(1-3\alpha^2-\alpha^3)(-1-3\alpha+\alpha^3)Z.$

Relations c): same as for b) except that the relations marked in b) by an * should be replaced with the following ones

2.1 $[Z_1, Z_1] = 2[Z, [Z, [X, X]]] - 2\alpha^2 Y + 4(2+\alpha)Z;$
2.2 $(\mathrm{ad}\,[X, X])^2 Z_1 = (-2-\alpha)[Z, [Z, [X, X]]] - 8(1+\alpha)Z + 4\alpha^2(2+\alpha)Y.$

§7. Remarks and problems

7.1. On proof

For the exceptional Lie algebras and superalgebras $\mathfrak{osp}_\alpha(4|2)$ the proof is direct: the quotient of the free Lie algebra generated by x, y and z modulo our relations is the needed finite dimensional one. For rank $\mathfrak{g} \leq 12$ we similarly computed relations for $\mathfrak{g} = \mathfrak{sl}(n)$, $\mathfrak{o}(2n+1)$ and $\mathfrak{sp}(2n)$; as Post pointed out, together with the result of [PH] on deformation (cf. 2.7) this completes the proof for Lie algebras. The results of [PH] on deformations can be directly extended for the case of $\mathfrak{sl}(2)$ replaced with $\mathfrak{osp}(1|2)$; this proves Theorem 4.3.

Our Theorem 2.6 elucidates Proposition 2 of [F]; we just wrote relations explicitly. Feigin claimed [F] that for $\mathfrak{SL}(\lambda)$ the relations of type 3 follow from the decomposition of $L^{2k_1} \wedge L^{2k_2} \subset L^{2k_1} \wedge L^{2k_1} \wedge L^{2k_1}$. We verified that this is so not only in Feigin's case but for all the above-considered algebras except $\mathfrak{e}_6$, $\mathfrak{e}_7$ and $\mathfrak{e}_8$: for the latter one should consider the whole $L^{2k_1} \wedge L^{2k_1} \wedge L^{2k_1}$, cf. [GL1]. Theorem 4.3 is a direct superization of Theorem 2.6.

7.2. Problems

1) How to present $\mathfrak{o}(2n)$ and $\mathfrak{osp}(2m|2n)$? One can select z as suggested in Sec. 1.1. Clearly, the form of z (hence, relations of type 1) and the number of relations of type 3 depend on n in contradistinction with the algebras considered above. Besides, the relations are not as neat as for the above algebras. We should, perhaps, have taken the generators as for $\mathfrak{o}(2n-1)$ and add a generator from L^{2n-2}. We have no guiding idea; to try at random is frustrating, cf. the relations we got for $\mathfrak{osp}_\alpha(4|2)$.

2) We could have similarly realized the Lie algebra $\mathfrak{SL}(\lambda)$ as the quotient of $U(\mathfrak{vect}(1))$, where $\mathfrak{vect}(1) = \mathfrak{der}\mathbb{C}[u]$. However, $U(\mathfrak{vect}(1))$ has no center except the constants. What are the generators of the ideal — the analogue of (2.0) — modulo which we should factorize $U(\mathfrak{vect}(1))$ in order to get $\mathfrak{SL}(\lambda)$? (Observe that in case $U(\mathfrak{g})$, where $\mathfrak{g}$ is a simple finite dimensional Lie superalgebra such that $Z(U(\mathfrak{g}))$ is not noetherian, the ideal — the analog of (5.0) — is, nevertheless, finitely generated, cf. [GL2].)

3) Feigin realized $\mathfrak{SL}(*)$ on the space of functions on the open cell of $\mathbb{C}P^1$, a hyperboloid, see [F]. Examples of [DGS] are similarly realized. Give any realization of $\mathfrak{o}/\mathfrak{sp}(*)$ and its superanalogues.

4) Other problems are listed in Sec. 8.1–8.3 below.

alg	N_{GB}	N_{comm}	D_{GB}	Space	Time
$\mathfrak{sl}(3)$	23 (24)	21 (21)	9 (4)	1300 (1188)	¡1 sec (< 1 sec)
$\mathfrak{sl}(4)$	69 (84)	70 (60)	17 (6)	3888 (3612)	¡1 sec (< 1 sec)
$\mathfrak{sl}(5)$	193 (218)	220 (126)	25 (8)	13556 (8716)	¡1 sec (< 1 sec)
$\mathfrak{sl}(6)$	444 (473)	476 (225)	33(10)	34692 (18088)	2 sec (< 1 sec)
$\mathfrak{sl}(7)$	893 (908)	937 (363)	41(12)	80272 (33700)	10 sec (1 sec)
$\mathfrak{sl}(8)$	1615(1594)	1632 (546)	49(14)	162128 (57908)	34 sec (3 sec)
$\mathfrak{sl}(9)$	2705(2614)	2714 (780)	57(16)	314056 (93452)	109 sec (6 sec)
$\mathfrak{sl}(10)$	4263 (4063)	4138 (1071)	65 (18)	534684 (143456)	336 sec (10 sec)
$\mathfrak{sl}(11)$	6405 (6048)	6224 (1425)	73 (20)	921972 (211428)	1058 sec (19 sec)

7.3. Serre relations are more convenient than ours

The following Table represents results of V. Kornyak's computations. N_{GB} is the number of relations in Groebner basis, N_{comm} is the number of non-zero commutators in the multiplication table, D_{GB} is a maximum degree of relations in GB, Space is measured in in bytes. The corresponding values for Chevalley generators/Serre relations are given in brackets.

For the other Lie algebras, especially exceptional ones, the comparison is even more unfavourable. Nevertheless, for $\mathfrak{SL}(\lambda)$ with noninteger λ there are only the Jacobson generators and we have to use them.

§8. Lie algebras of higher ranks. The analogs of the exponents and W-algebras

The following Tables 8.1 and 8.2 introduce the generators for the Lie algebras $U_{\mathfrak{g}}(\lambda)$ and the analogues of the exponents that index the generalized W-algebras (for their definition in the simplest cases from different points of view see [FFr] and [KM]; we will follow the lines of [KM]).

Recall that (see 0.1) one of the definitions of $U_{\mathfrak{g}}(\lambda)$ is as the associative algebra generated by $\tilde{\mathfrak{g}}$; we loosely denote it by $\tilde{S}^{\cdot}(\tilde{\mathfrak{g}})$. For the generators of $LU_{\mathfrak{g}}(\lambda)$ we take the Chevalley generators of $\mathfrak{g}$ (since by 7.3 they are more convenient) and the lowest weight vectors of the irreducible $\mathfrak{g}$-modules that constitute $\tilde{S}^2(\tilde{\mathfrak{g}})$.

8.1. The exponents

This section is just part of Table 1 from [OV] reproduced here for the convenience of the reader. Recall that if $\mathfrak{g}$ is a simple (finite dimensional) Lie algebra, $W = W_{\mathfrak{g}}$ is its Weyl group, $l = \text{rk}\ \mathfrak{g}$, $\alpha_1, \dots, \alpha_l$ the simple roots, α_0 the lowest root; the n_i the coefficients of linear relation among the α_i normed so that $n_0 = 1$; let $c = r_1 \cdot \dots \cdot r_l$, where r_i are the reflections from W associated with the simple roots, be the Killing–Coxeter element. The order h of c (the Coxeter number) is equal to $\sum_{i>0} n_i$. The eigenvalues of c are $\varepsilon^{k_1}, \dots, \varepsilon^{k_l}$, where ε is a primitive h-th root of unity. The numbers k_i are called the *exponents*. Then

The exponents k_i are the respective numbers k_i from Table 1.1, e.g., $k_1 = 1$. The number of roots of $\mathfrak{g}$ is equal to $l\sum_{i>0} n_i = 2\sum_{i>0} k_i$. The order of W is equal to

$$zl!\prod n_i = \prod_{i>0}(k_i + 1),$$

where z is the number of 1's among the n_i's for $i > 0$ (the number z is also equal to the order of the centrum $Z(G)$ of the simply connected Lie group G with the Lie algebra $\mathfrak{g}$). The algebra of W-invariant polynomials on the maximal diagonalizable (Cartan) subalgebra of $\mathfrak{g}$ is freely generated by homogeneous polynomials of degrees $k_i + 1$.

We will use the following notation:

For a finite dimensional irreducible representations of finite dimensional simple Lie algebras $R(\lambda)$ denotes the irreducible representation with highest weight λ and $V(\lambda)$ the space of this representation; $\rho = \frac{1}{2}\sum\limits_{\alpha>0}\alpha$ or ρ is a weight such that $\rho(\alpha_i) = A_{ii}$ for each simple root α_i.

The weights of the Lie algebras $\mathfrak{o}(2l)$ and $\mathfrak{o}(2l+1)$, $\mathfrak{sp}(2l)$ and $\mathfrak{f}_4$ $(l = 4)$ are expressed in terms of an orthonormal basis $\varepsilon_1, \dots, \varepsilon_l$ of the space $\mathfrak{h}^*$ over $\mathbb{Q}$. The weights of the Lie algebras $\mathfrak{sl}(l+1)$ as well as $\mathfrak{e}_7$, $\mathfrak{e}_8$ and $\mathfrak{g}_2$ $(l = 7, 8$ and 2, respectively) are expressed in terms of vectors $\varepsilon_1, \dots, \varepsilon_{l+1}$ of the space $\mathfrak{h}^*$ over $\mathbb{Q}$ such that $\sum\varepsilon_i = 0$. For these vectors $(\varepsilon_i, \varepsilon_i) = \frac{l}{l+1}$ and $(\varepsilon_i, \varepsilon_j) = \frac{1}{l+1}$ for $i \neq j$. The indices in the expression of any weight are assumed to be different.

The analogues of the exponents for $LU_{\mathfrak{g}}(\lambda)$ are the highest weights of the representations that constitute $\tilde{S}^k(\tilde{\mathfrak{g}})$.

Problem . *Interprete these exponents in terms of the analog of the Weyl group of $LU_{\mathfrak{g}}(\lambda)$ in the sence of [PS] and invariant polynomials on $LU_{\mathfrak{g}}(\lambda)$.*

Columns 2 and 3 the Table 8.2 are derived from Table 5 in [OV]. Columns 4 and 5 are results of a computer-aided study. To fill in the gaps is a research problem, cf. [GL2] for the Lie algebras different from $\mathfrak{sl}$ type.

The generators of $LU_{\mathfrak{g}}(\lambda)$ are the Chevalley generators $X_i^{\pm}$ of $\mathfrak{g}$ AND the lowest weight vectors from $\tilde{S}^2$. Denote the latter by z_1, z_2 (sometimes there is a third one, z_3). Then the relations are (recall that $h_i = [X_i^+, X_i^-]$):

(type 0) the Serre relations in $\mathfrak{g}$

(type 1) The relations between $X_i^{\pm}$ and z_j, namely:

$$X_i^-(z_j) = 0;$$
$$h_i(z_j) = \text{weight}_i(z_j);$$
$$(adX_i^+)^{\text{the power determined by the weight of } z_j}(z_j) = 0.$$

Problem . *Give an explicit form of the relations of higher types.*

8.2. Table. The Lie algebras $U_{\mathfrak{g}}(\lambda)$ as $\mathfrak{g}$-modules

$\mathfrak{g}$	ad	$\tilde{S}^2(\tilde{\mathfrak{g}})$	$\tilde{S}^3(\tilde{\mathfrak{g}})$	$\tilde{S}^k(\tilde{\mathfrak{g}})$
$\mathfrak{sl}(2)$	$R(2\pi)$	$R(4\pi)$	$R(6\pi)$	$R(2k\pi)$
$\mathfrak{sl}(3)$	$R(\pi_1+\pi_2)$	$R(2\pi_1+2\pi_2)$ $R(\pi_1+\pi_2)$	$R(3\pi_1+3\pi_2)$ $R(2\pi_1+2\pi_2)$	$R(k\pi_1+k\pi_2)$ $R((k-1)\pi_1+(k-1)\pi_2)$
$\mathfrak{sl}(4)$	$R(\pi_1+\pi_3)$	$R(2\pi_1+2\pi_3)$ $R(\pi_1+\pi_3)$ $R(2\pi_2)$	$R(3\pi_1+3\pi_3)$ $R(2\pi_1+2\pi_3)$ $R(2\pi_1+\pi_2)$ $R(\pi_2+2\pi_3)$ $R(\pi_1+\pi_3)$ $R(\pi_1+2\pi_2+\pi_3)$	
$\mathfrak{sl}(n+1)$ $n\geq 4$	$R(\pi_1+\pi_n)$	$R(2\pi_1+2\pi_n)$ $R(\pi_1+\pi_n)$ $R(\pi_2+\pi_{n-1})$	$R(3\pi_1+3\pi_n)$ $R(2\pi_1+2\pi_n)$ $R(2\pi_1+\pi_{n-1})$ $R(\pi_2+\pi_{n-1})$ $R(\pi_2+2\pi_n)$ $R(\pi_1+\pi_n)$ $R(\pi_1+\pi_2+\pi_{n-1}+\pi_n)$	

8.3. Tough problems

Even if the explicit realization of the exceptional Lie algebras by differential operators on the base affine space were known at the moment, it is, nevertheless, a difficult computer problem to fill in the blank spaces in the above table and similar tables for Lie superalgebras. To make plausible conjectures we have to compute $\tilde{S}^k(\tilde{\mathfrak{g}})$ to, at least, $k=4$.

Observe that due to a theorem of Kostant, every irreducible $\mathfrak{g}$-module containing the zero weight enters $U_{\mathfrak{g}}(\lambda)$ with multiplicity equal to that of the zero weight in this module; so only the column with second symmetric square and the row for $\mathfrak{sl}(2)$ are complete in Table 8.2.

§9. A connection with integrable dynamical systems

We recall the basic steps of the Khesin–Malikov construction and then superize them.

9.1. The Hamilton reduction

Let (M^{2n},ω) be a symplectic manifold with an action *act* of a Lie group G on M by symplectomorphisms (i.e., G preserves ω). The derivative of the G-action gives rise to a Lie algebra homomorphism $a:\mathfrak{g}=Lie(G)\longrightarrow\mathfrak{h}(2n)$. The action *act*, or rather, a is called a *Poisson* one, if a can be lifted to a Lie algebra homomorphism $\tilde{a} : \mathfrak{g} \longrightarrow \mathfrak{po}(2n)$, where the Poisson algebra $\mathfrak{po}(2n)$ is the nontrivial central extension of $\mathfrak{h}(2n)$.

For any Poisson G-action on M there arises a G-equivariant map $p : M \longrightarrow \mathfrak{g}^*$, called the *moment map*, given by the formula

$$\langle p(x), g\rangle = \tilde{a}(g)(x) \quad \text{for any} \quad x \in M, g \in \mathfrak{g}.$$

Fix $b \in \mathfrak{g}^*$; let $G_b \subset G$ be the stabilizer of b. Under certain regularity conditions (see [Ar]) $p^{-1}(b)/G_b$ is a manifold. This manifold is endowed with the symplectic form

$$\begin{gathered}\omega(\bar{v}, \bar{w}) = \omega(v, w) \quad \text{for arbitrary preimages} \quad v, w \text{ of } \bar{v}, \bar{w}, \text{ respectively}\\ \text{wrt the natural projection} \quad T(p^{-1}(b)) \longrightarrow T(p^{-1}(b)/G_b).\end{gathered}$$

The passage from M to $p^{-1}(b)/G_b$ is called *Hamilton reduction.* In the above picture M can be the *Poisson manifold*, i.e., ω is allowed to be nondegenerate not on the whole M; the submanifolds on which ω is nondegenerate are called *symplectic leaves.*

Example. Let $\mathfrak{g} = \mathfrak{sl}(n)$ and $M = \mathfrak{g}^*$, let G be the group N of uppertriangular matrices with 1 on the diagonal. The coadjoint N-action on $\mathfrak{g}^*$ is a Poisson one, the moment map is the natural projection $\mathfrak{g}^* \longrightarrow \mathfrak{n}^*$ and $\mathfrak{g}^*/N$ is a Poisson manifold.

9.2. The Drinfeld–Sokolov reduction

Let $\mathfrak{g} = \hat{\mathfrak{a}}^{(1)}$, where $\mathfrak{a}$ is a simple finite dimensional Lie algebra (the case $\mathfrak{a} = \mathfrak{sl}(n)$ is the one considered by Gelfand and Dickey), hat denotes the Kac–Moody central extension. The elements of $M = \mathfrak{g}^*$, can be identified with the $\mathfrak{a}$-valued differential operators:

$$(f(t)dt, az^*) \mapsto (tf(t) + at\frac{d}{dt}))\frac{dt}{t}.$$

Let N be the loop group with values in the group generated by positive roots of $\mathfrak{a}$. For the point b above take the element $y \in \mathfrak{a} \subset \hat{\mathfrak{g}}^*$ described in §3. If $\mathfrak{a} = \mathfrak{sl}(n)$, we can represent every element of $p^{-1}(b)/N$ in the form

$$t\frac{d}{dt} + y + \begin{pmatrix} b_1(t) & \dots & b_n(t)\\ 0 & \dots & 0\\ 0 & \dots & 0\end{pmatrix} \longleftrightarrow \frac{d^n}{d\varphi^n} + \tilde{b}_1(\varphi)\frac{d^{n-1}}{d\varphi^{n-1}} + \dots + \tilde{b}_n(\varphi),$$

To generalize the above to $\mathfrak{sl}(\lambda)$, Khesin and Zakharevich described the Poisson–Lie structure on symbols of pseudodifferential operators, see [KM] and references therin. Let us recall the main formulas.

9.3.1. The Poisson bracket on symbols of ΨDO

Set $D = d/dx$; define

$$D^\lambda \circ f = fD^\lambda + \sum_{k\geq 1}\binom{\lambda}{k} f^{(k)} D^{(\lambda-k)}, \quad \text{where } \binom{\lambda}{k} = \frac{\lambda(\lambda-1)\dots(\lambda-k+1)}{k!}.$$

Set

$$G = \left\{ D^\lambda (1 + \sum_{k\geq 1} u_k(x) D^{(-k)} \right\}$$

and

$$T_\lambda G = \left\{ \sum_{k\geq 1} v_k(x) D^{(-k)} \right\} \circ D^\lambda, \quad T_\lambda^* G = D^{-\lambda} \circ DO.$$

For $X = D^{-\lambda} \circ \sum_{k\geq 0} u_k(x) D^{(k)} \in T_\lambda^* G$ and $L = \left(\sum_{k\geq 1} v_k(x) D^{(-k)} \right) \circ D^\lambda \in T_\lambda G$ define the pairing $\langle X, L\rangle$ to be

$$\langle X, L\rangle = \mathrm{Tr}(L \circ X), \quad \text{where} \quad \mathrm{Tr} \sum w_k(x) D^{(k)}) = \mathrm{Res}|_{x=0} w_{-1}.$$

The Poisson bracket on ΨDS is defined on linear functionals by the formula

$$\{X, Y\}(L) = X(H_Y(L)), \quad \text{where} \quad H_Y(L) = (LY)_+ L - L(YL)_+.$$

Theorem. (Khesin–Malikov) *For $\mathfrak{a} = \mathfrak{sl}(\lambda)$ in the Drinfeld-Sokolov picture, the Poisson manifolds $p^{-1}(b)/N_b$ and ΨDS_λ are isomorphic. Each element of the Poisson leaf has a representative in the form*

$$t\frac{d}{dt} + y + \begin{pmatrix} b_1(t) & \dots & b_n(t) & \dots \\ 0 & \dots & 0 & \dots \\ 0 & \dots & 0 & \dots \end{pmatrix} \longleftrightarrow D^\lambda \left(1 + \sum_{k\geq 1} \tilde{b}_k(\varphi) D^{(-k)} \right).$$

The Drinfeld–Sokolov construction [DS], as well as its generalization to $\mathfrak{SL}(\lambda)$ and $\mathfrak{o}/\mathfrak{sp}(\lambda)$ ([KM]), hinges on a certain element that can be identified with the image of $X^+ \in \mathfrak{sl}(2)$ under the principal embedding. For the case of higher ranks this is the image in $U_\mathfrak{g}(\lambda)$ of the element $y \in \mathfrak{g}$ described in §3 for Lie algebras. In $\mathfrak{SL}(\lambda)$ and $\mathfrak{o}/\mathfrak{sp}(\lambda)$ this image is just d/dx (or the matrix whose only nonzero entries are the 1's under the main diagonal in the realization of $\mathfrak{SL}(\lambda)$ and $\mathfrak{o}/\mathfrak{sp}(\lambda)$ by matrices).

9.4. Superization

9.4.1 Basics

Further facts from Linear Algebra in Superspaces. The *tensor algebra* $T(V)$ of the superspace V is naturally defined: $T(V) = \bigoplus_{n\geq 0} T^n(V)$, where $T^0(V) = k$ and $T^n(V) = V \otimes \dots \otimes V$ (n factors) for $n > 0$.

The *symmetric algebra* of the superspace V is $S(V) = T(V)/I$, where I is the two-sided ideal generated by $v_1 \otimes v_2 - (-1)^{p(v_1)p(v_2)} v_2 \otimes v_1$ for $v_1, v_2 \in V$.

The *exterior algebra* of the superspace V is $E(V) = S(\Pi(V))$. Clearly, both the exterior and symmetric algebras of the superspace V are supercommutative superalgebras. It is worthwhile to mention that if $V_{\bar 0} \neq 0$ and $V_{\bar 1} \neq 0$, then both $E(V)$ and $S(V)$ are infinite dimensional.

A *Lie superalgebra* is defined with the Sign Rule applied to the definition of a Lie algebra. Its multiplication is called *bracket* and is usually denoted by $[\cdot, \cdot]$ or

$\{\cdot,\cdot\}$. If, however, we try to use this definition in attempts to apply the standard group-theoretical methods to differential equations on supermanifolds we will find ourselves at a loss: the supergroups and their modules are objects from different categories! Accordingly, the following (equivalent to the conventional, "naive" one, see [L]) definition becomes useful: a *Lie superalgebra* is a superalgebra $\mathfrak{g}$ (defined over a field or, more generally, a supercommutative superalgebra k); the bracket should satisfy the following conditions: $[X,X]=0$ and $[Y,[Y,Y]]=0$ for any $X\in(C\otimes\mathfrak{g})_{\bar 0}$ and $Y\in(C\otimes\mathfrak{g})_{\bar 1}$ and any supercommutative superalgebra C (the bracket in $C\otimes\mathfrak{g}$ is defined via C-linearity and Sign Rule).

With an associative (super)algebra A we associate Lie (super)algebras (1) A_L with the same (super)space A and the multiplication $(a,b)\mapsto[a,b]$ and (2) $\mathfrak{der}A$, the algebra of derivations of A, defined via the Sign and Leibniz Rules.

From a Lie superalgebra $\mathfrak{g}$ we construct an associative superalgebra $U(\mathfrak{g})$, called the *universal enveloping algebra* of the Lie superalgebra $\mathfrak{g}$ by setting $U(\mathfrak{g})=T(\mathfrak{g})/I$, where I is the two-sided ideal generated by the elements $x\otimes y-(-1)^{p(x)p(y)}y\otimes x-[x,y]$ for $x,y\in\mathfrak{g}$.

The *Poincaré–Birkhoff–Witt theorem* for Lie algebras extends to Lie superalgebras with the same proof (beware Sign Rule) and reads as follows:

If $\{X_i\}$ is a basis in $\mathfrak{g}_{\bar 0}$ and $\{Y_j\}$ is a basis in $\mathfrak{g}_{\bar 1}$, then the monomials $X_{i_1}^{n_1}\dots X_{i_r}^{n_r}Y_{j_1}^{\varepsilon_1}\dots Y_{j_s}^{\varepsilon_s}$, where $n_i\in\mathbb{Z}^+$ and $\varepsilon_j=0,1$, constitute a basis in the space $U(\mathfrak{g})$.

A superspace M is called a *left module* over a superalgebra A (or a *left A-module*) if there is given an even map *act*: $A\otimes M\to M$ such that if A is an associative superalgebra with unit, then $(ab)m=a(bm)$ and $1m=m$ and if A is a Lie superalgebra, then $[a,b]m=a(bm)-(-1)^{p(a)p(b)}b(am)$ for any $a,b\in A$ and $m\in M$. The definition of a *right A-module* is similar.

Convention. We endow every module M over a supercommutative superalgebra C with a two-sided module structure: the left module structure is recovered from the right module one and vice versa according to the formula $cm=(-1)^{p(m)p(c)}mc$ for any $m\in M$ and $c\in C$. Such modules will be called *C-modules.* (Over C, there are *two* canonical ways to define a two-sided module structure, see [L]; the meaning of such an abundance is obscure.)

The functor Π is, actually, tensoring by $\Pi(\mathbb{Z})$. So there are two ways to apply Π to C-modules: to get $\Pi(M)=\Pi(\mathbb{Z})\otimes_{\mathbb{Z}}M$ and $(M)\Pi=M\otimes_{\mathbb{Z}}\Pi(\mathbb{Z})$. The two-sided module structures on $\Pi(M)$ and $(M)\Pi$ are given via the Sign Rule.

Sometimes, instead of the map *act* a morphism $\rho: A\to\mathrm{End}M$ is defined if A is an associative superalgebra (or $\rho: A\longrightarrow(\mathrm{End}M)_L$ if A is a Lie superalgebra); ρ is called a *representation* of A in M.

The simplest (in a sense) modules are those which are *irreducible.* We distinguish *irreducible modules of G-type* (general); these do not contain invariant subspaces different from 0 and the whole module; and their "odd" counterparts, *irreducible modules of Q-type*, which do contain an invariant subspace which, however, is not a subsuperspace. Consequently, *Schur's lemma* states that *over $\mathbb{C}$ the* *centralizer of a set of irreducible operators is either* $\mathbb{C}$ *or* $\mathbb{C} \otimes \mathbb{C}^s = Q(1;\mathbb{C})$, see the definition of the superalgebras Q below.

The next in terms of complexity are *indecomposable* modules, which cannot be split into the direct sum of invariant submodules.

A C-module is called *free* if it is isomorphic to a module of the form $C \oplus \dots \oplus C \oplus \Pi(C) \oplus \dots \oplus \Pi(C)$ (C occurs r times, $\Pi(C)$ occurs s times).

The *rank* of a free C-module M is the element $\mathrm{rk} M = r + s\varepsilon$ from the ring $\mathbb{Z}[\varepsilon]/(\varepsilon^2 - 1)$. Over a field, $C = k$, we usually write just $\dim M = (r, s)$ or $r|s$ and call this pair the *superdimension* of M.

The module $M^* = \mathrm{Hom}_C(M, C)$ is called *dual* to a C-module M. If $(\cdot,\cdot)$ is the pairing of modules M^* and M, then to each operator $F \in \mathrm{Hom}_C(M, N)$, where M and N are C-modules, there corresponds the dual operator $F^* \in \mathrm{Hom}_C(N^*, M^*)$ defined by the formula

$$(F(m), n^*) = (-1)^{p(F)p(m)}(m, F^*(n^*)) \quad \text{for any } m \in M,\ n^* \in N^*.$$

Over a supercommutative superalgebra C a *supermatrix* is a supermatrix with entries from C, the parity of the matrix with only (i,j)-th nonzero element c is equal to $p(i) + p(j) + p(c)$. Denote by $\mathrm{Mat}(Par; C)$ the set of $Par \times Par$ matrices with entries from a supercommutative superalgebra C.

The even invertible elements from $\mathrm{Mat}(Par; C)$ constitute the *general linear group* $GL(Par; C)$. Put $GQ(Par; C) = Q(Par; C) \cap GL(Par; C)$.

On the group $GL(Par; C)$ an analogue of the determinant is defined; it is called the *Berezinian* (in honour of F. A. Berezin who discovered it). In the standard format the explicit formula for the Berezinian is:

$$\mathrm{Ber}\begin{pmatrix} A & B \\ D & E \end{pmatrix} = \det(A - BE^{-1}D)\det E^{-1}.$$

For the matrices from $GL(Par; C)$ the identity $\mathrm{Ber}\, XY = \mathrm{Ber}\, X \cdot \mathrm{Ber}\, Y$ holds, i.e., $Ber : GL(Par; C) \to GL(1|0; C) = GL(0|1; C)$ is a group homomorphism. Set $SL(Par; C) = \{X \in GL(Par; C) : Ber X = 1\}$. The *orthosymplectic* group of automorphisms of the bilinear form with the even canonical matrix is denoted (in the standard format) by $Osp(n|2m; C)$.

9.4.2. Pseudodiferential operators on the supercircle. Residues

Let V be a superspace. For $\theta = (\theta_1, \dots, \theta_n)$ set

$$\begin{gathered} V[x,\theta] = V \otimes \mathbb{K}[x,\theta];\ V[x^{-1}, x, \theta] = V \otimes \mathbb{K}[x^{-1}, x, \theta]; \\ V[[x^{-1}, \theta]] = V \otimes \mathbb{K}[[x^{-1}, \theta]]; \\ V((x,\theta)) = V \otimes \mathbb{K}[[x^{-1}]][x,\theta]. \end{gathered}$$

We call $V((x,\theta))x^\lambda$ the space of *pseudodifferential symbols.* Usually, V is a Lie (super)algebra. Such symbols correspond to pseudodifferential operators (pdo) of the form

$$\sum_{i=-\infty}^{n} \sum_{k_0+\dots+k_n=i} a_i(\partial_x)^{k_0}\theta_1^{k_1}\dots\theta_n^{k_n},$$

Here $k_i = 0$ or 1 for $i > 0$ and $a_i(x,\theta) \in V$. This is clear.

For any $P = \sum\limits_{i \leq m} P_i x^i \theta_0^k \theta^j \in V((x,\theta))$ we call $P_+ = \sum\limits_{i,j,k \geq 0} P_i x^i \theta_0^k \theta^j$ the *differential part* of P and $P_- = \sum\limits_{i,k<0} P_i x^i \theta_0^k \theta^j$ the *integral* part of P.

The space ΨDO of pdos is, clearly, the left module over the algebra $\mathcal{F}$ of functions. Define the left ΨDO-action on $\mathcal{F}$ from the Leibniz formula thus making ΨDO into a superalgebra.

Define the involution in the superalgebra ΨDO setting

$$(a(t,\theta)D^i\tilde{D}^j)^* = (-1)^{jip(\tilde{D})p(D)}\tilde{D}^j D^i a^*(x,\theta).$$

The following fact is somewhat unexpected. If D is an odd differential operator, then D^2 is well-defined as $\frac{1}{2}[D,D]$. Hence, we can consider the set $V((x,\theta))x^\lambda$ for an *odd* x! Therefore, there are two types of pdos: *contact* ones, when $D^2 \neq 0$ for odd D's and general ones, when all odd D's are nilpotent.

For the definition of distinguished stringy superalgebras, crucial in what follows, see [GLS].

Conjecture. There exists a residue for all distinguished dimensions, i.e. for the contact type pdos in dimensions $1|n$ for $n \leq 4$ and for the general pdos in dimensions $1|n$ for $n \leq 2$.

So far, however, the residue was defined only for contact type pdos, of $\mathfrak{k}^L$ type, and only for $n = 1$ at that.

Extend [MR] and define the *residue* of $P = \sum\limits_{i \leq m} P_i x^i \theta_0^k \theta^j \in V((x,\theta_0,\theta))$ for $n = 1$. We can do this thanks to the following exceptional property of $\mathfrak{k}^L(1|1)$ and $\mathfrak{k}^M(1|1)$. Indeed, over $\mathfrak{k}^L(1|1)$, the volume form "$dt\frac{\partial}{\partial\theta}$" is, more or less, $d\theta$: consider the quotient $\Omega^1/\mathcal{F}\alpha$, where α is the contact form preserved by $\mathfrak{k}^L(1|1)$; similarly, over $\mathfrak{k}^M(1|1)$, the transformation rules of $dt\frac{\partial}{\partial\theta}$" and $\tilde{\alpha}$, where $\tilde{\alpha}$ is the contact form preserved by $\mathfrak{k}^M(1|1)$, are identical. Therefore, define the residue by the formula

$$\text{Res } P = \text{ coefficient of } \frac{\theta}{x} \text{ in the expansion of } P_{-1}.$$

Remark. Manin and Radul [MR] considered the Kadomtsev–Petviashvili hierarchy associated with $\mathfrak{ns}$, i.e., for $D = K_\theta$. The formula for the residue allows one to directly generalize their result and construct a similar hierarchy associated with $\mathfrak{r}$, i.e., for $D = \tilde{K}_\theta$.

This new phenomenon — an invertible odd symbol — doubles the old picture: let θ_0 be the symbol of D, and let x be the symbol of the differential operator D^2. We see that the case of the odd D reduces to either $V((x,\theta_0,\theta))x^\lambda$ or $V((x,\theta_0^{-1},\theta))x^\lambda$.

9.5. Continuous Toda lattices

Khesin and Malikov [KM] considered straightforward generalizations of the Toda lattices — the dynamical systems on the orbits of the coadjoint representation of a simple finite diminsional Lie group G defined as follows. Let $\mathcal{X}$ be the image of $X^+ \in \mathfrak{sl}(2)$ in $\mathfrak{g} = Lie(G)$ under the principal embedding. Having identified $\mathfrak{g}$ with $\mathfrak{g}^*$ with the help of the invariant nondegenerate form, consider the orbit $\mathcal{O}_{\mathcal{X}}$. On $\mathcal{O}_{\mathcal{X}}$, the traces $H_i(A) = \operatorname{tr}(A + \mathcal{X})^i$ are the commuting Hamiltonians.

In our constructions we only have to consider in $LU_{\mathfrak{g}}(\lambda)$ either (for Lie algebras) the image of $\mathcal{X}$ or (for Lie superalgebras) the image of $\nabla^+ \in \mathfrak{osp}(1|2)$ under the superprincipal embedding of $\mathfrak{osp}(1|2)$. For superalgebras we also have to replace trace with the supertrace.

For the general description of dynamical systems on the orbits of the coadjoint representations of Lie supergroups see [LST]. A possibility of odd mechanics is pointed out in [LST] and in the subsequent paper by R. Yu. Kirillova in the same Procedings. To take such a possibility into account, we have to consider analogs of the principal embeddings for $\mathfrak{sq}(2)$. This is a full-time job; its results will be considered elsewhere.

References

[Ar] Arnold V., *Mathematical methods of classical mechanics*, Springer, 1989

[BO] Yu. A. Bakhturin, A. Yu. Olshansky, The approximations and characteristic subalgebras of free Lie algebras, In: *Proc. I. G. Petrovsky seminar*, 2, 1976, 145–150 (in Russian)

[BWV] Bergshoeff E., de Wit B., Vasiliev M., The structure of the super-$W_\infty(\lambda)$ algebra. *Nucl. Phys* B366, 1991, 315–346

[BMP] Bouwknegt P., McCarthy J., Pilch K., Quantum group structure in the Fock space resolutions of $\widehat{\mathfrak{sl}(n)}$ representations, *Commun. Math. Phys.*, 131, 1990, 339–368

[B] Burdík Č., Realizations of real semisimple Lie algebras: a method of construction, *J. Phys. A*: Math. Gen., 18, 1985, 3101–3111

[BGLS] Burdík Č., Grozman P., Leites D., Sergeev A., Realization of simple Lie algebras via differential operators, (to appear)

[Di] Dixmier J. *Algèbres envellopentes*, Gautier–Villars, Paris, 1974

[DGS] Donin J., Gurevich D., Shnider S., Quantization of function algebras on semisimple orbits in $\mathfrak{g}^*$. Talk at Internat. Conf. "Quantum groups and integrable systems" June 19 – 21, 1997, Prague.

[DS] Drinfeld V., Sokolov V., Lie algebras and equations of Korteveg–de Vries type, JOSMAR, 1984

[D] Dynkin E. B., Semi-simple subalgebras of semi-simple Lie algebras, Mat. Sbornik, 30, 1952, 111–244 (AMS Transl., v.6, ser. 2, 1957)

[E] Egorov G., How to superize $\mathfrak{gl}(\infty)$. In: Mickelsson J., Peckonnen O. (eds.) *Diff. Geometric Methods in Theoretical Physics* (Proc. conf 1991, Turku, Finland) World Sci., 1992, 135–146

[F] Feigin B. L., The Lie algebras $\mathfrak{gl}(\lambda)$ and cohomologies of Lie algebra of differential operators, Russian Math. Surveys, v. 43, 2, 1988, 157–158

[FFr] Feigin B. L., Frenkel E., Integrals of motion and quantum groups. In: Donagi R. e.a. (eds) *Integrable systems and quantum groups* LN in Math 1620, 1996, 349–418

[FO] Feigin B. L., Odessky A., Elliptic Sklyanin algebras, Funkt. Anal. Appl., 1989, v.23, n. 3, 45–54

[Go] Golod P., A deformation of the affine Lie algebra $A_1^{(1)}$ and hamiltonian systems on the orbits of its subalgebras. In: *Group-theoretical methods in physics* Proc. of the 3rd seminar, Yurmala, 1985, v.1, Moscow, Nauka, 1986, 368–376

[GL1] Grozman P., Leites D., Defining relations associated with the principal $\mathfrak{sl}(2)$-subalgebras. In: Dobrushin R., Minlos R., Shubin M., Vershik A. (eds.) *Contemporary mathematical physics* (F. A. Berezin memorial volume), Amer. Math. Soc. Transl. Ser.2, vol. 175, Amer. Math. Soc., Providence, RI, 1996, 57–68

[GL2] Grozman P., Leites D., Lie superalgebras of supermatrices of complex size: a closer view (to appear)

[GL3] Grozman P., Leites D., Defining Relations for Lie superalgebras with Cartan matrix, hep-th 9702073

[GLS] Grozman P., Leites D., Shchepochkina I., Lie superalgebras of string theories, hep-th 9702120

[KS] Kac V., Lie superalgebras, Adv. Math., 1976, 55–110

[K] Kac V., *Infinite dimensional Lie algebras*, Cambridge Univ. Press, Cambridge, 1991

[KR] Kac V., Radul A., Quasifinite highest weight modules over the Lie algebra of differential operators on the circle, Commun. Math. Phys., v. 157, 1993, 429–457

[Ka] Kashivara M., Representation theory and D-modules on flag varieties, Asterisque, 173–174, 1989, 55–110

[KM] Khesin B., Malikov F., Universal Drinfeld–Sokolov reduction and the Lie algebras of matrices of complex size, Comm. Math. Phys., v. 175, 1996, 113–134

[KV] Konstein S., Vasiliev M., Supertraces on the algebras of observables of the rational Calogero model with harmonic potential. J. Math. Phys 37(6), 1996, 2872–2891

[L1] Leites D., Quantization and supermanifolds. In: F. Berezin, M. Shubin *Schrödinger equation*, Kluwer, Dordrieht, 1991

[L2] Leites D. (ed.) *Seminar on supermanifolds*. Reports of Stockholm Univ., ##1–34, 1987–92

[LM] Leites D., Montgomery S., New simple filtered Lie superalgebras: a construction (to appear)

[LP] Leites D., Poletaeva E., Defining relations for Lie algebras of polynomial vector fields, Math. Scand., **81**, 1997, No. 1, 5–19.

[LSS] Leites D., Saveliev M., Serganova V., Embeddings of $\mathfrak{osp}(n|2)$ and the associated nonlinear supersymmetric equations. In: (Markov, M.A., Man'ko, V.I. and Dodonov, V.V., eds.) *Group Theoretical Methods in Physics* (Yurmala 1985), VNU Science Press, Utreht, v.1, 1986, 255–297

[LST] Leites D. , Semenov-Tian-Shansky M., Integrable systems and Lie superalgebras. In: L. D. Faddeev (ed.) *Proceeding of LOMI Seminars*, Nauka, Leningrad, v. 123, 1983, 92–97.

[LS] Leites D., Serganova, V., Defining relations for classical Lie superalgebras. I., In: Mickelsson J., Peckonnen O. (eds.) *Diff. Geometric Methods in Theoretical Physics* (Proc. conf 1991, Turku, Finland) World Sci., 1992, 194–201

[LAS] Leites D., Sergeev A., The automorphisms and real forms of Lie superalgebras of supermatrices of complex size (to appear)

[LSc] Leites D., Shchepochkina I., Classification of simple Lie superalgebras of polynomial growth (to appear)

[MR] Manin Yu., Radul A., Commun. Math. Phys., 98, 1985, 65–77

[M] Montgomery S., Constructing simple Lie superalgebras from associative graded algebras, J. Algebra 195 (1997), no. 2, 558–579

[OV] Onishchik A. L., Vinberg É. B., *Seminar on algebraic groups and Lie groups*, Springer, 1990

[PS] Penkov I., Serganova V., Generic irreducible representations of finite dimensional Lie superalgebras, International J. Math., v. 5, 1994, 389–419

[PH] Post G., Hijligenberg N. van den, $\mathfrak{gl}(\lambda)$ and differential operators preserving polynoials. Acta Appl. Math., v. 44, 1996, 257–268

[SV] Saveliev M., Vershik A., Continuum analogues of contragredient Lie algebras, Commun. Math. Phys., 126, 1989, 367–378

[SH] Sheinman O.M., Heighest weight modules over certain quasigraded Lie algebras over elliptic curves, Funct. Anal. Appl., 26:3, 1992, 203–208

[S] Sergeev A., Invariant polynomials on Lie superalgebras. In [L2], v. 32.

[Sh] Shoikhet B., Certain topics on the Lie algebra $\mathfrak{gl}(\lambda)$ representation theory, q-alg/9703029

[vJ] Van der Jeugt J., Principal five-dimensional subalgebras of Lie superalgebras, J. Math. Phys, v.27 (12), 1986, 2842–2847

Dept. of Math.,
Univ. of Stockholm,
Roslagsv. 101, Kräftriket hus 6,
S-106 91, Stockholm,
E-mail address: mleites@matematik.su.se

Operator Theory:
Advances and Applications, Vol. 114

A new local variant of the Hausdorff-Young inequality

Amir Kamaly

Abstract. A Carlson-type inequality is proved and it is applied to show a Babenko-Beckner type of the Hausdorff-Young inequality on n-dimensional torus. The relation between the sharp forms of the Hausdorff-Young inequality for functions (with small supports) in the space $L^p(\mathbb{T}^n)$ and Young's inequality for convolution (with even exponent) is also explored. The Babenko-Beckner constant B_p is applied to improve the constant of a norm inequality of an oscillatory integral.

1. Introduction

Fritz Carlson's inequality (1934) states [1] that

$$\sum_{n=1}^{\infty} a_n < \sqrt{\pi}\left(\sum_{n=1}^{\infty} a_n^2\right)^{\frac{1}{4}}\left(\sum_{n=1}^{\infty} n^2 a_n^2\right)^{\frac{1}{4}}$$

holds for any positive sequence $(a_n)_{n=1}^{\infty}$ such that not all a_n are 0. Let $a_n := \widehat{f}(n)$ for a periodic function f. Then there can be equality only if f is a multiple of f', and therefore an exponential function $C_0 e^{bx}$. This is plainly impossible.

Note that the sums $\sum_{n=1}^{\infty} a_n^2$ and $\sum_{n=1}^{\infty} n^2 a_n^2$ are supposed to be finite.

The corresponding integral inequality [2] is

$$\int_0^{\infty} f(x)\,dx \le \sqrt{\pi}\left(\int_0^{\infty} f^2(x)\,dx\right)^{\frac{1}{4}}\left(\int_0^{\infty} x^2 f^2(x)\,dx\right)^{\frac{1}{4}}.$$

Here there is equality when $f(x) := \frac{1}{\alpha+\beta x^2}$, for any positive α, β.

The other expression of Carlson's inequality is

$$\|f\|_{A(\mathbb{T})} \le C\Big(\|f\|_2\|f'\|_2\Big)^{\frac{1}{2}} \tag{1}$$

for $f \in A(\mathbb{T})$ and $\widehat{f}(0) = 0$.

1991 *Mathematics Subject Classification.* 42AXX, 42BXX, 47A30, 47A63.
Key words and phrases. Carlson's inequality, the Hausdorff-Young inequality, locally compact unimodular groups, nilpotent Lie groups, generic coadjoint orbits, adjoint operator, the Riesz potential, the Riesz interpolation theorem, Gaussian curvature.

Here $\|f\|_{A(\mathbb{T})} := \sum_{m\in\mathbb{Z}} |\widehat{f}(m)|$ and $A(\mathbb{T})$ is the space of continuous functions on $\mathbb{T}$ having an absolutely convergent Fourier series. The constant C in (1) depends on the definitions of $\mathbb{T}$ and the Fourier series of f.

The well-known classical Hausdorff-Young inequality (1912-1923) states that for any complex-valued function g in the Banach space $L^p(\mathbb{T})$,

$$\|\widehat{g}\|_{p'} \leq \|g\|_p \tag{2}$$

holds for $1 \leq p \leq 2$. Here and throughout the paper, p' is the dual exponent of p. Also, $\|\widehat{g}\|_{p'} := \left(\sum_{n\in\mathbb{Z}} |\widehat{g}(n)|^{p'}\right)^{\frac{1}{p'}}$ and $\|g\|_p := \left(\int_{\mathbb{T}} |g(x)|^p \, dx\right)^{\frac{1}{p}}$ are supposed to be finite.

Titchmarsh [5] proved (2) for the space $L^p(\mathbb{R})$ in 1924. In fact, (2) is true for locally compact unimodular groups [14]. This result is due to R. A. Kunze (1957). Hardy and Littlewood [7] showed that (2) is sharp and there is equality if and only if $g = C_0 e^{2\pi mix}$ for $m \in \mathbb{Z}$.

For the space $L^p(\mathbb{R}^n)$ and for the even integer p' the improvement [7] is due to K. I. Babenko (1961) and for all p due to W. Beckner [8] (1975). That is,

$$\|\widehat{f}\|_{p'} \leq B_p{}^n \|f\|_p, \quad B_p = \sqrt{\frac{p^{1/p}}{p'^{1/p'}}} \tag{3}$$

where $\widehat{f}(\xi) := \int_{\mathbb{R}^n} f(x) e^{-2\pi i<\xi,x>} dx$ is the Fourier transform of f and $<\xi, x> := \sum_{\nu=1}^n \xi_\nu x_\nu$. B_p is called the Babenko-Beckner constant.

B. Russo (1974) [15] and J. J. F. Fournier (1977) [16] proved (3) for certain classes of locally compact unimodular groups.

The extension of (3) is due to J. Inoue (1992) [13]. For certain classes of nilpotent Lie groups he improved (3) and obtained the constant

$$B_p^{\dim(G)-\frac{m}{2}}.$$

Here $G := \exp(\mathbf{g})$ and $\mathbf{g}$ is a Lie algebra with the dual space $\widehat{\mathbf{g}}$, $\dim(G)$ is the dimension of nilpotent Lie groups G and m is the dimension of generic coadjoint orbits of G in $\widehat{\mathbf{g}}$.

For the even integer p' [9] M. E. Andersson (1992) and for all p P. B. Sjölin [10] (1994) proved a Babenko-Beckner type of the inequality (3) for functions in the space $L^p(\mathbb{T})$, with small supports.

Define

$$H_{p,a} := \sup\left\{\frac{\|\widehat{g}\|_{p'}}{\|g\|_p} : \; g \in L^p(\mathbb{T}^n), \; \text{supp } g \subset \overline{B}(0,a), \; \|g\|_p \neq 0\right\}$$

and let $H_p := \lim_{a\to 0+} H_{p,a}$. Here and everywhere in the paper a obeys the restriction $0 < a < \frac{1}{2}$ and $\overline{B}(0,a)$ is a closed ball of radius a, centered at the origin. Also, the convolution of functions $g_1,\ g_2 \in L^p(\mathbb{T}^n)$ is $g_1 * g_2(x) := \int_{\mathbb{T}^n} g_1(x-t)g_2(t)dt$, where $\mathbb{T}^n := \left\{x \in \mathbb{R}^n : |x_\nu| \le \frac{1}{2},\ 1 \le \nu \le n\right\}$.

Let the multi-indices β and γ be vectors in $\mathbb{R}^n$ with components β_k and γ_k in $\mathbb{N}_0$ such that $\gamma \le \beta$ is equivalent to $\gamma_k \le \beta_k$ for all $1 \le k \le n$. Also, $m^\beta := \prod_{k=1}^n {m_k}^{\beta_k}$ for $m \in \mathbb{Z}^n$ and $0^0 := 1$.

Throughout this paper, we have $|\beta| := \sum_{k=1}^n \beta_k$, $\binom{\beta}{\gamma} := \prod_{k=1}^n \binom{\beta_k}{\gamma_k}$ and the operator $D^\beta := \prod_{k=1}^n \frac{\partial^{\beta_k}}{\partial x_k^{\beta_k}}$.

The purpose of this paper is to prove Carlson's inequality of type (1) on the n-dimensional torus and applying it to find the best (in some sense) upper bound for $H_{p,a}$. Different methods are used in estimating of $H_{p,a}$. We show that $H_p = {B_p}^n$ for $p \in [1,2]$. Also, we prove a Babenko-Beckner type of the Hausdorff-Young inequality on the n-dimensional torus and give one application of the Babenko-Beckner constant B_p. The relation between the sharp forms of the Hausdorff-Young inequality for functions (with small supports) in the space $L^p(\mathbb{T})$ and Young's inequality for convolution (with even exponent) is also explored. □

Theorem 1 (Generalization of Carlson's inequality). *Let $f \in A(\mathbb{T}^n)$ and $\widehat{f}(0) = 0$. Let the absolute value of the multi-index β be equal to the positive integer α such that $\alpha \ge 1$ and $\alpha > \frac{n}{q}$ where $1 < q \le 2$. Then we get*

$$\|f\|_{A(\mathbb{T}^n)} \le K_{n,q}^{(\alpha)} {\|f\|_q}^{1-\frac{n}{q\alpha}} \left(\sum_{|\beta|=\alpha} \|D^\beta f\|_q\right)^{\frac{n}{q\alpha}}. \tag{4}$$

In the case $\widehat{f}(0) \neq 0$, we obtain

$$\|f\|_{A(\mathbb{T}^n)} \le \|f\|_1 + K_{n,q}^{(\alpha)} {\|f\|_q}^{1-\frac{n}{q\alpha}} \left(\sum_{|\beta|=\alpha} \|D^\beta f\|_q\right)^{\frac{n}{q\alpha}},$$

where the positive constant $K_{n,q}^{(\alpha)}$ depends only on n, α and q.

Proof of Theorem 1. The technique is analogous to the case $n = 1$, due to Hardy [2]. Let $\widehat{f}(0) = 0$ and q' be the dual exponent of q. Define

$$S := \sum_{m \in \mathbb{Z}^n} |\widehat{f}(m)|^{q'}$$
$$T := \sum_{|\beta|=\alpha} \sum_{m \in \mathbb{Z}^n} |(2\pi i m)^\beta|^{q'} |\widehat{f}(m)|^{q'},$$

then $S = \|\widehat{f}\|_{q'}^{q'}$ and $T = \sum_{|\beta|=\alpha} {\|\widehat{D^\beta f}\|_{q'}}^{q'} \le \left(\sum_{|\beta|=\alpha} \|\widehat{D^\beta f}\|_{q'}\right)^{q'}$, because the inequality

$$\sum_m {a_m}^{q'} \le \left(\sum_m a_m\right)^{q'} \tag{4.1}$$

holds for $q' \geq 1$ and $a_m \geq 0$. Also, for t_1, $t_2 > 0$ define

$$P := \sum_{|\beta|=\alpha} \left(t_1 + t_2 |(2\pi m)^\beta|^{q'}\right).$$

Then, by Hölder's inequality we get

$$\begin{aligned}
\|f\|_{A(\mathbb{T}^n)} &= \sum_{m\in\mathbb{Z}^n} |\widehat{f}(m)| \\
&= \sum_{|m|>0} |\widehat{f}(m)| P^{\frac{1}{q'}} P^{-\frac{1}{q'}} \\
&\leq \left(\sum_{|m|>0} |\widehat{f}(m)|^{q'} P\right)^{\frac{1}{q'}} \left(\sum_{|m|>0} P^{-\frac{q}{q'}}\right)^{\frac{1}{q}} \\
&\leq \left(t_1 c_{n,\alpha} S + t_2 T\right)^{\frac{1}{q'}} \left[\sum_{|m|>0} \left(\sum_{|\beta|=\alpha} \left(t_1 + t_2 |(2\pi m)^\beta|^{q'}\right)\right)^{-\frac{q}{q'}}\right]^{\frac{1}{q}} \\
&\leq t_1{}^{-\frac{1}{q'}} \left(t_1 c_{n,\alpha} S + t_2 T\right)^{\frac{1}{q'}} \left[\sum_{|m|>0} \left(1 + \frac{t_2 C_{n,\alpha}}{t_1} |m|^{q'\alpha}\right)^{-\frac{q}{q'}}\right]^{\frac{1}{q}},
\end{aligned} \tag{4.2}$$

because

$$\sum_{|\beta|=\alpha} \left(t_1 + t_2 |(2\pi m)^\beta|^{q'}\right) = c_{n,\alpha} t_1 + t_2 \sum_{|\beta|=\alpha} |(2\pi m)^\beta|^{q'} \geq t_1 + t_2 C_{n,\alpha} |m|^{q'\alpha}.$$

Here and everywhere in this paper $c_{n,\alpha} := \sum_{|\beta|=\alpha} 1$ and $C_{n,\alpha}$ depends on n and α. It is not hard to see that the sum $\left(\sum_{|m|>0} \frac{1}{(1+|m|^{q'\alpha})^{\frac{q}{q'}}}\right)^{\frac{1}{q}}$ is finite for $\alpha > \frac{n}{q}$ and

$$\int_0^\infty \frac{dx}{(1 + x^{\frac{q'\alpha}{n}})^{\frac{q}{q'}}} = \frac{\Gamma\left(\frac{n(q-1)}{q\alpha}\right) \Gamma\left(\frac{(q-1)(q\alpha-n)}{q\alpha}\right)}{\frac{q\alpha}{n(q-1)} \Gamma(q-1)}. \tag{4.3}$$

See [23], p. 175, formula 14.

Now, by (4.2) and (4.3) we obtain

$$\|f\|_{A(\mathbb{T}^n)} \le c_0 t_1^{-\frac{1}{q'}} \left(t_1 c_{n,\alpha} S + t_2 T\right)^{\frac{1}{q'}} \left(\int_{\mathbb{R}^n} \frac{dx}{\left(1+\frac{t_2 C_{n,\alpha}}{t_1}|x|^{q'\alpha}\right)^{\frac{q}{q'}}}\right)^{\frac{1}{q}}$$

$$= c_0 t_1^{-\frac{1}{q'}} \left(\frac{t_1}{C_{n,\alpha} t_2}\right)^{\frac{n}{qq'\alpha}} (t_1 c_{n,\alpha} S + t_2 T)^{\frac{1}{q'}} \left(\int_{\mathbb{R}^n} \frac{dx}{(1+|x|^{q'\alpha})^{\frac{q}{q'}}}\right)^{\frac{1}{q}}$$

$$= c_0 t_1^{-\frac{1}{q'}} \left(\frac{t_1}{C_{n,\alpha} t_2}\right)^{\frac{n}{qq'\alpha}} (t_1 c_{n,\alpha} S + t_2 T)^{\frac{1}{q'}} \left(\int_0^\infty \int_{\{x\in\mathbb{R}^{n-1}:|x|=1\}} \frac{r^{n-1}drdx}{(1+r^{q'\alpha})^{\frac{q}{q'}}}\right)^{\frac{1}{q}}$$

$$= c_0 \left(\frac{w_{n-1}}{n}\right)^{\frac{1}{q}} t_1^{-\frac{1}{q'}} \left(\frac{t_1}{C_{n,\alpha} t_2}\right)^{\frac{n}{qq'\alpha}} (t_1 c_{n,\alpha} S + t_2 T)^{\frac{1}{q'}} \left(\int_0^\infty \frac{dx}{\left(1+x^{\frac{q'\alpha}{n}}\right)^{\frac{q}{q'}}}\right)^{\frac{1}{q}}$$

$$= c_0 A_{n,q}^{(\alpha)} \left\{ \left(\frac{t_1}{t_2}\right)^{\frac{n}{q\alpha}} c_{n,\alpha} S + \left(\frac{t_1}{t_2}\right)^{\frac{n}{q\alpha}-1} T \right\}^{\frac{1}{q'}}$$

$$= c_0 A_{n,q}^{(\alpha)} \left\{ \left(\frac{t_1}{t_2}\right)^{\frac{n}{q\alpha}} c_{n,\alpha} S + \left(\frac{t_1}{t_2}\right)^{\frac{n}{q\alpha}-1} T \right\}^{\frac{1}{q'}}.$$

Here

$$A_{n,q}^{(\alpha)} := \sqrt[q]{\frac{(q-1)w_{n-1}\Gamma\left(\frac{n(q-1)}{q\alpha}\right)\Gamma\left(\frac{(q-1)(q\alpha-n)}{q\alpha}\right)}{q\alpha\Gamma(q-1)}}\,(C_{n,\alpha})^{\frac{n(1-q)}{q\alpha}},$$

and w_{n-1} is the surface area of the unit sphere in $\mathbb{R}^{n-1}$.

Choose $t_1 = T$ and $t_2 = S$; then by the classical Hausdorff-Young inequality (2), two times, we get

$$\begin{aligned} \|f\|_{A(\mathbb{T}^n)} &\le c_0 A_{n,q}^{(\alpha)} \sqrt[q]{(c_{n,\alpha}+1)^{q-1}} \|\widehat{f}\|_{q'}^{1-\frac{n}{q\alpha}} \left(\sum_{|\beta|=\alpha} \|\widehat{D^\beta f}\|_{q'}\right)^{\frac{n}{q\alpha}} \\ &\le c_0 A_{n,q}^{(\alpha)} \sqrt[q]{(c_{n,\alpha}+1)^{q-1}} \|f\|_{q}^{1-\frac{n}{q\alpha}} \left(\sum_{|\beta|=\alpha} \|D^\beta f\|_{q}\right)^{\frac{n}{q\alpha}} \\ &= c_0 A_{n,q}^{(\alpha)} \sqrt[q]{(c_{n,\alpha}+1)^{q-1}} \|f\|_{q}^{1-\frac{n}{q\alpha}} \left(\sum_{|\beta|=\alpha} \|D^\beta f\|_{q}\right)^{\frac{n}{q\alpha}} \\ &= K_{n,q}^{(\alpha)} \|f\|_{q}^{1-\frac{n}{q\alpha}} \left(\sum_{|\beta|=\alpha} \|D^\beta f\|_{q}\right)^{\frac{n}{q\alpha}}. \end{aligned} \qquad (5)$$

Note that $\widehat{D^\beta f}(m) = (2\pi i m)^\beta \widehat{f}(m)$. □

Remark 1. The inequality (5) can also be expressed as in the following:

$$\|f\|_{A(\mathbb{T}^n)} \leq K_{n,q}^{(\alpha)} \|\widehat{f}\|_{q'}^{1-\frac{n}{q\alpha}} \left(\sum_{|\beta|=\alpha} \|D^\beta f\|_q \right)^{\frac{n}{q\alpha}} . \square \tag{5.1}$$

2. Application of Theorem 1 to obtain an estimate of $H_{p,a}$.

Theorem 2 (An upper bound for $H_{p,a}$)). *For a fixed $n \in \mathbb{N}$, there exists a positive constant C_0 which does not depend on a, such that*

$$H_{p,a} \leq \left(1 + C_0 a\right) B_p{}^n, \quad 1 \leq p \leq 2, \tag{6}$$

for sufficiently small a.

Proof of Theorem 2. The technique is analogous to the case $n = 1$, due to Y. Domar, [6]. Let α be the positive integer defined in Theorem 1. Define

$$\varphi(x) := \begin{cases} 1 & |x| \leq \frac{1}{2} \\ 0 & |x| \geq 1 \end{cases}$$

such that $\varphi \in C_0^\infty(\mathbb{R}^n)$, $0 \leq \varphi \leq 1$ and $\varphi_a := \varphi\left(\frac{x}{a}\right)$. Let $\Psi(x) := (e^{-2\pi i <b,x>} - 1)\varphi_a(x)$ where $b := (b_1, b_2, \ldots, b_n)$ and $|b_k| \leq \frac{1}{2}$, $1 \leq k \leq n$. Then $\Psi \in C_0{}^\infty(\mathbb{R}^n)$ and for $m \in \mathbb{Z}^n$ we get

$$\begin{aligned} |\widehat{\Psi}(m)| &= \left| \int_{|x|\leq a} \Psi(x) e^{-2\pi i <m,x>} dx \right| \leq a^n \int_{|y|\leq 1} \left| e^{-2\pi i a <b,y>} - 1 \right| dy \\ &\leq \pi \sqrt{n} a^{n+1} \int_{|y|\leq 1} dx = \Omega_n a^{n+1}, \end{aligned}$$

because

$$\left| e^{-2\pi i a <b,y>} - 1 \right| \leq 2\pi a |<b,y>| \leq \pi \sqrt{n} a.$$

Here $\Omega_n := \sqrt{n} \pi w_{n-1}$ and $w_{n-1} = \frac{2\pi^{\frac{n}{2}}}{\Gamma(\frac{n}{2})}$ is the surface area of the unit sphere in $\mathbb{R}^n$.

Furthermore, by Leibniz's formula, together with Minkowski's inequality we obtain

$$\begin{aligned}\sum_{|\beta|=\alpha}\|D^\beta\Psi\|_q &\le \pi\sqrt{n}a\sum_{|\beta|=\alpha}\|D^\beta\varphi_a\|_q+\sum_{|\beta|=\alpha}\sum_{\substack{\gamma\le\beta\\|\gamma|\neq 0}}\binom{\beta}{\gamma}\pi^{|\gamma|}\|D^{\beta-\gamma}\varphi_a\|_q\\
&\le \pi\sqrt{n}a^{1-\alpha+\frac{n}{q}}\sum_{|\beta|=\alpha}\|D^\beta\varphi\|_q+\sum_{|\beta|=\alpha}\sum_{\substack{\gamma\le\beta\\|\gamma|\neq 0}}\binom{\beta}{\gamma}\pi^{|\gamma|}a^{|\gamma|-|\beta|+\frac{n}{q}}\|D^{\beta-\gamma}\varphi\|_q\\
&\le a^{1-\alpha+\frac{n}{q}}\left\{\pi\sqrt{n}\sum_{|\beta|=\alpha}\|D^\beta\varphi\|_q+\sum_{|\beta|=\alpha}\sum_{\substack{\gamma\le\beta\\|\gamma|\neq 0}}\binom{\beta}{\gamma}\pi^{|\gamma|}\|D^{\beta-\gamma}\varphi\|_q\right\}\\
&= A_{n,q,\alpha}a^{1-\alpha+\frac{n}{q}},\end{aligned}\tag{6.1}$$

because

$$\sum_{|\beta|=\alpha}\left\|D^\beta\varphi_a\right\|_q=a^{\frac{n}{q}-\alpha}\sum_{|\beta|=\alpha}\left\|D^\beta\varphi\right\|_q.$$

Now, by Theorem 1 and invoking (6.1), we get

$$\begin{aligned}\|\widehat{\Psi}\|_1 &= \sum_{m\in\mathbb{Z}^n}|\widehat{\Psi}(m)|\le \Omega_n a^{n+1}+\sum_{|m|>0}|\widehat{\Psi}(m)|\\
&\le \Omega_n a^{n+1}+K_{n,q,\alpha}\|\Psi\|_q^{1-\frac{n}{q\alpha}}\left(\sum_{|\beta|=\alpha}\|D^\beta\Psi\|_q\right)^{\frac{n}{q\alpha}}\\
&\le \Omega_n a^{n+1}+\left[K_{n,q,\alpha}\left(\pi\sqrt{n}\|\varphi\|_q\right)^{1-\frac{n}{q\alpha}}A_{n,q,\alpha}^{\frac{n}{q\alpha}}\right]a\\
&\le \left\{\Omega_n+K_{n,q,\alpha}A_{n,q,\alpha}{}^{\frac{n}{q\alpha}}\left(\pi\sqrt{n}\|\varphi\|_q\right)^{1-\frac{n}{q\alpha}}\right\}a\\
&= C_0 a,\end{aligned}\tag{6.2}$$

because

$$\|\Psi\|_q\le\pi\sqrt{n}\|\varphi\|_q a^{1+\frac{n}{q}}.$$

Note that $\|\widehat{\Psi}\|_1:=\|\Psi\|_{A(\mathbb{T}^n)}$.

Choose $f\in L^p(\mathbb{R}^n)$, $g\in L^p(\mathbb{T}^n)$, such that $f=g$ on the ball $\overline{B}(0,a)$ and zero outside of the ball. Define $g_b(x):=e^{-2\pi i<x,b>}g(x)$. Then

$$\begin{aligned}\widehat{g_b}(m)&=\widehat{g}(m+b)\\ \|f\|_p&=\|g\|_p\\ \widehat{f_b}(m)&=\widehat{g_b}(m).\end{aligned}$$

Also, we get

$$\begin{aligned}\widehat{g_b}(m)-\widehat{g}(m) &= \int e^{-2\pi i<m,x>}\left(e^{-2\pi i<b,x>}-1\right)g(x)dx\\ &= \int_{\overline{B}(0,a)} \Psi(x)g(x)e^{-2\pi i<m,x>}dx\\ &= \int_{\overline{B}(0,a)} g(x)\Big(\sum_{m'\in\mathbb{Z}^n}\widehat{\Psi}(m')e^{2\pi i<m',x>}\Big)e^{-2\pi i<m,x>}dx\\ &= \sum_{m'\in\mathbb{Z}^n}\widehat{\Psi}(m')\widehat{g}(m-m').\end{aligned}$$

Thus we obtain

$$\|\widehat{g_b}-\widehat{g}\|_{p'} \leq \Big(\sum_{m'\in\mathbb{Z}^n}|\widehat{\Psi}(m')|\Big)\Big(\sum_{m\in\mathbb{Z}^n}|\widehat{g}(m)|^{p'}\Big)^{\frac{1}{p'}} = \|\widehat{g}\|_{p'}\,\|\widehat{\Psi}\|_1.$$

By the triangle inequality we have

$$\|\widehat{g}\|_{p'}-\|\widehat{g_b}\|_{p'} \leq \|\widehat{g_b}-\widehat{g}\|_{p'} \leq \|\widehat{g}\|_{p'}\,\|\widehat{\Psi}\|_1.$$

Also, for $t\in\mathbb{R}^n$, we obtain

$$\|\widehat{g}\|_{p'}\left(1-\|\widehat{\Psi}\|_1\right) \leq \|\widehat{g_b}\|_{p'} = \Big(\sum_{m\in\mathbb{Z}^n}|\widehat{g_b}(m)|^{p'}\Big)^{\frac{1}{p'}} = \Big(\sum_m\left|\widehat{f_b}(m)\right|^{p'}\Big)^{\frac{1}{p'}}.$$

Hence

$$\begin{aligned}\|\widehat{g}\|_{p'}\left(1-\|\widehat{\Psi}\|_1\right) &\leq \Big(\sum_m\int_{\{b:|b_k|\leq\frac{1}{2}\}}\left|\widehat{f_b}(m)\right|^{p'}db\Big)^{\frac{1}{p'}}\\ &= \Big(\sum_m\int_{\{t:\ |t_k-m_k|\leq\frac{1}{2}\}}\left|\widehat{f}(t)\right|^{p'}dt\Big)^{\frac{1}{p'}}\\ &= \|\widehat{f}\|_{p'},\end{aligned}$$

or

$$\|\widehat{g}\|_{p'} \leq \frac{\|\widehat{f}\|_{p'}}{1-\|\widehat{\Psi}\|_1} \leq \frac{\|\widehat{f}\|_{p'}}{1-C_0a}. \tag{7}$$

By (7), thus, we get

$$H_{p,a}=\sup_g\frac{\|\widehat{g}\|_{p'}}{\|g\|_p} \leq \frac{{B_p}^n}{1-C_0a}.$$

Because

$$\sup_f\frac{\|\widehat{f}\|_{p'}}{\|f\|_p}={B_p}^n, \tag{8}$$

(see [8], p. 160). Choose a such that $C_0a<\frac{1}{2}$, then

$$\frac{1}{1-C_0a}\approx 1+C_0a,$$

because $\frac{1}{1-C_0 a} = 1 + C_0 a + O\left({C_0}^2 a^2\right)$. Hence

$$H_{p,a} \le \left(1 + C_0 a\right) {B_p}^n.$$

□

3. Other methods to estimate $H_{p,a}$ and the $A(\mathbb{T}^n)$-norm of Ψ.

Theorem 3. *Let $f \in A(\mathbb{T}^n)$ and $\widehat{f}(0) = 0$. Let $1 < q \le 2$ and $|\beta| := \alpha$, where $\alpha \ge 1$ is a positive integer such that $\alpha > \frac{n}{q}$. Then we get*

$$\|f\|_{A(\mathbb{T}^n)} \le B^{(\alpha)}_{n,q} \sum_{|\beta|=\alpha} \|D^\beta f\|_q, \tag{9}$$

where $B^{(\alpha)}_{n,q} = C_{n,\alpha}^{\frac{1-q}{q}} \left(\sum_{|m|>0} |m|^{-q\alpha}\right)^{\frac{1}{q}}$. In the case $\widehat{f}(0) \ne 0$, we obtain

$$\|f\|_{A(\mathbb{T}^n)} \le \|f\|_1 + B^{(\alpha)}_{n,q} \sum_{|\beta|=\alpha} \|D^\beta f\|_q.$$

Proof of Theorem 3. Let $\widehat{f}(0) = 0$, then by Hölder's inequality, invoking (4.1) we get

$$\begin{aligned}
\|f\|_{A(\mathbb{T}^n)} &= \sum_{|m|>0} |\widehat{f}(m)| \left(\sum_{|\beta|=\alpha} |(2\pi m)^\beta|^{q'}\right)^{\frac{1}{q'}} \left(\sum_{|\beta|=\alpha} |(2\pi m)^\beta|^{q'}\right)^{-\frac{1}{q'}} \\
&\le \left(\sum_{|\beta|=\alpha} \sum_{|m|>0} |(2\pi i m)^\beta \widehat{f}(m)|^{q'}\right)^{\frac{1}{q'}} \left\{\sum_{|m|>0} \left(\sum_{|\beta|=\alpha} |(2\pi m)^\beta|^{q'}\right)^{-\frac{q}{q'}}\right\}^{\frac{1}{q}} \\
&\le C_{n,\alpha}^{\frac{1-q}{q}} \left(\sum_{|m|>0} |m|^{-q\alpha}\right)^{\frac{1}{q}} \sum_{|\beta|=\alpha} \|\widehat{D^\beta f}\|_{q'}.
\end{aligned}$$

By (2) we obtain

$$\|f\|_{A(\mathbb{T}^n)} \le B^{(\alpha)}_{n,q} \sum_{|\beta|=\alpha} \|D^\beta f\|_q$$

Because there exists a positive constant $C_{n,\alpha}$ such that

$$\sum_{|\beta|=\alpha} |(2\pi m)^\beta|^{q'} \ge C_{n,\alpha} |m|^{q'\alpha}. \tag{10}$$

It is obvious that the sum $\left(\sum_{|m|>0} |m|^{-q\alpha}\right)^{\frac{1}{q}}$ is finite for $\alpha > \frac{n}{q}$ and $|\widehat{f}(0)| \le \|f\|_1$. □

Remark 2. 1. An example of f in Lemma 1 is the function Ψ in the proof of Theorem 2.

2. The inequality (9) can be compared with Wirtinger's inequality which states

$$\|f\|_{L^2(\mathbb{T})} \leq \pi^{-2}\|f'\|_{L^2(\mathbb{T})},$$

for the 1-periodic function f such that $f \in C^1(\mathbb{T})$ and $f = 0$ on the boundary of $\mathbb{T}$. □

Second estimate of $\|\widehat{\Psi}\|_1$ and $H_{p,a}$ Let $1 < q \leq 2$, $n \in \mathbb{N}$, then, for a positive integer $\alpha \geq 1$ such that $\alpha > \frac{n}{q}$, there exists a positive constant C_0 which does not depend to a such that

$$\|\widehat{\Psi}\|_1 \leq C_0 a^{1-\alpha+\frac{n}{q}}.$$

As a consequence

$$H_{p,a} \leq \left(1 + C_0 a^{1-\alpha+\frac{n}{q}}\right) B_p{}^n. \tag{11}$$

Proof. Recall the functions f, g and Ψ from the proof of Theorem 2. By Theorem 3 and because of (6.1), we get

$$\begin{aligned}\|\widehat{\Psi}\|_1 &\leq \|\Psi\|_1 + B_{n,q}^{(\alpha)} \sum_{|\beta|=\alpha} \|D^\beta \Psi\|_q \\ &\leq \Omega_n a^{n+1} + A_{n,q,\alpha} B_{n,q}^{(\alpha)} a^{1-\alpha+\frac{n}{q}} \\ &\leq \left(\Omega_n + A_{n,q,\alpha} B_{n,q}^{(\alpha)}\right) a^{1-\alpha+\frac{n}{q}} \\ &= C_0 a^{1-\alpha+\frac{n}{q}}.\end{aligned} \tag{12}$$

Recall the first part of the inequality (7) from the proof of Theorem 2. Then we derive

$$H_{p,a} \leq \frac{B_p{}^n}{1 - C_0 a^{1-\alpha+\frac{n}{q}}} \approx \left(1 + C_0 a^{1-\alpha+\frac{n}{q}}\right) B_p{}^n.$$

□

Third estimates of $\|\widehat{\Psi}\|_1$ and $H_{p,a}$. Let $n \in \mathbb{N}$ and the positive integer $\alpha \geq 1$ such that $\alpha > \frac{n}{q}$, $1 < q \leq 2$. Assume $\epsilon > 0$ is a small number. We shall prove that α and q can be chosen under the above conditions so that $1 - \alpha + \frac{n}{q} = 1 - \epsilon$. Accordingly the estimate

$$\|\widehat{\Psi}\|_1 \leq C_0 a^{1-\epsilon}$$

can be obtained and as a consequence

$$H_{p,a} \leq \left(1 + C_0 a^{1-\epsilon}\right) B_p{}^n. \tag{13}$$

Proof. Since $1 - \alpha + \frac{n}{q} = 1 - \left(\alpha - \frac{n}{q}\right)$ it is sufficient to prove that α and q can be chosen under the above conditions so that $\alpha - \frac{n}{q} = \epsilon$. For $n = 1$ we take $q = \frac{1}{1-\epsilon}$ and $\alpha = 1$. Then $\alpha - \frac{1}{q} = 1 - (1 - \epsilon) = \epsilon$.

For $n = 2$ we choose $q = \frac{2}{2-\epsilon}$ and $\alpha = 2$. In the case $n \geq 3$ we set $q = \frac{n}{\alpha-\epsilon}$ and let α be an integer satisfies $\frac{n}{2} < \alpha < n$. Then $\alpha - \frac{n}{q} = \epsilon$ for $n \geq 2$. Now, by (6.1) we obtain

$$\sum_{|\beta|=\alpha} \|D^\beta \Psi\|_q \leq A_{n,q,\alpha} a^{1-\epsilon}$$

and

$$\|\widehat{\Psi}\|_1 \leq C_0 a^{1-\epsilon},$$

with the same C_0 as in (12). Also

$$H_{p,a} \leq (1 + a^{1-\epsilon} C_0) B_p^n.$$

□

Another estimate of $H_{p,a}$

For a fixed $n \in \mathbb{N}$, there exists a positive constant C_0 which does not depend to a, such that

$$H_{p,a} \leq \left(1 + C_0 \sqrt{a}\right) B_p{}^n. \tag{14}$$

Proof. The technique is suggested by Sjölin. Recall the function φ and the inequality

$$H_{p,a} \leq \left(\frac{1+\delta}{1-\delta} + \epsilon(a)\right)^n B_p{}^n,$$

from the proof of Lemma 1 (see below). By the mean value theorem we obtain

$$\begin{aligned}
\epsilon(a) &= \sup_{t\in\mathbb{R}} \left| \|\widehat{\varphi}\|_1 - \sum_m |\widehat{\varphi}\,(a(m-t))| \right| \\
&\leq \sup_t \sum_{m\in\mathbb{Z}} \int_{a(m-t)}^{a(m+1-t)} |\widehat{\varphi}(y) - \widehat{\varphi}(am - at)|\, dy \\
&\leq a \|h\|_1,
\end{aligned}$$

where $h(\xi) := \sup_{|x-\xi|\leq a} |\widehat{\varphi}'(x)|$. But

$$\widehat{\varphi}'(x) = \frac{1}{1-\delta}\left(\frac{1}{\delta^2} F'\left(\frac{x}{\delta}\right) - \delta F'(x)\right)$$

with $F(x) := \left(\frac{\sin(\pi x)}{\pi x}\right)^2$ (see Fig. 1) and

$$|F'(x)| = \pi \left| \frac{\sin(2\pi x)}{(\pi x)^2} - \frac{2(\sin(\pi x))^2}{(\pi x)^3} \right| \leq \frac{1}{\pi x^2} + \frac{2}{\pi^2 x^3} \leq \frac{1}{2}\left(\frac{1}{|x|^2} + \frac{1}{|x|^3}\right).$$

Furthermore

$$|F'(x)| \leq \begin{cases} A_0 & \text{on } [-1,1] \\ |x|^{-2} & \text{otherwise} \end{cases}$$

with $A_0 < 1$. (See Fig. 2 and Fig. 3.)

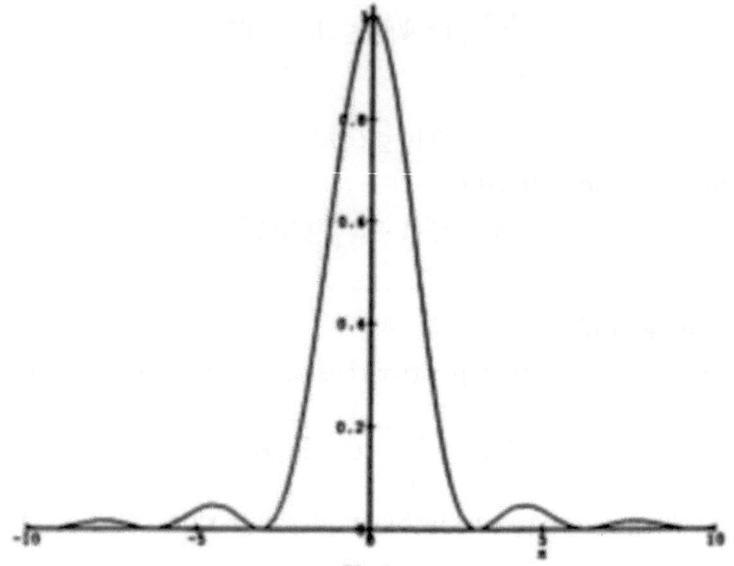

Fig. 1

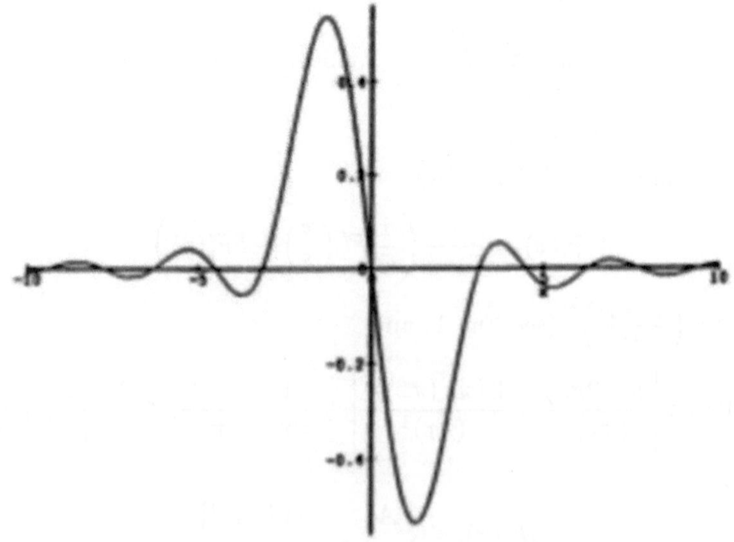

Fig. 2

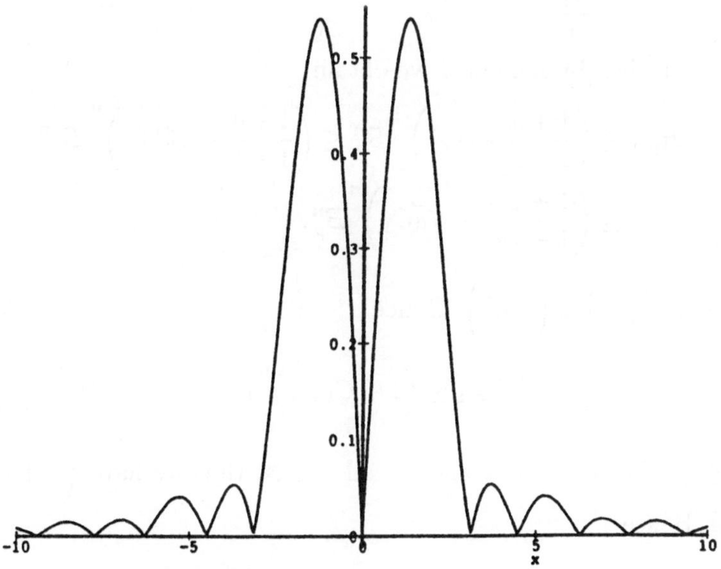

Fig. 3

Choose $a < \delta$ and define

$$h_1(\xi) := \sup_{|x-\xi|\leq a} |F'(x)|$$

$$h_\delta(\xi) := \sup_{|x-\xi|\leq a} \left|F'\left(\frac{x}{\delta}\right)\right|.$$

Then we get

$$h_1(\xi) \leq \begin{cases} A_0 & \text{on } [-1,1] \\ |\xi|^{-2} & \text{otherwise} \end{cases}$$

$$h_\delta(\xi) \leq \begin{cases} A_0 & \text{on } [-\delta,\delta] \\ \delta^2|\xi|^{-2} & \text{otherwise.} \end{cases}$$

But

$$\|h_1\|_1 \leq 2A_0 + \int_{|\xi|\geq 1} |\xi|^{-2} d\xi = 2(A_0+1)$$

and

$$\|h_\delta\|_1 \leq 2A_0\delta + \delta^2 \int_{|\xi|\geq\delta} |\xi|^{-2} d\xi = 2\delta(A_0+1).$$

Hence

$$\|h\|_1 \le \frac{1}{1-\delta}\Big(\frac{1}{\delta^2}\|h_\delta\|_1 + \delta\|h_1\|_1\Big) \le 2(A_0+1)\Big(\frac{1}{\delta}+\delta\Big)$$
$$\le \frac{C_1}{\delta}.$$

Choose $\delta = \sqrt{a}$, then by Lemma 1 we obtain

$$H_{p,a} \le \left(\frac{1+\delta}{1-\delta} + \epsilon(a)\right)^n B_p^n \le \left(\frac{1+\delta}{1-\delta} + a\,\|h\|_1\right)^n {B_p}^n$$
$$\le \left(\frac{1+\sqrt{a}}{1-\sqrt{a}} + \sqrt{a}C_1\right)^n B_p^n.$$

But $\frac{1}{1-\sqrt{a}} = 1 + \sqrt{a} + O\Big(\sqrt{a}^2\Big)$. Hence

$$H_{p,a} \le \Big(1 + C_1\sqrt{a}\Big)^n B_p^n.$$

Since $a^{\frac{n}{2}} \le \sqrt{a}$ for $0 < a < \frac{1}{2}$ and for all $n \in \mathbb{N}$, thus we have $\Big(1 + C_1\sqrt{a}\Big)^n \approx 1 + nC_1\sqrt{a}$. Hence

$$H_{p,a} \le \Big(1 + C_0\sqrt{a}\Big)B_p^n.$$

Here $C_0 := 4n(A_0+1)$. □

Remark 3. The upper bound (11) is better than (14) for the case $n = \alpha = 1$ because $\frac{1}{2} \le \frac{1}{q} < 1$ implies that $\sqrt{a} \ge a^{\frac{1}{q}} > a$. □

4. The Babenko-Beckner type of the Hausdorff-Young inequality

Theorem 4. $H_p = B_p^n$.

The proof is a consequence of the following two lemmas:

Lemma 1. $H_p \le B_p^n$.

Proof of Lemma 1. We get the result by letting $a \to 0^+$ in the formula (6) in Theorem 2. Here we give another proof. The technique is analogous to the case $n = 1$, due to Sjölin, [10].

For the case $p = 1$ it is obvious that $B_1^n = 1$ and $\|\widehat{f}\|_\infty \le \|f\|_1$. Let $1 < p \le 2$. For positive integer k, $1 \le k \le n$, and for $0 < \delta < \frac{1}{2}$, define the even functions

$$\varphi_k := \varphi(x_k) = \begin{cases} 1 & |x_k| \le 1 \\ \frac{1-\delta|x_k|}{1-\delta} & 1 \le |x_k| \le \frac{1}{\delta} \\ 0 & |x_k| > \frac{1}{\delta} \end{cases}$$

and $\phi_k := \varphi\left(\frac{x_k}{a}\right)$, then

$$\widehat{\varphi_k} = \widehat{\varphi}(\xi_k) = \frac{\delta}{1-\delta}\left\{\left(\frac{\sin\frac{\pi\xi_k}{\delta}}{\pi\xi_k}\right)^2 - \left(\frac{\sin\pi\xi_k}{\pi\xi_k}\right)^2\right\}$$

and

$$\|\widehat{\phi_k}\|_1 = \|\widehat{\varphi_k}\|_1 \le \frac{1+\delta}{1-\delta}. \tag{15}$$

Define the functions $\Phi := \prod_{k=1}^n \varphi_k$ and $\Phi_a := \prod_{k=1}^n \phi_k$, then we have

$$\|\widehat{\Phi}\|_1 = \|\widehat{\Phi_a}\|_1 \le \left(\frac{1+\delta}{1-\delta}\right)^n.$$

Recall the functions f and g from the proof of Theorem 2. We get

$$\|f\|_p = \|g\|_p, \quad \Phi_a f = \begin{cases} g & \text{on } \overline{B}(0,a) \\ 0 & \text{otherwise} \end{cases}$$

and for $m \in \mathbb{Z}^n$ we get

$$\widehat{g}(m) = \widehat{\Phi_a f}(m) = \widehat{\Phi_a} * \widehat{f}(m).$$

By Hölder's inequality we obtain

$$|\widehat{g}(m)| \le \|\widehat{\Phi_a}\|_1^{\frac{1}{p}} \left(\int_{\mathbb{R}^n} |\widehat{f}(\xi)|^{p'} |\widehat{\Phi_a}(m-\xi)| d\xi\right)^{\frac{1}{p'}}.$$

Hence

$$\|\widehat{g}\|_{p'} = \left(\sum_m |\widehat{g}(m)|^{p'}\right)^{\frac{1}{p'}} \le \|\widehat{\Phi_a}\|_1^{\frac{1}{p}} \left(\int_{\mathbb{R}^n} \left|\widehat{f}(\xi)\right|^{p'} \sum_m \left|\widehat{\Phi_a}(m-\xi)\right| d\xi\right)^{\frac{1}{p'}}.$$

But $\sum_m |\widehat{\Phi_a}(m-\xi)| = \sum_m a^n |\widehat{\Phi}(a(m-\xi))|$, is the Riemann sum of $\|\widehat{\Phi}\|_1$ that is

$$\sum_m \left|\widehat{\Phi_a}(m-\xi)\right| = \prod_{k=1}^n \left(\sum_{m_k \in \mathbb{Z}} a\,|\widehat{\varphi_k}\,(a(m_k-\xi_k))|\right) \le \left(\|\widehat{\varphi}\|_1 + \epsilon(a)\right)^n$$

where the positive function $\epsilon(a) \to 0$ as $a \to 0^+$, furthermore

$$\epsilon(a) := \sup_{\xi \in \mathbb{R}} \left| \|\widehat{\varphi}\|_1 - \sum_{m \in \mathbb{Z}} a\,|\widehat{\varphi}\,(a(m-\xi))| \right|.$$

Thus we obtain

$$\|\widehat{g}\|_{p'} \le \|\widehat{\Phi}\|_1^{\frac{1}{p}} \left\{ \int_{\mathbb{R}^n} |\widehat{f}(\xi)|^{p'} \left(\|\widehat{\varphi}\|_1 + \epsilon(a) \right)^n d\xi \right\}^{\frac{1}{p'}} \tag{16}$$
$$= \|\widehat{\varphi}\|_1^{\frac{n}{p}} (\|\widehat{\varphi}\|_1 + \epsilon(a))^{\frac{n}{p'}} \|\widehat{f}\|_{p'}$$
$$\le \left(\|\widehat{\varphi}\|_1 + \epsilon(a) \right)^{n\left(\frac{1}{p}+\frac{1}{p'}\right)} \|\widehat{f}\|_{p'}$$
$$= \left(\|\widehat{\varphi}\|_1 + \epsilon(a) \right)^n \|\widehat{f}\|_{p'}.$$

Invoking (15), (16) and because of (8), we get

$$\sup_g \frac{\|\widehat{g}\|_{p'}}{\|g\|_p} \le \left(\|\widehat{\varphi}\|_1 + \epsilon(a) \right)^n B_p^n$$

that is

$$H_{p,a} \le \left(\|\widehat{\varphi}\|_1 + \epsilon(a) \right)^n B_p^n \le \left(\frac{1+\delta}{1-\delta} + \epsilon(a) \right)^n B_p^n.$$

The proof is complete by letting δ and $a \to 0^+$. □

Lemma 2. $H_p \ge B_p^n$.

Proof of Lemma 2. The technique is analogous to the case $n = 1$, due to Andersson, [9]. Choose $f \in L^p(\mathbb{R}^n) \bigcap C_0{}^\infty(\mathbb{R}^n)$, with support in $\overline{B}\left(0, \frac{1}{2}\right)$, such that $f = g$ on $\overline{B}(0, a)$. Define $g_a := g\left(\frac{x}{a}\right)$, then $\widehat{g_a}(m) = a^n \widehat{f}(am)$ for $m \in \mathbb{Z}^n$, and

$$\begin{aligned} \|f\|_p &= \|g\|_p \\ a^{\frac{n}{p}} \|f\|_p &= \|g_a\|_p . \end{aligned} \tag{17.1}$$

By the Riemann sum of $\|\widehat{f}\|_{p'}$ and by invoking (17.1), we obtain

$$\frac{\|\widehat{f}\|_{p'}}{\|f\|_p} = \lim_{a \to 0+} \frac{\left(\sum_m a^n \left| \widehat{f}(am) \right|^{p'} \right)^{\frac{1}{p'}}}{\|f\|_p} = \lim_{a \to 0+} \frac{\left(\sum_m \left| a^n \widehat{f}(am) \right|^{p'} \right)^{\frac{1}{p'}}}{a^{\frac{n}{p}} \|f\|_p} \tag{17.2}$$
$$= \lim_{a \to 0+} \frac{\left(\sum_m |\widehat{g_a}(m)|^{p'} \right)^{\frac{1}{p'}}}{\|g_a\|_p} = \lim_{a \to 0+} \frac{\|\widehat{g_a}\|_{p'}}{\|g_a\|_p}$$
$$\le \lim_{a \to 0+} \sup_{g_a} \frac{\|\widehat{g_a}\|_{p'}}{\|g_a\|_p} \le \lim_{a \to 0+} H_{p,a} = H_p.$$

The proof is complete by (17.2) and by (8) because

$$B_p^n = \sup_f \frac{\|\widehat{f}\|_{p'}}{\|f\|_p} \le H_p$$

and the set of all functions in $C_0{}^\infty(\mathbb{R}^n)$ is dense in $L^p(\mathbb{R}^n)$. □

As a consequence of the previous lemmas

$$H_p = B_p{}^n. \square$$

5. An upper bound for B_p

It is known that $B_p{}^n \le 1$. By application of M. Riesz interpolation theorem, Andersson [9] proved that, given $1 \le p \le 2$ there are positive constants C_p and m_p such that every complex-valued function on $\mathbb{T}$ supported in an interval of length m_p satisfies

$$\|\widehat{f}\|_{p'} \le C_p \|f\|_p.$$

As a consequence

$$B_p \le C_p < 1 \quad \text{for } 1 < p < 2.$$

For completeness we include a proof that $B_p{}^n < 1$ for $1 < p < 2$ by the following elementary technique :

It is obvious that $B_1{}^n = B_2{}^n = 1$. Let $1 < p < 2$. Since $p^{\frac{1}{p}} < p'^{\frac{1}{p'}}$ is equivalent to

$$\left(\frac{1}{q}\right)^q < \left(\frac{1}{1-q}\right)^{(1-q)}$$

for $\frac{1}{2} < q < 1$, so it is enough to show that $f(q) > 0$ for $q \in [\frac{1}{2}, 1]$, where $f(q) := q \ln q - (1-q)\ln(1-q)$. But $f(\frac{1}{2}) = f(1) = 0$ and $f'(q) = 2 + \ln q(1-q)$ have only one zero in the interval $[\frac{1}{2}, 1]$, that is $q_0 = \frac{1}{2} + \frac{1}{4e}\sqrt{4e^2-1}$ which is a local maximum because $f''(q_0) = -8e\sqrt{4e^2-1} < 0$. Hence $f(q)$ is a non-negative function on $[\frac{1}{2}, 1]$. □

6. A duality between the sharp forms of the Hausdorff-Young inequality and Young's inequality for convolution

W. Beckner, [8], pp. 169-176, proved the sharp form of Young's inequality for convolution, namely :

$$\|f_1 * f_2 * \cdots * f_m\|_r \le B_{r'}^n \prod_{j=1}^m B_{p_j}^n \|f_j\|_{p_j}, \tag{18}$$

holds for (i) $1 < p_j < \infty$, $1 \le j \le m$, $1 \le r \le \infty$, or for (ii) $1 \le r', p_j \le 2$, $1 \le j \le m$. Here $1/r' = \sum_{j=1}^m 1/p_j'$ and $f_j \in L^{p_j}(\mathbb{R}^n)$. Note that $B_{p_j} \to 1$ as $p_j \to 1$ or ∞.

Fournier [16] proved that (18) is true for locally compact unimodular groups with no compact open subgroups.

We ask: is there a duality between the inequality (18) for the case (ii) and the sharp form of the Hausdorff-Young inequality

$$\|\widehat{g}\|_{p'} \le B_p^n \|g\|_p, \tag{3.1}$$

for an even integer p' and for $g \in L^p(\mathbb{T}^n)$ with a small support? Here we explore the relation between these inequalities.

Lemma 3. *Let $f \in L^r(\mathbb{R}^n)$, $g \in L^r(\mathbb{T}^n)$ for the positive even integer r ($2k$, say) and $f = g$ on $\overline{B}(0, \frac{1}{2k})$. Let supp f = supp $g \subset \overline{B}(0, \frac{1}{2k})$. Then*

$$\|\widehat{g}\|_r = \|\widehat{f}\|_r.$$

Proof of Lemma 3. For the case $n = 1$, the technique is due to Andersson, [9]. It is obvious that

$$(i)\ \widehat{g}(\mu) = \widehat{f}(\mu),\ \mu \in \mathbb{Z}^n.$$

$$(ii)\ f^{*k} := \underbrace{f * f * \cdots * f}_{k\text{-times}} = \begin{cases} g^{*k} & x \in \overline{B}(0, \frac{1}{2}) \\ 0 & \text{otherwise} \end{cases}.$$

$$(iii)\ \widehat{f^{*k}} = \widehat{f}^k, \text{and } \widehat{g^{*k}} = \widehat{g}^k.$$

By $(i), (ii), (iii)$, and invoking the Plancherel theorem, we get

$$\begin{aligned}
\|\widehat{g}\|_r &= \Big(\sum_{\mu \in \mathbb{Z}^n} |\widehat{g}(\mu)|^{2k}\Big)^{\frac{1}{r}} = \Big(\sum_{\mu} |\widehat{g^{*k}}(\mu)|^2\Big)^{\frac{1}{r}} \\
&= \Big(\int_{\mathbb{T}^n} |g^{*k}(x)|^2 dx\Big)^{\frac{1}{r}} = \Big(\int_{\mathbb{R}^n} |f^{*k}(x)|^2 dx\Big)^{\frac{1}{r}} \\
&= \Big(\int_{\mathbb{R}^n} |\widehat{f^{*k}}(\xi)|^2 d\xi\Big)^{\frac{1}{r}} = \Big(\int_{\mathbb{R}^n} |\widehat{f}(\xi)|^{2k} d\xi\Big)^{\frac{1}{r}} \\
&= \|\widehat{f}\|_r.
\end{aligned}$$

□

Theorem 5. *For an even integer p', the sharp form of the Hausdorff-Young inequality* (3.1) *can be obtained from the convolution inequality (18).*

Proof of Theorem 5. Let $p' = 2k$ for $k \in \mathbb{N}$. Recall the functions f and g from the proof of Theorem 2, then by Lemma 3 and (18), we get

$$\begin{aligned}
\|\widehat{g}\|_{p'}^k &= \|\widehat{g}\|_{2k}^k = \|\widehat{f}\|_{2k}^k = \|\widehat{f}^k\|_2 = \|\widehat{f^{*k}}\|_2 \\
&= \|f^{*k}\|_2 \le \left(B_p^n \|f\|_p\right)^k = \left(B_p^n \|g\|_p\right)^k.
\end{aligned}$$

Hence

$$\|\widehat{g}\|_{p'} \le {B_p}^n \|g\|_p.$$

□

To explore the opposite direction, namely, to answer the question whether the inequality (18) for the case (ii) can be obtained from the inequality (3.1) or not, we need to prove that (3.1) is valid for the group $\mathbb{Z}^n$. In the orther words: Does the expression

$$\left\|\widehat{\widehat{g_1 * g_2}}\right\|_r \leq B_{r'}{}^n \left\|\widehat{g_1 * g_2}\right\|_{r'} \tag{19}$$

hold for the even integer r and for $g_j,\ j = 1,2$ defined on $\mathbb{T}^n$ with small supports which depend on r? Unfortunately we can not apply the result of Russo and Fournier (see Introduction, p. 2) because the locally compact unimodular group $\mathbb{Z}^n$ has a compact open subgroup, namely the group $\{0\}$. Here 0 is the identity element of the group. But if (19) is true, then the inequality (3.1) implies (18) for $1 \leq r', p_j \leq 2$. First, for smooth functions in the space $L^p(\mathbb{R}^n)$ with small supports and then, by the approximation theory, for the whole space $L^p(\mathbb{R}^n)$. With this circumstance, the proof is easy by mathematical induction if we show it for the case $m = 2$. Let $p_1 = p$ and $p_2 = q$, $\frac{1}{r} = \frac{1}{p} + \frac{1}{q} - 1$. Define $\widetilde{g_j} := g_j(-x),\ j = 1,2$ then by (3.1) and by (19), invoking Hölder's inequality we obtain

$$\begin{aligned} \|g_1 * g_2\|_r &= \left\|\widetilde{g_1 * g_2}\right\|_r = \left\|\widehat{\widehat{g_1 * g_2}}\right\|_r \leq B_{r'}{}^n \left\|\widehat{g_1 * g_2}\right\|_{r'} \\ &= B_{r'}{}^n \left\|\widehat{g_1}\widehat{g_2}\right\|_{r'} \leq B_{r'}{}^n \left\|\widehat{g_1}\right\|_{p'} \left\|\widehat{g_2}\right\|_{q'}, \end{aligned}$$

where $\frac{1}{r'} = \frac{1}{p'} + \frac{1}{q'}$. Again, by (3.1) and Lemma 3 we get

$$\begin{aligned} \|f_1 * f_2\|_r = \|g_1 * g_2\|_r &\leq B_{r'}{}^n B_p{}^n B_q{}^n \|g_1\|_p \|g_2\|_q \\ &= B_{r'}{}^n B_p{}^n B_q{}^n \|f_1\|_p \|f_2\|_q . \ \square \end{aligned}$$

7. Questions

1. Can one improve the upper bound (6) in Theorem 2? At least, by our technique it can be possible only if $\sum_{|\beta|=\alpha} \|D^\beta f\|_q < C\|\widehat{f}\|_{q'}$. This is plainly impossible because of the Hausdorff-Young inequality. By the inequality (9) we know that $B_{n,q}^{(\alpha)} \sum_{|\beta|=\alpha} \|D^\beta f\|_q \geq \|\widehat{f}\|_1$. But this, see the proof of Theorem 3, leads to the upper bound (11) which is worse than (6). So, in this sense, the upper bound (6) is the best.

2. Since $H_p = B_p{}^n$, there must also exist a lower bound for $H_{p,a}$. Can one find a lower bound for $H_{p,a}$? $\square$

8. Application of the Babenko-Beckner constant B_p

Many norm inequalities in Fourier analysis which are related to the Hausdorff-Young inequality can be improved in the sense of the constant of the inequalities. For instance in [12], one can find the following norm inequality

$$\|T_\lambda f\|_{L^q(\mathbb{R}^n)} \leq A\lambda^{-\frac{n}{q}} \|f\|_{L^p(\mathbb{R}^{n-1})}$$

where $1 \leq p \leq 2$ and $q = \left(\frac{n+1}{n-1}\right)p'$. The oscillatory integral

$$T_\lambda f(\xi) := \int_{\mathbb{R}^{n-1}} e^{i\lambda\Phi(x,\xi)}\psi(x,\xi)f(x)dx,$$

mapping functions on $\mathbb{R}^{n-1}$ to functions on $\mathbb{R}^n$, [12], pp. 379-380. With this result, one can then prove the restriction theorem for the Fourier transform

$$\left(\int_{S_0} |\widehat{f}(\xi)|^q d\sigma(\xi)\right)^{\frac{1}{q}} \leq C_{p,q}(S_0)\|f\|_{L^p(\mathbb{R}^n)} \tag{20}$$

for $1 \leq p \leq \frac{2n+2}{n+3}$ and $q = \left(\frac{n-1}{n+1}\right)p'$. Here $S \subset \mathbb{R}^n$ is a manifold of dimension $n-1$ whose Gaussian curvature is nowhere zero, and S_0 is a compact subset of S, [12], pp. 352–365.

For $n = 2$, the sub-manifold S_0 is a smooth curve in the plane and $1 \leq p < \frac{4}{3}$ and $q = \frac{p'}{3}$. For this case, we want to give an estimate for the constant $C_{p,q}(S_0)$ in (20).

In the following theorem we apply the sharp form of the Hardy-Littlewood-Sobolev inequality proved by E.H. Lieb, [21], namely

$$\||x|^{-\lambda} * f\|_s \leq N_{s,\lambda,n}\|f\|_r$$

for $\frac{1}{r} + \frac{\lambda}{n} = 1 + \frac{1}{s}$ and $1 < s, r, \frac{\lambda}{n} < \infty$ where

$$N_{s,\lambda,n} = \pi^{\frac{\lambda}{2}} \frac{\Gamma(\frac{n}{2} - \frac{\lambda}{2})}{\Gamma(n - \frac{\lambda}{2})} \left(\frac{\Gamma(\frac{n}{2})}{\Gamma(n)}\right)^{-1+\frac{\lambda}{n}}$$

is the so-called

Riesz potential constant. Here s is the dual exponent of r. In our case, Theorem 6, we take $\lambda := \frac{2(p-1)}{2-p}$, $n := 1$ and $r := \frac{2-p}{3-2p}$ which give

$$C_R = \pi^{\frac{5p-6}{2(2-p)}} \frac{\Gamma\left(\frac{4-3p}{2(2-p)}\right)}{\Gamma\left(\frac{3-2p}{2-p}\right)},$$

the so-called sharp form of the Riesz potential constant [21].

Assume that S_0 is parameterized by $S_0 := \{(t, \phi(t)) : t \in [0,1]\}$ where $\phi''(t) \neq 0$. With the previous notation, we improve the constant in (20).

Theorem 6. *Let $B_{\frac{p}{2-p}}$ be the Babenko-Beckner constant for $1 \leq p < \frac{4}{3}$. Assume that $q = \frac{p}{3(p-1)}$. Then*

$$C_{p,q}(S_0) = C_{S_0}\left(\sqrt{C_R}\right)^{\frac{2-p}{p}} B_{\frac{p}{2-p}}$$

where C_{S_0} depends only on ϕ.

Note: An explicit expression for C_{S_0} will follow from the proof.

Proof of Theorem 6. Define the operator R by

$$Rf(\xi) = \int_{\mathbb{R}^2} e^{-2\pi i<x,\xi>} f(x)dx, \quad \xi \in S_0,$$

for any smooth function f on $\mathbb{R}^2$. Rf is the restriction of the Fourier transform of f on S_0. We are then concerned with the inequality

$$\|Rf\|_{L^q(S_0)} \leq C_{p,q}(S_0)\|f\|_{L^p(\mathbb{R}^2)},$$

for $1 \leq p < \frac{4}{3}$ and $q = \frac{p'}{3}$. Since we know that an operator and its adjoint have the same norm [17], p. 712) let us verify the equivalent inequality

$$\|Sf\|_{L^{p'}(\mathbb{R}^2)} \leq C_{p,q}(S_0)\|f\|_{L^{q'}(S_0)}$$

where $S = R^*$, the adjoint operator of R, that is

$$Sf(x) = \int_{S_0} e^{2\pi i<x,\xi>} f(\xi)d\sigma(\xi), \ x \in \mathbb{R}^2.$$

Using the parameterization of S_0, [11], we may rewrite

$$Sf(x_1, x_2) = \int_0^1 e^{2\pi i[x_1 t + x_2\phi(t)]} F(t)dt$$

where $F(t) = f(t, \phi(t))\,[1 + \phi'(t)^2]^{\frac{1}{2}}$. Define $\frac{1}{2}C_0 := \max_{0\leq t\leq 1}[1 + \phi'(t)^2]^{\frac{1}{2}}$. Then consider

$$(Sf)^2(x_1, x_2) = \int_0^1 \int_0^1 e^{2\pi i\{x_1(t_1+t_2)+x_2[\phi(t_1)+\phi(t_2)]\}} F(t_1)F(t_2)dt_1dt_2.$$

Now make the change of variables: $u_1 = t_1 + t_2$ and $u_2 = \phi(t_1) + \phi(t_2)$, with the Jacobian

$$J = \frac{\partial(u_1, u_2)}{\partial(t_1, t_2)} = \begin{vmatrix} 1 & 1 \\ \phi'(t_1) & \phi'(t_2) \end{vmatrix} = \phi'(t_2) - \phi'(t_1).$$

For $t_1 \neq t_2$, there exists a positive constant C_1 such that

$$\left|\frac{\phi'(t_2) - \phi'(t_1)}{t_2 - t_1}\right| = |\phi''(t_0)| \geq C_1,$$

and since $\phi''(t) \neq 0, \ \forall t \in [0, 1]$ we have

$$|\phi'(t_2) - \phi'(t_1)| \geq C_1|t_2 - t_1|.$$

Since the mapping $(t_1, t_2) \mapsto (u_1, u_2)$ is two-to-one, we obtain

$$(Sf)^2(x) = \int_D e^{2\pi i<x,u>} G(u)du,$$

for some region $D \subset \mathbb{R}^2$, with

$$G(u_1, u_2) = \frac{2F(t_1)F(t_2)}{|\phi'(t_2) - \phi'(t_1)|}, \ t_1 = t_1(u_1, u_2), \ t_2 = t_2(u_1, u_2).$$

Thus we find that

$$(Sf)^2 = \widetilde{(\widehat{G_\circ})},$$

where $G_\circ(u) = G(u)$ if $u \in D$ and zero if $u \notin D$.

By the sharp form of the Hausdorff-Young inequality, we yield

$$\left\|\widehat{G_\circ}\right\|_{L^{r'}(\mathbb{R}^2)} \leq B_r^2 \left\|G_\circ\right\|_{L^r(\mathbb{R}^2)},$$

for $1 \leq r \leq 2$. Let $p' := 2r'$, then $r = \frac{p}{2-p}$ and it is clear that

$$\|Sf\|^2_{L^{p'}(\mathbb{R}^2)} = \left\|(Sf)^2\right\|_{L^{r'}(\mathbb{R}^2)} = \left\|\widetilde{\widehat{G_\circ}}\right\|_{L^{r'}(\mathbb{R}^2)} = \left\|\widehat{G_\circ}\right\|_{L^{r'}(\mathbb{R}^2)}.$$

But

$$\begin{aligned}\|G_\circ\|^r_{L^r(\mathbb{R}^2)} &= \int_0^1\int_0^1 |F(t_1)|^r|F(t_2)|^r|J|^{1-r}dt_1dt_2\\ &\leq C_0^{2r}C_1^{1-r}\int_0^1\int_0^1 |f\left(t_1,\phi(t_1)\right)|^r|f\left(t_2,\phi(t_2)\right)|^r|t_2-t_1|^{1-r}dt_1dt_2.\end{aligned}$$

By Hölder's inequality

$$\|G_\circ\|^r_{L^r(\mathbb{R}^2)} \leq C_{S_0}^{2r}\left(\int_0^1\left(\int_0^1|f(t_1)|^r|t_2-t_1|^{1-r}dt_1\right)^{s'}dt_2\right)^{\frac{1}{s'}}\left(\int_0^1|f(t_2)|^{rs}dt_2\right)^{\frac{1}{s}},$$

with $s = \frac{2}{3-r}$ and $C_{S_0} := C_0C_1^{\frac{1-r}{2r}}$ (see the remark below). Now, by the Hardy-Littlewood-Sobolev inequality, [12], p. 354,

$$\left(\int_0^1\left(\int_0^1|f(t_1)|^r|t_2-t_1|^{1-r}dt_1\right)^{s'}dt_2\right)^{\frac{1}{s'}} \leq C_R\left(\int_0^1|f(t_1)|^{rs}dt_1\right)^{\frac{1}{s}},$$

whenever $1 \leq r < 2$. Thus we obtain

$$\|G_\circ\|^r_{L^r(\mathbb{R}^2)} \leq {C_{S_0}}^{2r}C_R\left(\int_0^1|f(t)|^{rs}dt\right)^{\frac{2}{s}} \leq {C_{S_0}}^{2r}C_R\,\|f\|^{2r}_{L^{q'}(S_0)},$$

where $q' = rs$. This gives

$$\left\|(Sf)^2\right\|_{L^{r'}(\mathbb{R}^2)} \leq {C_{S_0}}^2{C_R}^{\frac{1}{r}}{B_r}^2\,\|f\|^2_{L^{q'}(S_0)},$$

with $q' = \frac{2r}{3-r}$. For $1 \leq r < 2$, we have $1 \leq q' < 4$. Also note that $2r' = 3q$. Thus we have

$$\|Sf\|_{L^{p'}(\mathbb{R}^2)} \leq C_{S_0}C_R^{\frac{1}{2r}}B_r\,\|f\|_{L^{q'}(S_0)},$$

for $1 \leq q' < 4$ and $p' = 3q$, which is equivalent to

$$\|Rf\|_{L^q(S_0)} \leq C_{S_0}\left(\sqrt{C_R}\right)^{\frac{2-p}{p}}B_{\frac{p}{2-p}}\,\|f\|_{L^p(\mathbb{R}^2)},$$

for $1 \leq p < \frac{4}{3}$ and $3q = p'$, as the theorem says.

We would like to remark that the constant C_R depends only on $|t|^{1-r}$ so-called the kernel of Riesz potential. □

Remark 4. A lower and an upper bound of the constant C_{S_0} can easily be found. Redefine $C_1 := \min_{0\leq t\leq 1} |\phi''(t)|$. Then

$$C_0 \leq C_{S_0} \leq C_0 C_1^{-\frac{1}{4}}, \quad \text{if } 0 < C_1 \leq 1 \text{ and}$$

$$C_0 C_1^{-\frac{1}{4}} \leq C_{S_0} \leq C_0, \quad \text{if } C_1 > 1. \ \square$$

Acknowledgment It is a pleasure to thank professor Sjölin who suggested this paper. Also, many thanks to professor Domar for helping me to find references about Carlson's inequality.

References

[1] F. Carlson, *Une inégalité,* Arkiv för matematik, astronomi och fysik, **25 B,** No. 1, (1934) 1–5.

[2] G. H. Hardy, *A note on two inequalities,* J. London Math. Soc., **11,** (1936) 167–170.

[3] N. J. Krugljak, L. Maligranda and L. E. Persson, *A Carlson type inequality with blocks and interpolation,* Studia Math. **104,** No. 2 (1993) 161–180.

[4] G. H. Hardy, J. E. Littlewood, G. Pólya, Inequalities, Cambridge University Press, London (1934) 182–196.

[5] E. C. Titchmarsh, Introduction to the theory of Fourier integrals, Oxford, at Clarendon Press, (1937).

[6] Y. Domar, private communication with P. Sjölin.

[7] K. I. Babenko, *An inequality in theory of Fourier integrals,* Amer. Math. Soc. Transl. (2) **44,** (1965) 115–128.

[8] W. Beckner, *Inequalities in Fourier integrals,* Annals of Math., **102,** (1975) 159–182.

[9] M. E. Andersson, *Local variants of the Hausdorff-Young inequality,* M. Gyllenberg, L. E. Persson (Eds.) Analysis, Algebra and Computers in Mathematical Research: Proceedings of the 21st Nordic congress of mathematicians, (1994) Marcel-Dekker, New York.

[10] P. Sjölin, *A remark on the Hausdorff-Young inequality.* Proceedings of the Amer. Math. Soc. **123,** No. 10 (1995) 3085–3088.

[11] P. Sjölin, A lecture on Fourier analysis and singular integrals, (1994) Royal Inst. of Tech. Stockholm, Sweden.

[12] E. M. Stein, Harmonic Analysis, Princeton University Press, (1993).

[13] J. Inoue, *L^p-Fourier transforms on nilpotent Lie groups and solvable Lie groups action on Siegel domains,* Pacific J. of Math., **155,** No. 2, (1992) 295–318.

[14] R. A. Kunze, *L^p-Fourier transforms on locally compact unimodular groups,* Trans. Amer. Math. Soc., **89,** (1958) 519–540.

[15] B. Russo, *The norm of the L^p-Fourier transform on unimodular groups,* Trans. Amer. Math. Soc. **192,** (1974) 293–318.

[16] J. J. Fournier, *Sharpness in Young's inequality for convolution,* Pacific J. of Math., **72,** No. 2, (1977) 383–397.

[17] E. Hewitt and K. A. Ross, Abstract Harmonic Analysis, I, II, Springer-Verlag, Berlin, (1970).

[18] C. Fefferman, *Inequalities for strongly singular convolution operators,* Acta. Math., **124.** (1970) 9–36.

[19] A. Zygmund, *On Fourier coefficients and transforms of functions of two variables,* Studia Math., **50,** (1974) 189–202.

[20] H. J. Brascamp and E. H. Lieb, *Best constants in Young's inequality, its converse and its generalization to more than three functions,* Advances in Math., **20,** (1976) 151–173.

[21] E. H. Lieb, *Sharp constants in the Hardy-Littlewood-Sobolev and related inequalities,* Annals of Math. **118,** No. 1, (1983) 349–374.

[22] H. Dym and H. P. Mckean, Fourier series and integrals, Academic Press, London, (1972) 47–48.

[23] L. Råde and B. Westergren, Mathematics Handbook for Seience and Engineering, Studentlitteratur, Sweden, (1995).

Department of Electromagnetic Theory
Royal Institute of technology
SE-100 44 Stockholm,
Sweden
E-mail address: amir@tet.kth.se

Operator Theory:
Advances and Applications, Vol. 114

Spectral asymptotics of the N particle Schrödinger equation when $N \to \infty$ and normal forms of the quadratic boson operators

V. Kucherenko

Introduction

Quadratic boson operators have arise in the asymptotic approach to the N particle Schrödinger equation.

Let us consider the N particle stationary Schrödinger equation:

$$\sum_{i=1}^{N}\{-\frac{h^2}{2m}\Delta_{x_i} + U(x_i)\}\psi + \sum\sum_{1\le i<j\le N} \alpha_0 V(x_i - x_j)\psi = E\psi \tag{1}$$

$$\psi \in L_2(R^{N3}), \quad x \in R^3.$$

Here V is the potential of interaction and $U(x)$ is the external potential (V, U are real valued functions).

In many particle problems, physicists are interested in an approach which can give a classic limit when $N \to \infty$. For that goal physicists use the statistical limit.

Statistical limit

We suppose that the potentials $V(r)$, $U(r)$ have period L; the parameters L, N tend to ∞ but the density $\rho \stackrel{\text{def}}{=} N/L^3$ is preserved.

Of course, real systems do not have periodic potential. Therefore, we define periodic potentials V_L, U_L so that when $L \to \infty$: $V_L \to V$; $U_L \to U$. Let us define [2]

$$V_k \stackrel{\text{def}}{=} L^{-3} \int \exp(-i\frac{2\pi}{L}(k,y))V(y)dy,$$

$$V_L \stackrel{\text{def}}{=} \sum_k \exp(i\frac{2\pi}{L}(k,x))V_k,$$

$$U_L \stackrel{\text{def}}{=} \sum_k \exp(i\frac{2\pi}{L}(k,x))U_k.$$

We also suppose that the eigenfunction $\psi(x)$ of the Schrödinger equation (1) satisfies periodic conditions

$$\psi(\cdots x_j^\alpha + L \cdots) = \psi(\cdots x_j^\alpha \cdots).$$

Periodic conditions were introduced by Born, Karman to explain the classic effect in the case $N \to \infty$.

Let us introduce new dimensionless coordinates $x' \leftarrow x/L$. In the sequel we'll omit the sign' for the notation of the dimensionless coordinate and suppose that $x \in (R/\mathbb{Z})^3 \stackrel{\text{def}}{=} T$.

Let us assume that the potential $V_L(r)$ has the typical radius of interaction equal to $r_0 : V_L(r) = \Phi(r/r_0)$, $\int \Phi(\eta)d\eta = 1$, $\Phi \in C_0^\infty(T)$. Then $\alpha_0 V_L = \alpha_0 \delta^3 \times V_0(r)$, $V_0(r) = \delta^{-3}\Phi(r/\delta); \delta \stackrel{\text{def}}{=} r_0/L$ in the new dimensionless coordinates.

Dividing equation (1) by $N\alpha_0(r_0/L)^3$ we obtain in the dimensionless coordinate an equation on the torus T^N:

$$H_N\psi \stackrel{\text{def}}{=} \sum_{i=1}^{N}\{-\varepsilon^2\Delta_{x_i} + U_0(x_i)\}\psi + \\ + \frac{1}{N}\sum\sum_{1\le i<j\le N} V_0(x_i, x_j)\psi = \lambda\psi \tag{2}$$

$$\psi \in L_2(T^N), \quad \varepsilon^2 = N^{-2/3}(h^2/2m\alpha_0 r_0^3 \rho^{1/3}),\ \lambda = E/N\alpha_0(r/L)^3.$$

In equation (2) we have three parameters: $1/N \to 0$; and ε^2, δ depending on the physical properties of the system.

For some real physical systems applicable asymptotics are calculated for the equation (2) under the assumption that ε, δ are fixed but $N \to \infty$. So in the sequel we consider the case when $\varepsilon = 1$, $\delta = 1$, $N \to \infty$.

The open asymptotic problem, which arises under some relation between parameters ε, δ, N, is formulated in Section 3.

According to the principles of quantum mechanics all particles are divided in two different groups:

1. Bosons-which have symmetric eigenfunctions

$$\psi(\cdots x_i \cdots x_j \cdots) = \psi(\cdots x_j \cdots x_i \cdots).$$

2. Fermions –which have antisymmetric eigenfunctions.

Now and then we'll consider boson systems. There are many papers on the asymptotic approach to the equation (4) [1–9]. (On can find many references on this problem in [3]). I'll follow papers [4–6].

To justify the asymptotic approach by N. N. Bogolubov [1,2] we introduce in [4] the "calibrovochnoe" transformation of the potentials U, V which does not change equation (2). So, for any fixed $f \in L_2(T)\|f\| = 1$ let us put

$$C(V, N) \stackrel{\text{def}}{=} \iint_{T\times T} V_0(z_1, z_2)|f(z_1)|^2|f(z_2)|^2 dz_1 dz_2$$

$$\tilde{U}(x) = U_0(x) + V_h(x) + \frac{N-1}{2N}C(V, N); V_h(x) \stackrel{\text{def}}{=} \frac{N-1}{N}\int_T V_0(x, y)|f(y)|^2 dy; \tag{3}$$

$$\tilde{V}(x,y) \stackrel{\text{def}}{=} V_0(x,y) - \int_T V_0(x,z)|f(z)|^2 dz - \int_T V_0(y,z)|f(z)|^2 dz + C(V,N).$$

It is easy to prove that the operator $\tilde{H}_N$ from (2) with the new potentials $\tilde{U}, \tilde{V}$ will be identicaly equal to the operator H_N (from (2)) with the initial potentials U_0, V_0 because the transformation (3) does not change the function $\Phi(U,V)$:

$$\Phi(U,V) \stackrel{\text{def}}{=} \sum_{j=1}^{N} U_0(j) + \frac{1}{N} \sum\sum_{1 \le i < j \le N} V_0(x_i, x_j); \; V_0(x,y) = V_0(y,x).$$

Evidently, the potential $\tilde{V}$ satisfies the conditions:

$$\tilde{V}(x,y) = \tilde{V}(y,x); \quad \int_T \tilde{V}(x,y)|f(y)|^2 dy \equiv 0 \tag{4}$$

In the sequel we'll use the transformed potential $\tilde{U}, \tilde{V}$. From the properties (4) it follows that $\tilde{\tilde{V}} = \tilde{V}$, $\tilde{\tilde{U}} = \tilde{U}$.

1. Representation of secondary quantization for the Schrödinger equation (2)

Let us denote by $\mathcal{H}^N$ the Hilbert space of symmetric functions from $L_2(T^N)$ and let $U_0 \in C^\infty(T)$, $V_0 \in C^\infty(T^2)$. Evidently the H_N with the domain of definition $\mathcal{H}^N \cap C_0^\infty(T^N)$ is a symmetric operator on the Hilbert space $\mathcal{H}^N$ and its closure is the selfadjoint operator $\hat{H}_N$ in $\mathcal{H}^N$ with the domain of definition $\mathcal{D}(\hat{H}_N) = W_2^2(T^N) \cap \mathcal{H}^N$.

To obtain the spectral asymptotics of the operator $\hat{H}_N$ we introduce the sequence of the operators $\hat{H}_k$ on the spaces $\mathcal{H}^k : D(\hat{H}_k) \stackrel{\text{def}}{=} W_2^2(T^k) \cap \mathcal{H}^k$;

$$\hat{H}_k \stackrel{\text{def}}{=} \sum_{j=1}^{k} \hat{H}_1(x_j) + \frac{1}{N} \sum\sum_{1 \le i < j \le k} \tilde{V}(x_i, x_j); \tag{5}$$

$$\hat{H}_1(x) \stackrel{\text{def}}{=} -\frac{1}{2}\Delta_x + \tilde{U}(x).$$

Then we consider the Fock space $\mathcal{H}_B = \mathcal{H}^0 \oplus \mathcal{H}^1 \oplus \mathcal{H}^2 \oplus \cdots$ ($\mathcal{H}^0$ is a one dimensional space) and in $\mathcal{H}_B$ consider the operator $\hat{H}$ which is the direct sum of the operators $\hat{H}_k : \hat{H} = 0 \oplus \hat{H}_1 \oplus \hat{H}_2 \oplus \cdots$.

Let

$$\Phi = \{\Phi_0, \Phi_1(x_1), \Phi_2(x_1, x_2), \cdots\}^T \in \mathcal{H}_B$$

and

$$\Phi_j \equiv 0 \text{ when } j \ge N(\Phi),$$
$$\Phi_j \in \mathcal{H}^j \cap W_2^2(T^j). \tag{6}$$

Then

$$\hat{H}\Phi \stackrel{\text{def}}{=} \{0, \hat{H}_1\Phi_1, \hat{H}_2\Phi_2, \cdots, \hat{H}_N\Phi_N, 0, 0, \cdots\}^T.$$

In [9] was proved that the closure of $\hat{H}$ of the space of finite functions (6) is a selfajoint operator on $\mathcal{H}_B$.

The restriction of the operator $\hat{H}$ to the subspace $\mathcal{H}^{(N)} : \Phi^N = \{0, \cdots, 0, \Phi_N, 0, \cdots\}^T$ coincides with the operator $\hat{H}_N$.

To have a good calculation with the operator $\hat{H}$ in the Fock space $\mathcal{H}_B$ Dirak, Fock, Wigner, Iordan introduced so called representation of secondary quantization through the operators of creation $\psi^+(\xi)$ and annihilation $\psi(\xi)$. For the conjugate operator we'll use the symbol +; for the complex conjugate, the symbol $*$.

These operators are defined by the formulas

$$(\psi^+(\xi)\Phi)_k(x_1, \cdots, x_k) = k^{-1/2} \sum_{i=1}^{k} \delta(x_i - \xi) \times$$
$$\times \Phi_{k-1}(x_1, \cdots, x_{i-1}, x_{i+1}, \cdots, x_k) \tag{7}$$

$$(\psi(\xi)\Phi)_{k-1}(x_1, \cdots, x_k) = \sqrt{k}\Phi_k(x_1, \cdots, x_{k-1}, \xi)$$

$\psi(\xi)\Phi_0 = 0$ for all ξ, $\Phi_0 \stackrel{\text{def}}{=} \{1, 0, \cdots, 0, \cdots\}^T$.

The operators in the usual sense are the operators:

$$\psi_f \stackrel{\text{def}}{=} \int_T \psi(\xi) f(\xi) d\xi, \quad f \in L_2(T);$$
$$\psi_f^+ \stackrel{\text{def}}{=} \int_T \psi^+(\xi) f^*(\xi) d\xi. \tag{8}$$

The vector $\Phi_0 = \{1, 0, \cdots\}^T$ is called the vacuum vector: $\psi_f^+ \Phi_0 = \{0, f(x), 0, \cdots\}^T$. Φ_0 is the state without particles. In the state $\psi_f^+ \Phi_0$ there is one particle with the density of probability $|f(x)|^2$.

Operators $\psi^+(\xi)$, $\psi(\eta)$ satisfy commutator relations:

$$[\psi(\eta), \psi^+(\xi)] = \delta(\eta - \xi); [\psi(\eta), \psi(\xi)] = 0; \quad [\psi^+(\eta), \psi^+(\xi)] = 0. \tag{9}$$

Through the operators ψ^+, ψ the direct sum $\hat{H}$ on the Fock space $\mathcal{H}_B$ can be written in the form [2,3]:

$$\hat{H} = \int_T \psi^+(x) \hat{H}_1(x) \psi(x) dx +$$
$$+ \frac{1}{2N} \int \int_{T^2} \psi^+(x) \psi^+(y) \tilde{V}(x, y) \psi(y) \psi(x) dx dy,$$

$$\hat{H}_1 \stackrel{\text{def}}{=} -\frac{1}{2}\Delta_x + \tilde{U}(x). \tag{10}$$

It is useful to notice that the product $\psi^+(\xi)\psi(\eta)$ has the property

$$\psi^+(\xi)\psi(\eta)\{0 \cdots 0 \Phi_k 0 \cdots\}^T = \{0 \cdots 0 F_k 0 \cdots\}^T.$$

That means that a "k particle" function, $k \geq 1$ maps to a "k particle" function. Further the operator

$$\hat{N} \stackrel{\text{def}}{=} \int_T \psi^+(\xi)\psi(\xi)d\xi \tag{11}$$

has the property

$$\hat{N}\{0, \cdots, 0, \Phi_k, 0, \cdots\}^T = k\{0, \cdots, 0, \Phi_k, 0, \cdots\}^T$$

for all k, and is called the "number of particles" operator.

2. Asymptotic expantion

Following Bogolubov [1,2], let us introduce a unitary transformation U_f in the Fock space $\mathcal{H}_B$:

$$\begin{aligned} U_f \stackrel{\text{def}}{=} \exp\{\sqrt{N}\int_T dx(\psi^+(x)f(x) - \\ -\psi(x)f^*(x))\}, f \in C^\infty(T). \end{aligned} \tag{12}$$

As the operators ψ^+, ψ satisfy the commutator condition $[\psi(\xi), \psi^+(\eta)] = \delta(\xi - \eta)$, then [1–3,13],

$$U_f^+\psi^+U_f = \psi^+ + \sqrt{N}f^*, \quad U_f^+\psi U_f = \psi + \sqrt{N}f. \tag{13}$$

Therefore, the operator $U_f^+\hat{H}U_f$ one can get from the operator $\hat{H}$ substituting instead of ψ^+, ψ expressions (13). Thus we obtain expansion [1–3,4,7,8]

$$\begin{aligned} &U_f^+\hat{H}U_f = \hat{H}(\psi^+ + \sqrt{N}f^*, \psi + \sqrt{N}f) = \\ &NH(f^*, f) + \sqrt{N}\{\int_T \frac{\delta H(f^*, f)}{\delta f^*(x)}\psi^+(x)dx + \int_T \frac{\delta H(f^*, f)}{\delta f(x)}\psi(x)dx\} + \\ &H_2(\psi^+, \psi) + \frac{1}{\sqrt{N}}H_3 + \frac{1}{N}H_4. \end{aligned} \tag{14}$$

Here, according to (4):

$$\begin{aligned} H(f^*, f) = &\int_T f^*(x)(H_1 f)(x)dx \\ &+ \int_T \tilde{U}(x)|f|^2(x)dx \\ &+\frac{1}{2}\int_T |\nabla f|^2 dx, \end{aligned} \tag{15}$$

$$\frac{\delta H}{\delta f^*(x)} = -\frac{1}{2}\Delta f + \tilde{U}(x)f(x). \tag{16}$$

H_2 is a quadratic operator according to ψ^+, ψ; H_3, H_4 are operators of order 3 and 4 according to ψ^+, ψ.

We choose f as the solution of the Hartree equation

$$\frac{\delta H}{\delta f^*(x)} = \Omega f(x)$$

which with the initial potential U_0, V_0 has the form:

$$-\frac{1}{2}\Delta_x f + U_0(x)f(x) + \frac{N-1}{N}\int_T V_0(x,y)|f(y)|^2 dy = \{\Omega - C(V,N)\times\frac{N-1}{2N}\}f(x).$$

Then the term near $\sqrt{N}$ will has the form

$$\Omega(\int_T (\psi^+(x)f(x) + \psi(x)f^*(x))dx).$$

To calculate the asymptotics of eigenvalues for the N particle Hamiltonian H_N it is convenient to get rid of the linear terms in ψ^+, ψ in the expansion (14). For that purpose we can use instead of the operator $\hat{H}_N$ the following operator,

$$\tilde{H} \overset{\text{def}}{=} \hat{H}_N - \Omega\hat{N}$$

because on the N particle function the spectrum of the $\tilde{H}$ and the $\hat{H}_N$ differ by the constant ΩN. For the operator $\tilde{H}$ we obtain an expansion of type (14) but without the linear term in ψ^+, ψ:

$$U_f^+ \tilde{H} U_f = N\{H(f^*, f) - \Omega\} + +\tilde{H}_2(\psi^+, \psi) + \frac{1}{\sqrt{N}}H_3 + \frac{1}{N}H_4. \tag{17}$$

From the expansion (17) we see that the main *variable* term is the quadratic operator $\tilde{H}_2(\psi^+, \psi)$, and its spectrum should give the spectral asymptotics of the N particle Schrödinger operator (2).

Now we formulate the rigorous results. Let us consider the quadratic form

$$F(u,v) \overset{\text{def}}{=} \int_T \{\frac{1}{2}|\nabla_x u|^2 + \frac{1}{2}|\nabla_x v|^2 + (\tilde{U} - \Omega)(|u|^2(x) + |v|^2(x))\}dx \tag{18}$$

$$+ \iint_{T\times T} \tilde{V}(x,y)\{f^*(x)u(x) + f(x)v(x)\}\{f^*(y)u(y) + f(y)v(y)\}^* dxdy$$

on the intersection of the space $W_2^1(T) \oplus W_2^1(T)$ with the subspace $L \overset{\text{def}}{=} \{(u,v) : (u,f) = 0,\ (v,f^*) = 0\}$. In [4] the following theorem was proved.

Theorem 1. *Let $U, V \in L_\infty$ if dimension $T \le 3$ and let $U, V \in C^{2(n-3)}$ if $n \overset{\text{def}}{=}$ dimension $T \ge 4$. If ess min $|f| > 0$ and inf $\{F(u,v)/(\|u\|^2 + \|v\|^2)\}$ on the intersection $L \cap \{W_2^1(T) \oplus W_2^1(T)\}$ is positive;*

Then: 1) there is a proper linear canonical transformation G such that

$$G^+ H_2 G = \sum_{\lambda_j > 0} \lambda_j \xi_j^+ \xi_j + \gamma,\ \gamma \overset{\text{def}}{=} -\sum_{\lambda_j>0} \lambda_j \|v_j\|^2; [\xi_i, \xi_j] = 0, [\xi_i, \xi_j^+] = \delta_{ij}$$

where λ_j are the eigenvalue of the operator C and $\{u_j, v_j\}^T$ are the eigenfunctions:

$$C = \begin{pmatrix} A & B \\ -B^* & -A^* \end{pmatrix},\quad C\{u_j, v_j\}^T = \lambda_j\{u_j, v_j\}^T \tag{19}$$

(here $*$ is the symbol for the complex conjugate) and the kernels of the operators B, A are:

$$B(x,y) \stackrel{\text{def}}{=} \tilde{V}(x,y)f(y)f(x) \tag{20}$$

$$A(x,y) = -\frac{1}{2}\Delta_x\delta(x-y) + \delta(x-y)[\tilde{U}(x) - \Omega] + \tilde{V}(x,y)f(x)f^*(y).$$

2) The spectral series $\lambda(m,N) \stackrel{\text{def}}{=} N(\Omega + H(f,f^*)) + \gamma + \sum_{\lambda_j>0} \lambda_j m_m; \quad m_j \in \mathbb{Z}^+$

defines the asymptotics of the spectral series of the operator (2). This means that there are eigenvalues λ_N^m *of the operator (2) such that*

$$|\lambda_N^m - \lambda(m,N)| \leq N^{-1/2} g(\sum jm_j),$$

where a g is some nondecreasing function.

3. Unsolved problem

If the parameter ε^2 has the order $N^{-2/3}$; $\delta \sim N^{-1/4}$, wich is the typical case for a dense medium, then on the eigenfunctions of the operator $H_2(\psi^+,\psi) : H_2(\psi^+,\psi)\Phi_j = (\lambda_j + \gamma)\Phi_j$ we have $\|N^{-1/2}H_3(\psi^+,\psi)\Phi_j\| \gg \lambda_j$, but $\|N^{-1}H_4(\psi^+,\psi)\Phi_j\| \ll \lambda_j$. Therefore, for the dense medium, the spectral asymptotics of the N particle Schrödinger equation will be defined by the spectrum of the model cubic operator with some small parameter μ:

$$\mu \sum_{j=1}^{M} \lambda_j \xi_j^+ \xi_j + \sum_{i,j,k=1}^{M} \{a_{ijk}\xi_i^+\xi_j^+\xi_k + a_{ijk}^*\xi_k^+\xi_j\xi_i\}, \quad \mu \to +0.$$

That general problem is open.

4. Quadratic operators

Let us consider symmetric quadratic operator:

$$\begin{aligned} H_2(\psi^+,\psi) \stackrel{\text{def}}{=} \frac{1}{2}\int\int_{T^2} &\{B(x,y)\psi^+(x)\psi^+(y) + \\ &+2A(x,y)\psi^+(x)\psi(y) + \\ &+B^*(x,y)\psi(x)\psi(y)\}dxdy; \end{aligned} \tag{21}$$

$$B(x,y) = B(y,x); \quad A^*(y,x) = A(x,y).$$

Conditions on the potentials: $U, V \in L_\infty$; $V(x,y) = V(y,x)$ if $\dim T = n \leq 3$. If dimension $T = n \geq 4$ then $V \in C^{2(n-3)}(T^2)$, $U \in C^{2(n-3)}(T)$.

We consider the case when the operators B, A are defined by the formulas (20).

F. A. Berezin [9] proved that the operator H_2 which is defined on the finite function with component $\Phi_j \in W_2^2(T^j) \cap \mathcal{H}^j$ has selfadjoint closure.

The reduction of quadratic operator H_2 to simpler form can be achieved by a unitary transformation Q of the operators ψ^+, ψ:

$$\xi = Q^+\psi Q; \quad \xi^+ = Q^+\psi^+ Q;$$
$$Q^+Q = QQ^+ = E$$

Usually mathematicians and physicists use a linear proper canonical unitary transformation [1–3,9] of the form

$$\{\xi, \xi^+\} = \mathcal{K}\{\psi, \psi^+\}, \quad \mathcal{K} \stackrel{\text{def}}{=} \begin{pmatrix} \mathbb{U}^+ & -\mathbb{V}^+ \\ -\mathbb{V}^T & \mathbb{U}^T \end{pmatrix}. \tag{22}$$

We assume [4] that $\mathbb{U}, \mathbb{V}$ are integral operators with kernels

$$\mathbb{U}(x, y) = \sum u_j(x) f_j^*(y),$$
$$\mathbb{V}(x, y) = \sum v_j(x) f_j^*(y) \tag{23}$$

where $\{f_j\}$ is any fixed orthogonal basis in $L_2(T)$. The operator $\mathbb{U}$ must be bounded and the $\mathbb{V}$ must be the Hilbert–Schmidt operator [9].

Now let us introduce an operator $\mathcal{A}$

$$\mathcal{A} \stackrel{\text{def}}{=} \begin{pmatrix} \mathbb{U} & \mathbb{V}^* \\ \mathbb{V} & \mathbb{U}^* \end{pmatrix} \tag{24}$$

In the sequel we will use kernels $\mathbb{U}, \mathbb{V}$ such that

$$\mathcal{K}\mathcal{A} = \mathcal{A}\mathcal{K} = \begin{pmatrix} E & O \\ O & E \end{pmatrix}. \tag{25}$$

Therefore,

$$\{\psi, \psi^+\}^T = \mathcal{A}\{\xi, \xi^+\}^T. \tag{26}$$

N. N. Bogolubov [1,2] offered to choose functions $\{u_j, v_j\}^T$ for the kernels (23) from the eigenfunctions of the operator $\mathbb{C}$ (19):

$$Au_j + Bv_j = \lambda u_j$$
$$-B^* u_j - A^* v_j = \lambda v_j; \tag{27}$$

but in the general case he restricted himself [1,2] only to a finite dimensional operator $\mathbb{C}$ without adjoint vectors and complex eigenvalues. So, in this approach the key point is the investigation of the operator$\mathbb{C}$'s spectrum and its root subspaces. We do it in [4–6] using M.V. Keldish's theorems [10–12].

To apply Keldish's theorem, we note that our operator $\mathbb{C}$ can be brought to the form

$$\mathbb{C} = D + T; \quad D = \begin{pmatrix} -\frac{1}{2}\Delta + 1 & 0 \\ 0 & \frac{1}{2}\Delta - 1 \end{pmatrix},$$

$$Tf = \begin{pmatrix} -1 & 0 \\ 0 & +1 \end{pmatrix} f + \int \|t_{ij}(x,y)\| f(y) dy$$

with $t_{ij}(x,y) \in L_\infty$. Evidently, on $L_2 \oplus L_2$ the operator D is selfadjoint and has only discrete spectrum. We proved [5] that the operators D^{-1}, TD^{-1}, $D^{-1}TD^{-1}$ are compact, and

$$\sum s_j (D^{-1}TD^{-1})^{2p} < \infty, \quad \rho \geq [\frac{n+1}{4}] + 1;$$

where n is the dimension of torus. So, applying Keldish's theorem, we obtain the theorem [4,5].

Theorem 2. 1) *The spectrum of the operator $\mathbb{C}$ consists of isolated points with finite algebraical multiplicity.*

2) *The root space system is complete in $L_2 \oplus L_2$.*

3) *There is a constant R such that when $|\lambda_j| > R$ a) $Im\,\lambda_j = 0$; b) the operator $\mathbb{C}$ at the point λ_j does not have adjoint vectors; c) the eigenvectors $\{u_j^\alpha, v_j^\alpha\}$ corresponding to λ_j satisfy the inequality*

$$\lambda_j \int_{Tour} \{|u_j^\alpha|^2 - |v_j^\alpha|^2\} dx > 0 \quad \alpha = 1, ..., m_j; \tag{28}$$

m_j is the dimension of the root subspace corresponding to λ_j.

If we suppose that the spectrum of the operator $\mathbb{C}$ is real and $\mathbb{C}$ does not have adjoint vectors, then we can express the operators $\mathbb{U}, \mathbb{V}$ through the eigenvectors for which

$$\int_{Tour} \{|u_j^\alpha|^2 - |v_j^\alpha|^2\} dx = 1;$$

and bring the quadratic form H_2 to the form [4,5]:

$$H_2(\psi^+, \psi) = \sum_j \lambda_j \xi_j^+ \xi_j + \gamma E \tag{29}$$

where γ is some constant.

In the general case the operator $\mathbb{C}$ has adjoint vectors. Therefore, we need to work with root spaces and choose vectors $\{u_j^\alpha, v_j^\alpha\}^T$ in some special way from the root spaces. The next lemma will be helpful [5,6].

Lemma 1. 1) *If $\lambda \in \sigma(\mathbb{C})$ and $\lambda \neq 0$, then $-\lambda^* \in \sigma(\mathbb{C})$, and the dimension of the root space R_λ is equal to the dimension of the root space $R_{-\lambda}$.*

2) *If $\lambda \in \sigma(\mathbb{C})$, then $\lambda^* \in \sigma(\mathbb{C})$.*

3) *If* $\lambda_j + \lambda_k \neq 0$, *then the root spaces* $R_{\lambda_j}, R_{\lambda_k}$ *are orthogonal relative to the bilinear form*

$$\langle IX, Y\rangle \stackrel{\text{def}}{=} \int_T (X_2Y_1 - X_1Y_2)dx. \tag{30}$$

$$I \stackrel{\text{def}}{=} \begin{pmatrix} O & E \\ -E & O \end{pmatrix}$$

4) *Let us join the root spaces according the rule*

$$R_j \stackrel{\text{def}}{=} R_{\lambda_j} \oplus R_{-\lambda_j} \oplus R_{\lambda_j^*} \oplus R_{-\lambda_j^*} \text{ if } \operatorname{Im} \lambda_j \neq 0 \text{ and } \operatorname{Re} \lambda_j \neq 0;$$

$$R_j \stackrel{\text{def}}{=} R_{\lambda_j} \oplus R_{-\lambda_j} \text{ if } \operatorname{Im} \lambda_j = 0 \text{ or } \operatorname{Re} \lambda_j = 0, \text{ but } \lambda_j \neq 0;$$

$$R_j \stackrel{\text{def}}{=} R_{\lambda_j} \text{ if } \lambda_j = 0;$$

and let us introduce the operator $J \stackrel{\text{def}}{=} \begin{pmatrix} O & E \\ E & O \end{pmatrix}$.

Then: (i) The bilinear form $\langle IX, Y\rangle$ *is not degenerate on the subspaces* R_j.

(ii) The subspaces R_j *have even dimension* $2m_j < \infty$.

(iii) $R_{-\lambda_j^*} = J(R_{\lambda_j}^*)$; $J(R_j^*) = R_j$.

(iv) R_j, R_k *are orthogonal with respect to the bilinear form* $\langle IX, Y\rangle$.

(v) Let $L \stackrel{\text{def}}{=} \begin{pmatrix} -E & O \\ O & E \end{pmatrix}$ *and* P_j *be the orthogonal projector of the space* $L_2(T) \oplus L_2(T)$ *on* R_j.

Then in the subspace R_j *there are* m_j *eigenvectors* φ_j^α, $\alpha = 1, \cdots, m_j$ *of the operator* P_jLP_j *with positive eigenvalues* $\mathcal{H}_j^\alpha > 0$; *and there are* m_j *eigenvectors* $J(\varphi_j^{\alpha *})$ *with negative eigenvalues* $-\mathcal{H}_j^\alpha$. *These vectors make a basis for* R_j *and satisfy the relations:*

$$\langle I\varphi_j^\alpha, J(\varphi_j^{\beta *})\rangle = \delta\alpha\beta\mathcal{H}_\alpha; \langle I\varphi_j^\alpha, \varphi_j^\beta\rangle = 0$$

$$\langle IJ(\varphi_j^{\alpha *}), J(\varphi_j^{\beta *})\rangle = 0; \ 1 \leq \alpha, \beta \leq m_j.$$

(vi) There is a number R, *such that when* $|\lambda_j| > R$ *the following holds: a)* $Im\lambda_j = 0$; *b) vectors* $J(\varphi_j^{\alpha *})\alpha = 1, \cdots, m_j$ *are the eigenvectors of the operator* $\mathbb{C}$ *with the eigenvalue* $\lambda_j > 0$.

Thus, let $2m_j$ be the dimension of R_j; we split the orthogonal system $\{f_1(x), \cdots, f_N(x), \cdots\}$ into the blocks $\{f_j^\alpha\}$, $1 \leq \alpha \leq m_j$, and respectively will represent the kernel $\mathbb{U}, \mathbb{V}$ in block form:

$$\mathbb{U}(x, y) = \sum_{j=1}^{\infty} \mathbb{U}_j(x, y), \quad \mathbb{V}(x, y) = \sum_{j=1}^{\infty} \mathbb{V}(x, y) \tag{31}$$

$$\mathbb{U}_j(x, y) = \sum_{\alpha=1}^{m_j} u_j^\alpha(x) f_j^\alpha(y), \quad \mathbb{V}_j(x, y) = \sum_{\alpha=1}^{m_j} v_j^\alpha(x) f_j^\alpha(y).$$

The functions u_j^α, v_j^α are defined by the formula

$$\{u_j^\alpha, v_j^\alpha\}^T \stackrel{\text{def}}{=} J(\varphi_j^{\alpha *})/\sqrt{\mathcal{H}_j^\alpha},\ \alpha = 1, \cdots, m_j \tag{32}$$

where the functions φ_j^α and positive numbers $\mathcal{H}_j^\alpha$ are defined in the conclusion (v) of the Lemma 1.

For the operators defined in (31), the following theorem has been proved [5].

Theorem 3. *Let $U \in L_\infty(T)$, $V \in L_\infty(T^2)$ if dimension $T \leq 3$; and $U \in C^{2(n-3)}(T)$, $V \in C^{2(n-3)}(T^2)$ if $n \stackrel{\text{def}}{=}$ dimension $T \geq 4$.*

Then: 1) *The operator $\mathbb{U}$ is bounded on the space $L_2(T)$, and the operators $\mathbb{V}, \mathbb{V}^+$ are the Hilbert–Schmidt operators on the space $L_2(T)$.*

2) *The operators $\mathcal{K}$ of (22) and $\mathcal{A}$ of (24) satisfy condition (25).*

3) *If the basis $\{f_j^\alpha\}$ for $L_2(T)$ is real, then there are finitely many matrices*

$$\|A_j^{\alpha,\beta}\|, \quad \|B_j^{\alpha\beta}\|, \quad 1 \leq \alpha, \beta \leq j; \quad B_j^T = B_j; \quad A_j^+ = A_j$$

such that the canonical transformation (23), (31) brings the quadratic operator (21) to the form

$$H_2(\psi^+, \psi) = \sum_{j=1}^{M} H_j(\xi_j^+, \xi_j) + \sum_{j=M+1}^{\infty} \lambda_j \xi_j^+ \xi_j. \tag{33}$$

Here: $\xi_j = (\xi_j^\alpha, \cdots, \xi_j^{m_j})$, $\xi_j^\alpha \stackrel{\text{def}}{=} \int_T \xi(x) f_j(x) dx$;

$$H_j(\xi_j^+, \xi_j) = \frac{1}{2} \sum_{\alpha=1}^{m_j} \sum_{\beta=1}^{m_j} \{B_j^{\alpha\beta} \xi_j^{\alpha+} \xi_j^{\beta+} + 2A_j^{\alpha\beta} \xi_j^{\alpha+} \xi_j^\beta + B_j^{\alpha\beta *} \xi_j^\alpha \xi_j^\beta\}; \tag{34}$$

$[\xi_j^\alpha, \xi_j^\beta] = 0$, $[\xi_j^\alpha, \xi_k^{\beta+}] = \delta_{jk}\delta_{\alpha\beta}$ *and* $\lambda_j > 0$ *when* $j \geq M+1$.

$$\gamma \stackrel{\text{def}}{=} \frac{1}{2} \int\int_{T^2} B(x,y)\{\int_T \mathbb{V}(x,z)\mathbb{U}^*(y,z)dz\}dxdy +$$
$$\int\int_{T^2} A(x,y)\{\int_T \mathbb{V}(x,z)\mathbb{V}^*(y,z)dz\}dxdy +$$
$$\frac{1}{2} \int\int_{T^2} B^*(x,y)\{\int_T \mathbb{U}(x,z)\mathbb{V}^*(y,z)dz\}dxdy.$$

The matrices A_j, B_j [5] are calculated through the vectors $\{u_j^\alpha, v_j^\alpha\}$. Theorem 2 has reduced the problem of normal forms for the quadratic operator (21) to the problem of normal forms of the finite dimensional quadratic operators H_j (34).

5. Classification of finite dimensional quadratic operators $H^j(a_j^+, a_j)$ and its spectrum

Let us introduce finite dimension operators (35):

$$\mathcal{K}_j = \begin{pmatrix} \mathbb{U}_j^+ & -\mathbb{V}_j^+ \\ -\mathbb{V}_j^T & \mathbb{U}_j^T \end{pmatrix},$$

$$\mathcal{A}_j = \begin{pmatrix} \mathbb{U}_j & \mathbb{V}_j^* \\ \mathbb{V}_j & \mathbb{U}_j^* \end{pmatrix},$$

$$I_j \stackrel{\text{def}}{=} \begin{pmatrix} 0 & E_j \\ -E_j & 0 \end{pmatrix}. \tag{35}$$

From the properties of the canonical transformation (22), (24) it follows [9], [5] that

$$\mathcal{A}_j\mathcal{K}_j = \mathcal{K}_j\mathcal{A}_j = E;$$
$$\mathcal{K}_j^T I_j \mathcal{K}_j = I_j;$$
$$\mathcal{A}_j^T I_j \mathcal{A}_j = I_j. \tag{36}$$

Therefore, $\mathcal{A}_j, \mathcal{K}_j$ are complex symplectic matrices.

Let us set consider relation between the complex matrices $\mathcal{A}_j, \mathcal{K}_j$ making proper canonical transformations of operators $a_j^{\alpha+}, a_j^{\alpha}$ and a real symplectic matrix representing a canonical transformation of real phase space.

Let us set $z_\alpha = x^\alpha + ip^\alpha$, $z_\alpha^* = x^\alpha - ip^\alpha$, where x, p are real. There is a matrix K, such that

$$\{z, z^*\}^T = K\{x, p\}^T.$$

It is easy to prove that $(K^T)^{-1} I K^{-1} = (i/2)I$.

Statement 1. *If S is a symplectic matrix, then the matrix $\mathcal{A}_j \stackrel{\text{def}}{=} KSK^{-1}$ and its inverse satisfy condition (36) and $\mathcal{A}_j, \mathcal{K}_j$ are complex symplectic matrices of the form (35).*

Now let us associate with the quadratic operator

$$H^j(a_j^+, a_j) \stackrel{\text{def}}{=} \frac{1}{2}\{\langle B_j a_j^+, a_j^+\rangle + 2\langle A a_j, a_j^+\rangle + \langle B_j^* a_j, a_j\rangle\}$$

the real Hamiltonian

$$H_j(x,p) = \frac{1}{2}\langle B_j^* z, z\rangle + \langle A_j z^*, z\rangle + \frac{1}{2}\langle B_j z^*, z^*\rangle, \tag{37}$$

where $\langle a, b\rangle \stackrel{\text{def}}{=} \sum_{\alpha=1}^n a_\alpha b_\alpha$.

From the conditions $B_j^T = B_j$, $A_j^+ = A_j$ it follows that $H_j(x,p)$ is a real valued function.

For the real quadratic Hamiltonian, according to the paper of J. Willamson [14], D. Galin [15], [16] wrote a list of standard forms to which $H_j(x,p)$ can be reduced by linear real symplectic transformations. To use this list I need [5].

Lemma 2. 1) *The matrices* $\frac{i}{2}I$ *Hess* $H_j(x,p)$ *and* $\mathbb{C}_j$ *which is the restriction of the operator* $\mathbb{C}$ *to the subspace* R_j, *are similar.*

2) *The matrix* $\mathbb{C}_j$ *and hence the matrix* $\frac{i}{2}I$ *Hess* $H_j(x,p)$ *has spectrum:*

$$\lambda_j, -\lambda_j, \lambda_j^*, -\lambda_j^* \quad \text{if} \quad \text{Im } \lambda_j \neq 0; \text{Re } \lambda_j \neq 0;$$

$$\lambda_j, -\lambda_j \quad \text{if} \quad \lambda_j \text{ is real or imaginary .}$$

The classification of standard forms for the quadratic Hamiltonian is done in [14–17] according to its eigenvalues and the size of Jordan box of the matrix I *Hess* H_j.

Therefore, taking standard form for $H_j^\mu(x,p)$ by substituting $x = \frac{1}{2}(a_j^+ + a_j)$, $p = \frac{1}{2i}(a_j - a_j^+)$ we obtain normal forms

$$H_j^\mu(a_j^+, a_j) \stackrel{\text{def}}{=} \frac{1}{2}\{H_j^\mu(\frac{a_j^+ + a_j}{2}, \frac{1}{2i}(a_j - a_j^+)) + (H_j^\mu(\frac{a_j^+ + a_j}{2}, \frac{1}{2i}(a_j - a_j^+)))^+\} + \text{const} \tag{38}$$

to which H_j can be reduced by the linear finite dimensional canonical transformation (35) with the matrix $\mathcal{A}_j^\mu \stackrel{\text{def}}{=} KS_\mu K^{-1}$. Here S_μ is the real symplectic matrix which reduces $H_j(x,p)$ to $H_j^\mu(x,p)$.

The spectrum of normal forms

The operator H_j has discrete spectrum if and only if its eigenvalues λ_j are real and its matrix $\mathbb{C}_j$ does not have a Jordan box. In that case $H^j(a_j^+, a_j)$ can be reduced to the form [5]:

$$H^j(a_j^+, a_j) = \lambda_j \sum_{\alpha=1}^{m_j} \xi_\alpha^+ \xi_\alpha + \gamma_j E;$$

and has eigenvalues $(\lambda_j n + \gamma_j)$, $n \geq 0$.

In all other cases operator H_j has only continuous spectrum. Its spectrum is $\sigma(H^j) = R_1$ except for one case when $\lambda_j = 0$ and $\mathbb{C}$ has simple Jordan box; in that case $\sigma(H^j) = \pm R_1^+$.

As we can use the representation

$$a_j = \frac{1}{\sqrt{2}}(\frac{\partial}{\partial x_j} + x_j), \quad a_j^+ = (-\frac{\partial}{\partial x_j} + x_j)/\sqrt{2};$$

then from (38) and the foregoing it follow that the spectrum for the operator

$$H(x, -i\frac{\partial}{\partial x}) + H(x, -i\frac{\partial}{\partial x})^+$$

with quadratic symbol $H(x,p)$ has the same property as the spectrum of the operator (38).

6. The spectrum of quadratic operator $H_2(\psi^+, \psi)$

From Theorem 3 follows the existence of the unitary operator Q on the Fock space $\mathcal{H}_B$ such that $\xi = Q^+\psi Q$ and

$$QH_2(\psi^+, \psi)Q^+ = \sum_{j=1}^{M} H_j(a_j, a_j) + \sum_{j=M+1}^{\infty} \lambda_j a_j^+ a_j + \gamma E$$

$$a_j = (a_j^1, \cdots, a_j^{m_j}), \quad a_j^\alpha \overset{\text{def}}{=} \int \psi(x) f_j^\alpha(x) dx) \tag{39}$$

$$[a_j^\alpha, a_k^{\beta +}] = \delta_{\alpha\beta}\delta_{jk}.$$

The quadratic operators H_j are defined in Theorem 3.

It is evident that finite dimentional operators H_j are permutable. Therefore, from the spectral properties of the operators H_j and the spectral property of the operator $\mathbb{C}$ (Theorem 2) Theorem 4 follows.

Theorem 4. *Let the potentials U, V satisfy the conditions of Theorem 3. Then:* 1) *If for any eigenvalue $\lambda_j \neq 0$ the operator $\mathbb{C}$ has adjoint vector then $\sigma(H_2) = R^1$.* 2) *If the spectrum $\sigma(\mathbb{C})$ is real and the operator $\mathbb{C}$ does not have adjoint vectors then there is a unitary transformation Q of the Fock space $\mathcal{H}_B$ such that*

$$QH_2(\psi^+\psi)Q^+ = \sum_{j=1}^{M} \lambda_j a_j^+ a_j + \sum_{j=M+1}^{\infty} \lambda_j a_j^+ a_j + \gamma E \tag{40}$$

where $\lambda_j > 0$ if $j \geq M+1$.

Any eigenvector $\{u_j, v_j\}^T$ of the operator $\mathbb{C}$, $\mathbb{C}\{u_j, v_j\}^T = \lambda_j\{u_j, v_j\}^T$, which satisfies the condition $\|u_j\| > \|v_j\|$ corresponds to the term $\lambda_j a_j^+ a_j$ in the sum of (40).

The spectrum of the operator (40) consists of the points

$$\sum \lambda_j m_j, \quad m_j \in \mathbb{Z}^+$$

and their limit points.

References

[1] N. N. Bogolubov, *About the theory of superfluidity,* Izv. AN SSSR, Physica, **11,** N 1 (1947), 77–90.

[2] N. N. Bogolubov, N. N. Bogolubov (Jnr), An introduction to quantum statistical mechanics, Moscow, "Nauka" 1984.

[3] P. Ring, P. Schuck, *The nuclear many-body problem,* Springer-Verlag, 1980.

[4] V. V. Kucherenko, V. P. Maslov, *The spectrum of the N bosons system when $N \to \infty$,* Doklady Russian Akademii Nauk (Russia) **348,** N 2 (1996), 169–172.

[5] V. V. Kucherenko, V. P. Maslov, *The normal forms of the quadratic bosons operators,* Mathematical Notes, **61,** N 1 (1997), 69–90.

[6] V. V. Kucherenko, V. P. Maslov, *Reduction of the quadratic bosons operator to the normal forms,* Doklady Russian Akademii Nauk, **350,** N 2 (1996), 162–165.

[7] V. P. Maslov, O. U. Shvedov, *The spectrum of the N particles Hamiltonian for the big N and superfluidity,* Doklady Russian Akademii Nauk, **335,** N 1 (1994), 42–46.

[8] V. P. Maslov, O. U. Shvedov, *The method of the complex germ in the Fock space,* Theoretical and mathematical physic, **104,** N 2 (1995), 310–329.

[9] F. A. Berezin, *The method of secondary quantization,* Moscow, Nauka (1965).

[10] M. V. Keldish, *About eigenvalues and eigenfunctions for some class of non selfadjoint equations,* Doklady Akademii Nauk SSSR, **77,** N 1 (1951), 11–14.

[11] M. B. Keldish, V. B. Lidskii, *The problems of the spectral theory for the not selfadjoint operators,* Trudi IV Vsesouznogo matemat. Siezda SSSR. **I,** (1963), 101–120.

[12] I. Gohberg, M. G. Kreyn, *Introduction to the theory of linear non selfadjoint operators.* Moscow, Nauka, 1965.

[13] N. N. Bogolubov, D. V. Shirkov, *The quantum fields,* Moscow, Nauka, 1980.

[14] J. Williamson, *On an algebraic problem concerning the normal forms of linear dynamical systems,* Amer. J. of Math. **58,** N 1 (1936), 141-163.

[15] D. M. Galin, *Versal deformation of the linear Hamilton systems,* Trudy seminara Im, I.G. Petrovskogo. Moscow, Izdat MGU, V I (1975), 63–74.

[16] V. I. Arnold, Mathematical methods of classical mechanics, Moscow, Nauka, 1989.

[17] A. D. Bruno, The restricted 3–Body problem. Plane periodic orbits, Walter der Grerybez, Berlin, 1994.

Departamento de Matemáticas,
Instituto Politécnico Nacional,
Edificio 9, 3er piso ESFM,
U.P. Adolfo López Mateos, Col. Lindavista,
C.P. 07738, México D.F.
E-mail address: valeri@esfm.ipn.mx

Operator Theory:
Advances and Applications, Vol. 114

A survey of Q_p spaces

Peter Lappan

1. 1. Introduction and preliminaries

The subject of Q_p spaces has been developed only over the last three to four years, with the result that many of the basic papers dealing with it are just coming out. Our goal is to publicize some of the basic results in the subject, together with a variety of related ideas. It is not the purpose of this paper to announce previously unknown results, and everything mentioned here either has been published elsewhere, or will be shortly. No proofs are given. The list of references given at the end of the paper has deliberately been kept short, but substantial lists of further references can be found in any of the recent papers listed in the references given here. Thus, the casual reader may get some idea of what the basic ideas are and where they head, while the more serious reader can pursue the subject as completely as desired.

Let $D = \{z : |z| < 1\}$. All functions mentioned here are analytic in D until further notice. We say that a function f is in the class Q_p if

$$\sup_{a\in D} \int\int_D |f'(z)|^2 g^p(z,a)\, dx\, dy < \infty,$$

where $g(z,a)$ is the Green's function in D with singularity at a. We say that a function f is in the class $Q_{p,0}$ if

$$\lim_{|a|\to 1} \int\int_D |f'(z)|^2 g^p(z,a)\, dx\, dy = 0.$$

We define BMOA $= Q_1$ and VMOA $= Q_{1,0}$. Also, the Dirichlet space DA coincides with Q_0. The space DA consists of all analytic functions for which the area of the image, counted according to multiplicity, is finite.

We say that a function f is a *Bloch function*, and we write $f \in B$ if

$$\sup_{z\in D} |f'(z)|(1-|z|^2) < \infty,$$

and we say that f is in the "little Bloch space", and we write $f \in B_0$ if

$$\lim_{|z|\to 1} |f'(z)|(1-|z|^2) = 0.$$

Perhaps the result which began the study of Q_p spaces in earnest was the following.

Theorem 1.1 (Aulaskari and Lappan [2]). *For $p > 1$, $Q_p = B$ and $Q_{p,0} = B_0$.*

It had been known previously that $Q_2 = B$.

2. Q_p spaces

A basic result concerning Q_p spaces is the following.

Theorem 2.1 (Aulaskari, Xiao, and Zhao [3]). *For $0 < p < q \leq 1$, $Q_p \subset Q_q$ and $Q_{p,0} \subset Q_{q,0}$. Further, both of these containments are strict.*

It is also true that $AD \subset Q_{p,0} \subset Q_p$ for each p, $0 < p < 1$. However, $Q_{p,0} \subset$ VMOA for each p, $0 < p < 1$, while each space Q_p contains functions not in VMOA.

There are some interesting examples illustrating the ideas above. We make use of the following criterion for a function to be in the spaces Q_p and $Q_{p,0}$.

Theorem 2.2 (Aulaskari, Xiao, and Zhao [3]). *Let $0<p<1$ and let $f(z) = \sum a_n z^{2^n}$. If $\sum \frac{2^{n(1-p)}}{n}|a_n|^2 < \infty$, then $f \in Q_p$ and $f \in Q_{p,0}$.*

Example 2.3. Fix p, $0 < p < 1$, and let $f(z) = \sum \frac{2^{n(p-1)/2}}{n} z^{2^n}$. Then we have that $f \in Q_p$ and $f \notin \bigcup_{0<k<p} Q_k$, and thus $\bigcup_{0<k<p} Q_k$ is a proper subset of Q_p.

Example 2.4. Fix p, $0 < p < 1$, and let $f(z) = \sum n 2^{n(p-1)/2} z^n$. Then $f \in \bigcap_{p<k<1} Q_k$ but $f \notin Q_p$, and so Q_p is a proper subset of $\bigcap_{p<k<1} Q_k$.

Thus, unions and intersections of "blocks" of Q_p spaces form spaces which are not themselves Q_p spaces. This kind of result is also valid for the spaces $Q_{p,0}$, that is, for $0 < p < 1$, $\bigcup_{0<k<p} Q_{k,0}$ is a proper subset of $Q_{p,0}$, and $Q_{p,0}$ is a proper subset of $\bigcap_{p<k<1} Q_{k,0}$. This last result is valid for $p = 0$, as $AD = Q_0$ is a proper subset of $\bigcap_{0<k<1} Q_{k,0}$ (and thus, $AD = Q_0$ is also a proper subset of $\bigcap_{0<k<1} Q_k$).

The definition for $f \in Q_p$ is that

$$\sup_{a\in D} \int\int_D |f'(z)|^2 g^p(z,a)\, dx\, dy < \infty.$$

For a fixed point a, we can break D up into two sets,

$$D_1 = \{z : g(z,a) \leq 1\} \quad \text{and} \quad D_2 = \{z : g(z,a) > 1\}.$$

Whether or not the function f is in the space Q_p depends more on the behavior of the integral over the set D_1 rather than the behavior of the integral over the smaller set D_2. Thus, the behavior of the derivative $f'(z)$ on most of D is much more significant than some seemingly "bad" local behavior.

If a function is univalent, or even multivalent, then it is the case that

$$f \in B \iff f \in Q_p \quad \text{for each } p, 0 < p < 1.$$

(Under the same circumstances, it is also the case that

$$f \in B_0 \iff f \in Q_{p,0} \quad \text{for each } p, 0 < p < 1.)$$

As a consequence of this result, we can show that, for each p, $0 < p < 1$, the space $Q_{p,0}$ is a proper subspace of the space Q_p. As an example, which will work for all $p > 0$, let f be a conformal mapping of the unit disk D onto the region $W = \{z = x + iy : -1 < y < 1\}$. Then f is a Bloch function, so $f \in Q_p$ for each p, $0 < p < 1$. However, $f \notin B_0 = Q_{1,0}$, so f is not in any of the spaces $Q_{p,0}$, $0 < p < 1$. (For a conformal mapping to be in the space B_0, requires that, as $f(z)$ becomes large, the radius of any disk in $f(D)$ centered at $f(z)$ must approach zero. But in the example indicated, there are disks of radius 1 contained in $f(D) = W$ which are arbitrarily far from the origin.)

3. Some related spaces

Definition 3.1. We say that $f \in B^\alpha$ if

$$\sup\{|f'(z)|(1 - |z|^2)^\alpha : z \in D\} < \infty.$$

Similarly, we say that $f \in B_0^\alpha$ if

$$\lim_{|z|\to 1} |f'(z)|(1 - |z|^2)^\alpha = 0.$$

The spaces $\{B^\alpha\}$ are called *α-Bloch* spaces. Note that $B^1 = B$, the Bloch space, and $B_0^1 = B_0$, the little Bloch space. It is obvious that, for $0 < \alpha < \delta < 1$, $B_0^\alpha \subseteq B^\alpha \subseteq B_0^\delta \subseteq B^\delta$. Further, for $0 \le \alpha \le \frac{1}{2}$, it is clear that $B^\alpha \subset H^\infty \cap DA$, and, for $0 \le \alpha < 1$, $B^\alpha \subset H^\infty$. We note that the function $f(z) = \log(1 - z)$ is in the space Q_p for each p, $0 < p < 1$, but $f \notin B^\alpha$ for any $\alpha < 1$. Thus, $B^\alpha \neq Q_p$ for $p > 0$ and $\alpha < 1$. Further, $B^\alpha \subset H^\infty$ for each $\alpha < 1$, while DA contains unbounded functions, so $B^\alpha \neq DA$ for $\alpha < 1$. However, for $\frac{1}{2} < \alpha < 1$, it has been shown in [4] that $B^\alpha \subset Q_\delta$ for $\delta > 2\alpha - 1$ but B^α is not contained in $Q_{2\alpha-1}$. Thus, the spaces $\{B^\alpha : 0 < \alpha < 1\}$ form a collection of proper subsets of H^∞ but that $B^1 = B$, a class much larger that H^∞. Note that $H^\infty \subset \text{BMOA} \subset B$, where each containment here is proper, so the collection $\{B^\alpha\}$ skips right over H^∞ and over BMOA as α "jumps" from less than 1 to 1 itself. Further, the function $f(z) = \sum \frac{1}{n^2} z^{2^n}$ is a bounded function which is in none of the Q_p spaces since $\{\frac{2^{n(1-p)}}{n^4}\}$ is unbounded for $0 < p < 1$. This also means that f is in none of the spaces $\{B^\alpha\}$, for $\alpha < 1$.

Some variants of the α-Bloch spaces are the so called Besov spaces (see [1]). For $1 < p < \infty$, we say that

$$f \in (\text{Bes})_p \Longleftrightarrow \iint_D |f'(z)|^p (1 - |z|)^{p-2} dx\, dy < \infty.$$

To link $(\text{Bes})_p$ to the space B^p, we note that the integral in the definition of $(\text{Bes})_p$ is approximately

$$\iint_D (|f'(z)|^p (1 - |z|))^p d\sigma,$$

where $d\sigma = (1-|z|^2)^{-2}dx\,dy$ denotes the non-Euclidean hyperbolic element of area. Thus, for a function f to be in the space $(Bes)_p$, it is necessary that the expression $(|f'(z)|(1-|z|^2))^p$ be small for most of the hyperbolic area of the disk D. For $1<p<\infty$, $(Bes)_p \subseteq \text{VMOA} = Q_{1,0}$, while for $2<p<\infty$, $(\text{Bes})_p \subseteq Q_{q,0}$ for $(p-2)/p<q<1$. The lower bound on q here is best possible, since $(\text{Bes})_p \not\subset Q_{(p-2)/p,0}$. Further, we note that for $1<p<q$, we have $(\text{Bes})_p \subseteq (\text{Bes})_q$. Thus, if we apply the results above, we see that

$$(\text{Bes})_p \subset Q_{r,0} \text{ for } 0<r<1 \text{ and } 1<p<2.$$

It is worth noting that $(Bes)_2 = AD$.

One additional class of spaces fits into this circle of ideas. If $f(z) = \sum a_n z^n$, we say that $f \in D_\alpha$ if $\sum n^\alpha |a_n|^2 < \infty$. It is easily seen that

$$f \in D_\alpha \Longleftrightarrow \iint_D |f'(z)|^2(1-|z|^2)^{1-\alpha}dx\,dy < \infty.$$

We note here that $D_0 = H^2$, the Hardy space, while $D_1 = AD$ and $D_q \subseteq D_p$ for $0 \le p < q$. (Note that $\text{BMOA} \subset H^2$.) Here are some of the known results: $B^\beta \subset D_\alpha$ for $0<\beta<1-(\alpha/2)$ and $D_\alpha \subset B^\beta$ for $(3-\alpha)/2<\beta$. For $1-(\alpha/2)<\beta<(3-\alpha)/2$, neither B^β nor D_α is contained in the other.

Define $M(D_\alpha) = \{g : g \circ f \in D_\alpha \text{ whenever } f \in D_\alpha\}$. For $\alpha=0$, $D_\alpha = H^2$ and $M(H^2) \subset B$. We then have that $B^\beta \subset M(D_\alpha)$ for $0<\beta<1-\alpha/2$. Otherwise, there is no containment in either direction. Also, $M(D_\alpha) \subset Q_{1-\alpha}$ for $0 \le \alpha < 1$. Finally, we have mentioned already the situation where a function is univalent. If f is either multivalent or univalent, then

$$f \in B \Longleftrightarrow f \in Q_p \text{ for each } p, 0<p<1,$$

and

$$f \in B_0 \Longleftrightarrow f \in Q_{p,0} \text{ for each } p, 0<p<1.$$

As a framework for dealing with several of these spaces simultaneously, R. Zhao [6] introduced a space as follows: for $p>0$, $q>-2$, and $s \ge 0$, we say that $f \in F(p,q,s)$ if

$$\sup_{a\in D} \iint_D |f'(z)|^p(1-|z|)^q g^s(z,a)\,dx\,dy < \infty,$$

and we say that $f \in F_0(p,q,s)$ if

$$\lim{}_{|a|\to 1} \iint_D |f'(z)|^p(1-|z|)^q g^s(z,a)\,dx\,dy = 0.$$

Clearly, $F(2,0,s) = Q_s$, $F_0(2,0,s) = Q_{s,0}$, $F(p,p-2,0) = (\text{Bes})_p$ and $F(2,1-\alpha,0) = D_\alpha$. For a much less obvious equality, Zhao proved that for $\alpha>0$, $p>0$, $s>1$, it is true that $B^\alpha = F(p,p\alpha-2,s)$.

Using the fact that $\text{BMOA} = F(2,0,1)$, Zhao defined $\text{BMOA}(\alpha,p) = F(p, p\alpha-2,1)$. (Note that this means $\text{BMOA} = \text{BMOA}(1,2)$). Zhao also proved that, for $\alpha<\delta$, $p_1>0$, $p_2>0$, $\text{BMOA}(\alpha,p_1) \subset \text{BMOA}(\delta,p_2)$, and the inclusion is a strict inclusion. If $\alpha > 0$ and $p_1 < p_2$, then we also have $\mathrm{BMOA}(\alpha, p_1) \subset \mathrm{BMOA}(\alpha, p_2)$, where the inclusion is strict.

General containments for the spaces $F(p, q, s)$ are unclear when $s < 1$, since, for example, the Bloch space $B = F(3, 2, s)$ for $s > 1$, while the Hardy space $H^2 = D_0 = F(2, 1, 0)$ but neither of these spaces is contained in the other.

Since this is a survey and not a complete listing of results, we will now leave the analytic case and turn our attention to the meromorphic case.

4. Some classes of meromorphic functions

From this point forward, we let f be meromorphic. We say that $f \in Q_p^{\#}$ if

$$\sup_{a \in D} \iint_D (f^{\#}(z))^2 g^p(z, a)\, dx\, dy < \infty,$$

where $f^{\#}(z) = |f'(z)|/(1 + |f(z)|^2)$ denotes the spherical derivative of f. Also, we say that $f \in Q_{p,0}^{\#}$ if

$$\lim_{|a| \to 1} \iint_D (f^{\#}(z))^2 g^p(z, a)\, dx\, dy = 0.$$

Corresponding to the analytic space $\mathrm{BMOA} = Q_1$ is the class $Q_1^{\#} = UBC$, the so called class of meromorphic functions of uniformly bounded characteristic. Also, $Q_{1,0}^{\#}$ is known as UBC_0. UBC can be characterized as the collection of those meromorphic functions f such that

$$\sup_{a \in D} T(r, f_a) < \infty,$$

where $f_a(z) = f((a - z)/(1 - \bar{a}z))$ and $T(r, g)$ denotes the Nevanlinna characteristic function of g. By using the Ahlfors-Shimizu form of the Nevanlinna Characteristic, we get the equation

$$T(r, f) = \int_0^r \frac{A(t, f)}{t} dt,$$

where

$$A(t, f) = \iint_{|z| \le t} (f^{\#}(z))^2 dx\, dy.$$

To characterize BMOA by similar concepts, let

$$A^*(t, f) = \iint_{|z| \le t} |f'(z)|^2 dx\, dy$$

and let

$$T^*(r, f) = \int_0^r \frac{A^*(t, f)}{t} dt.$$

An analytic function, $f \in \text{BMOA} \Longleftrightarrow \sup_{a\in D} T^*(r, f_a) < \infty$. These characterizations of UBC and BMOA by using characteristic functions are due to S. Yamashita [5]. We say that a function f is a *normal function*, and write

$$f \in N \Longleftrightarrow \sup_{z\in D} f^\#(z)(1-|z|^2) < \infty.$$

Similarly, we say

$$f \in N_0 \Longleftrightarrow \lim_{|z|\to 1} f^\#(z)(1-|z|^2) = 0.$$

The class of normal functions has been extensively studied over the past 40 years.

We have already mentioned a version of the following result.

Theorem 4.1 (Aulaskari-Lappan [2]). *If f is meromorphic in D, then*

$$f \in N \Longleftrightarrow f \in Q_p^\# \text{ for each } p > 1,$$

and

$$f \in N_0 \Longleftrightarrow f \in Q_{p,0}^\# \text{ for each } p > 1.$$

It is worth noting that, although many of the results in the meromorphic case are basically the same as in the analytic case, in several cases the proofs are different due to the fact that the spherical derivative $f^\#$ is not additive in any reasonable sense—it does not satisfy the triangle inequality, for example. Thus, in the meromorphic case, in general, the collections of functions described are usually classes rather than spaces, since there is usually no additive closure.

Analogous to the analytic case, we have the following containments: For $0 \le p < q$, $Q_p^\# \subset Q_q^\#$ and $Q_{p,0}^\# \subset Q_{q,0}^\#$, and, for $p > 0$, $Q_{p,0}^\# \subset Q_p^\#$. Each of these containments is a strict containment. Also, for $0 \le p < 1$, $Q_p^\# \subset \bigcap_{q>p} Q_q^\#$ and, for $0 < p < 1$, $Q_{p,0}^\# \subset \bigcap_{q>p} Q_{q,0}^\#$, $\bigcup_{0<q<p} Q_q^\# \subset Q_p^\#$, and $\bigcup_{0<q<p} Q_{q,0}^\# \subset Q_{p,0}^\#$. All of these containments are strict. As in the analytic case, we note that, for $0 < p < 1$, $DS \subset Q_{p,0}^\# \subset Q_p^\# \subset UBC \subset N$, and $Q_{p,0}^\# \subset UBC_0 \subset N_0$, where DS is the spherical Dirichlet space

$$DS = \{f : \iint_D (f^\#)^2 dx\, dy < \infty\}.$$

Further, since $f^\#(z) \le |f'(z)|$, we have, for $0 < p < 1$, that $Q_p \subset Q_p^\#$ and $DA \subset Q_{p,0} \subset Q_{p,0}^\#$, and an example similar to those discussed before shows that, for each $p > 0$, there exists a function in $Q_p^\#$ which is not in UBC_0.

We can define the meromorphic Besov spaces by

$$f \in (\text{Bes})_p^\# \Longleftrightarrow \iint_D (f^\#(z))^p (1-|z|^2)^{p-2} dx\, dy < \infty.$$

As in the analytic case, we get, for $p \ge 2$,

$$(\text{Bes})_p^\# \subset \bigcap_{(p-2/p<q<1} Q_{q,0}^\#,$$

where the containment is strict and best possible. Also, if $f \in (\mathrm{Bes})_p^\#$ for some p with $1 < p \leq 2$, then we have

$$f \in N_0 \Longleftrightarrow f \in \bigcap_{0<q<1} Q_{q,0}^\#.$$

(If $p > 2$, then $(\mathrm{Bes})_p^\# \subset N_0$ is a consequence of a previous result of Aulaskari and Lappan.) Also, it is easy to see that $(\mathrm{Bes})_p \subset (\mathrm{Bes})_p^\#$ and that $(\mathrm{Bes})_2^\# = DS$.

We say that f is an *α-normal function* and denote

$$f \in N^\alpha \Longleftrightarrow \sup_{z\in D}(1-|z|^2)^\alpha f^\#(z) < \infty,$$

and

$$f \in N_0^\alpha \Longleftrightarrow \lim_{|z|\to 1}(1-|z|^2)^\alpha f^\#(z) = 0.$$

Here, $N^1 = N$, the class of all normal functions and $N_0^1 = N_0$. As in the analytic case, we have

$$\text{(1) for } 0 \leq \alpha < 1/2,\ N^\alpha \subset \bigcap_{0<p\leq 1} Q_{p,0}^\#,$$

and

$$\text{(2) for } 1/2 \leq \alpha < 1,\ N^\alpha \subset \bigcap_{2\alpha-1<p\leq 1} Q_{p,0}^\#.$$

These inclusions are both strict and best possible, in the sense that $N^\alpha \not\subset Q_{2\alpha-1,0}^\#$. In fact, $B^\alpha \subset N^\alpha$ and $B^\alpha \not\subset Q_{2\alpha-1,0}^\#$

There are a large number of open problems within the circle of ideas we have described. One direction in which there are a number of basic open problems is the study of Q_p spaces on Riemann surfaces.

References

[1] R. Aulaskari and G. Csordas, *Besov spaces and $Q_{q,0}$ spaces,* Acta Sci. Math. **60,** (1995), 31–48.

[2] R. Aulaskari and P. Lappan, *Criteria for an analytic function to be Bloch and a harmonic normal or meromorphic function to be normal,* in "Complex Analysis and its Applications," 136–146, Pitman Research Notes in Mathematics Series, **305,** Longman, Harlow, 1995.

[3] R. Aulaskari, J. Xiao, and R. Zhao, *On subspaces and subsets of BMOA and UBC,* Analysis **15,** (1995), 101–121.

[4] R. Aulaskari, P. Lappan, J. Xiao, and R. Zhao, *On α-Bloch spaces and multipliers of Dirichlet spaces,* J. Math. Anal. Appl. **209,** (1997), 103–121.

[5] S. Yamashita, *Functions of uniformly bounded characteristic,* Ann. Acad. Sci. Fenn. Ser. A I Math. **17,** (1982), 349–367.

[6] R. Zhao, *On a general family of function spaces,* Ann. Acad. Sci. Fenn. Math. Dissertationes, **105,** (1996), 1–56.

Department of Mathematics,
Michigan State University,
East Lansing, Michigan 48824, U.S.A.
E-mail address: mthchair@math.msu.edu

Operator Theory:
Advances and Applications, Vol. 114

Hurwitz-type and space-time-type duality theorems for Hermitian Hurwitz pairs

Julian Ławrynowicz and Osamu Suzuki

Abstract. The paper aims at proving general duality theorems of Hurwitz-type and space-time-type for Hermitian Hurwitz pairs (abbreviated as HHP). In particular, we can get dualities between the Minkowski space-time with signature $(+,---)$ and the neutral space with signature $(++,--)$ related to the Penrose theory.

Introduction

It is known that an Hermitian Hurwitz pair (HHP, for short) gives rise to two kinds of Clifford algebras with different signatures [3]. In this paper, we introduce a concept of duality for HHPs and compare these Clifford algebras by use of the duality theorems. We consider the duality theorems for the case of $(\mathbb{C}^4(I_{2,2}), \mathbb{R}^5(I_{2,3}))$ in a quite detailed manner because this pair gives the Minkowski space-time with signature $(+,---)$ and the neutral space with signature $(++,--)$ related to the Penrose theory, and the duality theorems describe relationships between these spaces. Also we may find that there exists a certain unicity of the 4-dimensional space-times with different signatures. In fact, we can see that these spaces can be transformed to each other by the use of dualities.

Here it seems reasonable to explain why we are concerned with HHPs. The Clifford algebra which is obtained from HHPs is a very special one, called Hurwitz algebra. The necessary and sufficient condition for a Clifford algebra to be a Hurwitz algebra is that it admits a #-self adjoint representation; cf. Definition 3 below. This condition is very important from the physical point of view. In fact, it is equivalent to the statement that the Dirac operator which is defined by a Clifford algebra becomes #-self adjoint with respect to some indefinite scalar product. Hence we can find all #-self adjoint operators with the use of HHPs.

In Section 1 we recall some facts on HHPs and in Section 2 we give some criterions on the irreducibility conditions. We see that two kinds of Hurwitz algebras,

1991 *Mathematics Subject Classification.* Primary 32C37; Secondary 46C20.
Key words and phrases. Clifford analysis, Hurwitz pair, duality theorem, space with an indefinite scalar product.
Research of the first author partially supported by the State Committee for Scientific Research (KBN) grant PB 2 P03A 016 10.

which are denoted by $\mathcal{H}_{\sigma-1,s}$ and $\mathcal{H}_{s-1,\sigma}$, can be obtained from an HHP:

$$\begin{array}{ccc} & (\mathbb{C}^p(I_{\sigma',s'}),\ \mathbb{R}^n(I_{\sigma,s})) & \\ \swarrow & & \searrow \\ \mathcal{H}_{\sigma-1,s} & & \mathcal{H}_{s-1,\sigma} \end{array} \tag{1}$$

Hence, we are led to consider relationships between these Hurwitz algebras. This is nothing but the duality theorems formulated and proved in Sections 3-5, connected with a characterization of HHPs in terms of the Fueter and Dirac equations. In fact, we can define a #-linear isomorphism (i.e., $\alpha^{(1)}_{(p)}\# = \#\alpha^{(1)}_{(p)}$) with respect to the metric of signature (σ, s):

$$\alpha^{(1)}_{(p)} : \mathcal{H}_{\sigma-1,s} \to \mathcal{H}_{s-1,\sigma},$$

which is called the *Hurwitz duality* (Theorem 3).

1. Hermitian Hurwitz pairs

In this section we recall some basic facts on Hermitian Hurwitz pairs (HHPs, for short).

1.1. Basic definitions

Definition 1. Let $\mathbb{C}^n(\kappa)$ be an n-dimensional Hermitian vector space with metric of signature (σ', s'), $\sigma' + s' = n$:

$$\langle\langle u, v\rangle\rangle_\kappa = u^*\kappa v, \quad u, v \in \mathbb{C}^n, \quad \kappa = I_{\sigma',s'} := \begin{pmatrix} I_{\sigma'} & 0 \\ 0 & -I_{s'} \end{pmatrix}$$

and $\mathbb{R}^p(\eta)$ a p-dimensional real vector space with symmetric metric of signature (σ, s), $\sigma + s = p$:

$$\langle x, y\rangle_\eta = x^t\eta y, \quad x, y \in \mathbb{R}^p, \quad \eta = I_{\sigma,s} = \begin{pmatrix} I_\sigma & 0 \\ 0 & -I_s \end{pmatrix}.$$

Definition 2. A pair $(\mathbb{C}^n(\kappa), \mathbb{R}^p(\eta))$ is called *Hermitian Hurwitz pair* (HHP, for short) whenever there exists a bilinear mapping $\phi : \mathbb{R}^p \times \mathbb{C}^n \to \mathbb{C}^n$ such that:

(i) $\langle\langle \phi(y, v), \phi(y, v)\rangle\rangle_\kappa = \langle y, y\rangle_\eta \langle v, v\rangle_\kappa$ for $y \in \mathbb{R}^n$, $v \in \mathbb{C}^n$ (*the Hurwitz condition*);

(ii) There is no subspace V of $\mathbb{C}^n$, $\{0\} \subsetneqq V \subsetneqq \mathbb{C}^n$ such that $\phi|_{\mathbb{R}^p\times V} : \mathbb{R}^p \times V \to V$ (*irreducibility condition*).

In this paper we are mainly concerned with the indefinite HHPs $(\mathbb{C}^n(\kappa), \mathbb{R}^p(\eta))$, $s \neq 0$. For $s = 0$ the situation is different and the corresponding results are stated as remarks.

1.2. Hurwitz algebra

We can get two kinds of Clifford algebras from an HHP in the following manner: Let $(\epsilon_1, \epsilon_2, \ldots, \epsilon_n)$ and $(e_1, e_2, \ldots, e_n)$ be the canonical bases of $\mathbb{R}^p$ and $\mathbb{C}^n$, respectively. We define the sequence (C_α) of matrices, $C_\alpha \in M_n(\mathbb{C})$, $\alpha = 1, 2, \ldots, p$, by

$$\phi(\epsilon_\alpha, e_j) = \sum_{k=1}^{p} C^k_{\alpha j} e_k.$$

Then we have

$$\phi(y, v) = (y_1 C_1 + \cdots + y_p C_p)v, \quad y^t = (y_1 \ldots y_p), \quad v \in \mathbb{C}^n.$$

In the following we call ϕ the *Hurwitz mapping* and $C_1, \ldots, C_p$ the *Hurwitz matrices.* Then the condition (i) in Definition 2 reads

$$C^\#_\alpha C_\beta + C^\#_\beta C_\alpha = 2\eta_{\alpha\beta} I_n, \tag{2}$$

where $C^\#_\alpha = \kappa C^*_\alpha \kappa^{-1}$, $(\eta_{\alpha\beta}) = \eta$, $I_n = I_{n,0}$.

R e m a r k 1. $C^\#_\alpha C_\alpha = I_n$, $\alpha = 1, 2, \ldots, \sigma$, $C^\#_\alpha C_\alpha = -I_n$, $\alpha = \sigma+1, \ldots, \sigma+s$.

R e m a r k 2. In the case where $s = 0$, we have $n = \sigma'$, $s' = 0$. Hence we can see that $C^\# = C^*$, $C^* = \bar{C}^t$.

We define $S^{(\rho)}_\alpha$, $\alpha \neq \rho$, $\alpha = 1, 2, \ldots, p$, for a fixed ρ by

$$S^{(\rho)}_\alpha = iC^{-1}_\rho C_\alpha.$$

Then (2) implies

$$\begin{aligned} &\text{(i)}\ \ S^{(\rho)}_\alpha S^{(\rho)}_\beta + S^{(\rho)}_\beta S^{(\rho)}_\alpha = 2\eta_{\rho\rho} I_n, \\ &\text{(ii)}\ \ S^{(\rho)\#}_\alpha = S^{(\rho)}_\alpha, \end{aligned} \qquad 1 \le \alpha, \beta \le p, \quad \alpha \neq \rho, \beta \neq \rho. \tag{3}$$

Hence we have obtained Clifford algebras $C^{(\sigma-1,s)}$ and $C^{(s-1,\sigma)}$ from the same HHP, which admit the special linear representation (3).

D e f i n i t i o n 3. A Clifford algebra which admits a representation with the condition (ii) in (3) is called a *Hurwitz algebra* and denoted by $\mathcal{H}_{\sigma-1,s}$ or $\mathcal{H}_{s-1,\sigma}$ whenever it is central (cf. Theorem 2 below).

1.3. Existence theorem

In analogy to [12], Theorem 1, we can prove

T h e o r e m 1. (i). *When a Hurwitz pair is given, we can construct the corresponding HHP. The converse is also true.*

(ii) $(\mathbb{C}^n(I_{\sigma',s'}), \mathbb{R}^p(I_{\sigma,s}))$ *is an HHP if and only if* $(\sigma, s) \neq$ *(odd, odd). In this case we have*

$$n = 2^{[\frac{1}{2}p - \frac{1}{2}]}, \qquad \sigma' = s' = \frac{1}{2}n,$$

where $\mathbb{R} \ni x \mapsto [x]$ *denotes the greatest integer function.*

R e m a r k 3. In the case where $s=0$, we can see that (i) still holds. Yet (ii) holds with the replacement of $\sigma' = s' = \frac{1}{2}n$ by $\sigma' = n$ and $s' = 0$.

1.4. Basic construction

We can give an explicit construction method in an inductive manner. Suppose that generators $S_1, \ldots, S_{\sigma+s}, \sigma+s \equiv 1 \mod 2)$ of $\mathcal{H}_{\sigma,s}$ are given. By the condition (ii) of (3), we have

$$S_\alpha = \begin{pmatrix} A_\alpha & iB_\alpha \\ iB_\alpha^* & -D_\alpha \end{pmatrix}, \quad \begin{array}{l} A_\alpha, B_\alpha, D_\alpha \in M_n(\mathbb{C})^n, \\ A_\alpha^* = A_\alpha, \; D_\alpha^* = D_\alpha, \quad \alpha = 1, 2, \ldots, \sigma+s. \end{array}$$

Then we can get the Hurwitz algebra $\mathcal{H}_{\sigma,s+2}$ as follows:

$$\widetilde{S}_\alpha = \left(\begin{array}{cc|cc} A_\alpha & 0 & 0 & iB_\alpha \\ 0 & D_\alpha & iB_\alpha^* & 0 \\ - & - & - & - \\ 0 & iB_\alpha & -A_\alpha & 0 \\ iB_\alpha^* & 0 & 0 & -D_\alpha \end{array}\right), \quad \alpha = 1, 2, \ldots, p, \tag{4}$$

$$\widetilde{S}_{p+1} = \begin{pmatrix} 0 & I_n \\ -I_n & 0 \end{pmatrix}, \quad \widetilde{S}_{p+2} = \begin{pmatrix} 0 & I_{\frac{1}{2}n} \otimes \sigma_3 \\ iI_{\frac{1}{2}n} \otimes \sigma_3 & 0 \end{pmatrix},$$

where $\sigma_3 = \left(\begin{smallmatrix} 1 & 0 \\ 0 & -1 \end{smallmatrix}\right)$. Another representation of (4) is given in

L e m m a 1. *If $\sigma \equiv 1 \pmod 2$ and $s \equiv 0 \pmod 2$, the Hurwitz algebra $\mathcal{H}_{\sigma,s}$ exists. Denote by $S_1, S_2, \ldots, S_{\sigma+s}$ its generators. Then we have:*

(i) $\{\tilde{S}_\alpha\}_{\alpha=1,2,\ldots,\sigma+s+1}$ *which is defined by*

$$\tilde{S}_\alpha = \begin{pmatrix} S_\alpha & 0 \\ 0 & -S_\alpha \end{pmatrix}, \qquad \alpha = 1, 2, \ldots, \sigma+s,$$
$$\tilde{S}_{\sigma+1} = \begin{pmatrix} 0 & I_n \\ -I_n & 0 \end{pmatrix}, \tag{5}$$

generates the Hurwitz algebra $\mathcal{H}_{\sigma,s+1}$ with respect to the metric $I'_{\frac{1}{2}n} \otimes I'_1$, $I'_k = \begin{pmatrix} I_k & 0 \\ 0 & -I_k \end{pmatrix}$.

(ii) $\{\widehat{S}_\alpha\}_{\alpha=1,2,\ldots,\sigma+s+1}$ *which is defined by*

$$\widehat{S}_\alpha = \begin{pmatrix} 0 & S_\alpha \\ -S_\alpha & 0 \end{pmatrix}, \qquad \alpha = 1, 2, \ldots, \sigma+s,$$

$$\widehat{S}_{\sigma+s+1} = \begin{pmatrix} I_n & 0 \\ 0 & -I_n \end{pmatrix},$$

generates $\mathcal{H}_{s+1,\sigma}$ with respect to the same metric.

P r o o f. Relations (3) may be checked by direct calculations and thus omitted here.

R e m a r k 4. In the case where $s=0$, we have also the basic construction:

$$\tilde{S}_\alpha = \begin{pmatrix} S_\alpha & 0 \\ 0 & -S_\alpha \end{pmatrix}, \qquad \alpha = 1,2,\ldots,\sigma,$$

$$\tilde{S}_{\sigma+1} = \begin{pmatrix} 0 & I_n \\ I_n & 0 \end{pmatrix}, \quad \tilde{S}_{\sigma+2} = \begin{pmatrix} 0 & iI_n \\ -iI_n & 0 \end{pmatrix},$$

where S_α are generators of $\mathcal{H}_{\sigma,0}$.

1.5. Irreducibility condition

We see that the Hurwitz algebras which are obtained above are isomorphic to the corresponding full matrix algebras. Hence the irreducibility condition (ii) in Definition 2 can be checked. Some criterions for the irreducibility will be given in Section 2.

1.6. Low-dimensional Hurwitz algebras

We give generators of low-dimensional Hurwitz algebras. We denote the Pauli matrices by

$$\sigma_1 = \begin{pmatrix} 0 & 1 \\ 1 & 0 \end{pmatrix}, \quad \sigma_2 = \begin{pmatrix} 0 & -i \\ i & 0 \end{pmatrix}, \quad \sigma_3 = \begin{pmatrix} 1 & 0 \\ 0 & -1 \end{pmatrix}. \tag{6}$$

E X A M P L E 1: $n=2$, $p=3$, $(\mathbb{C}^2(I_{1,1}), \mathbb{R}^3(I_{1,2}))$ with

$$\begin{aligned} \mathcal{H}_{1,1}: \ S_2^{(1)} &= \sigma_3, \quad S_3^{(1)} = i\sigma_2, \\ \mathcal{H}_{0,2}: \ S_1^{(3)} &= i\sigma_1, \quad S_2^{(3)} = i\sigma_2. \end{aligned} \tag{7}$$

E X A M P L E 2: $n=2$, $p=4$, $(\mathbb{C}^2(I_{1,1}), \mathbb{R}^4(I_{2,2}))$ with

$$\mathcal{H}_{1,2}: \ S_1^{(3)} = i\sigma_1, \qquad S_2^{(3)} = i\sigma_2, \qquad S_3^{(3)} = \sigma_3. \tag{8}$$

R e m a r k 5. In this case and, generally, in the case $\sigma = s$, we have only one Hurwitz algebra.

R e m a r k 6. $\mathcal{H}_{2,0}: S_2^{(1)} = \sigma_1$, $S_3^{(1)} = \sigma_2$, $S_4^{(1)} = \sigma_3$.

E x a m p l e 3: $n = 4$, $p = 5$, $(\mathbb{C}^4(I_{2,2}), \mathbb{R}^5(I_{2,3}))$ with

$$\mathcal{H}_{1,3} : S_2^{(1)} = \begin{pmatrix} I_2 & 0 \\ 0 & -I_2 \end{pmatrix}, \quad S_3^{(1)} = \begin{pmatrix} 0 & \sigma_1 \\ -\sigma_1 & 0 \end{pmatrix},$$

$$S_4^{(1)} = \begin{pmatrix} 0 & \sigma_2 \\ -\sigma_2 & 0 \end{pmatrix}, \quad S_5^{(1)} = \begin{pmatrix} 0 & \sigma_3 \\ -\sigma_3 & 0 \end{pmatrix},$$

$$\mathcal{H}_{2,2} : S_1^{(5)} = \begin{pmatrix} 0 & \sigma_3 \\ -\sigma_3 & 0 \end{pmatrix}, \quad S_2^{(5)} = \begin{pmatrix} 0 & i\sigma_3 \\ \sigma_3 & 0 \end{pmatrix}, \tag{9}$$

$$S_3^{(5)} = \begin{pmatrix} \sigma_2 & 0 \\ 0 & \sigma_2 \end{pmatrix}, \quad S_4^{(5)} = \begin{pmatrix} \sigma_1 & 0 \\ 0 & \sigma_1 \end{pmatrix},$$

E x a m p l e 4: $n = 4$, $p = 5$, $(\mathbb{C}^4(I_{2,2}), \mathbb{R}^5(I_{3,2}))$ with

$$\mathcal{H}_{3,1} : S_2^{(1)} = \begin{pmatrix} \sigma_1 & 0 \\ 0 & -\sigma_1 \end{pmatrix}, \quad S_3^{(1)} = \begin{pmatrix} \sigma_2 & 0 \\ 0 & -\sigma_2 \end{pmatrix},$$

$$S_4^{(1)} = \begin{pmatrix} \sigma_3 & 0 \\ 0 & -\sigma_3 \end{pmatrix}, \quad S_5^{(1)} = \begin{pmatrix} 0 & I_2 \\ -I_2 & 0 \end{pmatrix},$$

$$\mathcal{H}_{0,4} : S_1^{(5)} = \begin{pmatrix} 0 & I_2 \\ -I_2 & 0 \end{pmatrix}, \quad S_2^{(5)} = \begin{pmatrix} 0 & i\sigma_1 \\ i\sigma_1 & 0 \end{pmatrix}, \tag{10}$$

$$S_3^{(5)} = \begin{pmatrix} 0 & i\sigma_2 \\ i\sigma_2 & 0 \end{pmatrix}, \quad S_4^{(5)} = \begin{pmatrix} 0 & i\sigma_3 \\ i\sigma_3 & 0 \end{pmatrix}.$$

R e m a r k 7. Comparing $\mathcal{H}_{1,3}$ with $\mathcal{H}_{0,4}$ we see only the difference in the multiplication by an imaginary number.

R e m a r k 8. $\mathcal{H}_{0,4} : S_1^{(1)} = \begin{pmatrix} 0 & \sigma_1 \\ \sigma_1 & 0 \end{pmatrix}, \ S_2^{(1)} = \begin{pmatrix} 0 & \sigma_2 \\ \sigma_2 & 0 \end{pmatrix},$

$$S_3^{(1)} = \begin{pmatrix} 0 & \sigma_3 \\ \sigma_3 & 0 \end{pmatrix}, \quad S_4^{(1)} = \begin{pmatrix} I_2 & 0 \\ 0 & -I_2 \end{pmatrix}.$$

The examples of generators given above provide some representations. They are unique up to U, $U^{\#}U = 1$ or -1 [3]. Another representation is used for the space-time duality (see Theorem 4 below).

2. Irreducibility criterions

Here we give some criterions on the irreducibility conditions for Hurwitz mappings. This completes the discussion in [3], where the irreducibility conditions were not studied in detail. The results of this section are more or less known in several contexts, but we are going to give all the proofs for the sake of completeness.

At first we introduce the concept of a pre-Hurwitz algebra.

DEFINITION 4. Let $C^{(\sigma,s)}$ be a Clifford algebra with the signature (σ, s). $C^{(\sigma,s)}$ is called *pre-Hurwitz algebra*, which is denoted by $\widehat{\mathcal{H}}_{\sigma,s}$ if there is a representation $\psi : C^{(\sigma,s)} \to M_n(\mathbb{C})$ satisfying the conditions:

(i) $\psi(S_\alpha)^{\#} = \psi(S_\alpha)$, $\alpha = 1, 2, \ldots, \sigma + s$,

(ii) $\psi(S_\alpha)\psi(S_\beta) + \psi(S_\beta)\psi(S_\alpha) = 2\eta_{\alpha\beta} I_n$,

where S_α, $\alpha = 1, 2, \ldots, \sigma + s$, are generators of $C^{(\sigma,s)}$,

$$(\eta_{\alpha\beta}) = \begin{pmatrix} I_\sigma & 0 \\ 0 & -I_s \end{pmatrix},$$

and $\#$ implies an adjoint operator with respect to a certain (indefinite) metric on $\mathbb{C}^n$.

Moreover, under the condition:

(iii) $\psi(C_{\sigma,s}) \otimes \mathbb{C}$ is central, i.e.,

$$(\psi(C_{\sigma,s}) \otimes \mathbb{C})' = \{cI_n : c \in \mathbb{C}\},$$

$C^{(\sigma,s)}$ is called *Hurwitz algebra* and denoted by $\mathcal{H}_{\sigma,s}$. With the above notation, we prove the following.

Theorem 2. *Let $\widehat{\mathcal{H}}_{\sigma,s}$ be a pre-Hurwitz algebra with generators S_α, $\alpha = 1, 2, \ldots, p$, $p = \sigma + s$, having a representation $\psi : \widehat{\mathcal{H}}_{\sigma,s} \to M_n(\mathbb{C})$. Then the following conditions are equivalent:*

(i) $\phi : \mathbb{R}^{p+1} \times \mathbb{C}^n \to \mathbb{C}^n$, $\phi = x_0 I_n + \sum_{\alpha=1}^{p} x_\alpha \mathbf{S}_\alpha$,

where $\mathbf{S}_\alpha = \phi(S_\alpha)$, $\alpha = 1, 2, \ldots, p$, is irreducible;

(ii) *$\widehat{\mathcal{H}}_{\sigma,s}$ is the Hurwitz algebra $\mathcal{H}_{\sigma,s}$, i.e.,*

$$(\psi(\mathcal{H}_{\sigma,s}) \otimes \mathbb{C})' = \{cI_n : c \in \mathbb{C}\};$$

(iii) *$\widehat{\mathcal{H}}_{\sigma,s} \otimes \xrightarrow{\psi} \to M_n(\mathbb{C})$ is a bijection.*

Remark 9. The above equivalence does not hold when $\phi(\widehat{\mathcal{H}}_{\sigma,s}) \otimes \mathbb{C}$ is replaced with $\phi(\widehat{\mathcal{H}}_{\sigma,s})$ itself. The simplest example is as follows:

EXAMPLE 5. For

$$S_1 = \begin{pmatrix} 1 & 0 \\ 0 & -1 \end{pmatrix}, \ S_2 = \begin{pmatrix} 0 & 1 \\ -1 & 0 \end{pmatrix}; \ S_1' = \begin{pmatrix} 1 & 0 \\ 0 & -1 \end{pmatrix}, \ S_2' = \begin{pmatrix} 0 & i \\ i & 0 \end{pmatrix},$$

we see that both $\{S_1, S_2\}$ and $\{S_1', S_2'\}$ generate $\widehat{\mathcal{H}}_{1,1}$, where $\widehat{\mathcal{H}}_{1,1} = \mathcal{H}_{1,1}$. Although they do not constitute $M_2(\mathbb{C})$, $\mathcal{H}_{1,1} \otimes \mathbb{C} = M_2(\mathbb{C})$ holds. Also we notice that $(\mathcal{H}_{1,1})' = \mathbb{C}$.

The non-trivial part of the proof is (i) $\Longrightarrow$ (iii). We need two lemmas.

Lemma 2. *A representation $\psi : C_{\sigma,s} \to M_n(\mathbb{C})$ is reducible if and only if*

(i) $n = 2^{[\frac{1}{2}p-\frac{1}{2}]}$, $\sigma + s = p$, *and*

(ii) ψ *is equivalent to the basic construction.*

Proof. Let $\psi : \mathcal{H}_{\sigma,s} \to M_n(\mathbb{C})$ be an arbitrary representation, which may be, in general, reducible. When $s \geq 2$, we may assume that

$$S_{\sigma+s-1} = \begin{pmatrix} 0 & I_{\frac{1}{2}n} \\ -I_{\frac{1}{2}n} & 0 \end{pmatrix}, \quad S_{\sigma+s} = \begin{pmatrix} 0 & iI_{\frac{1}{2}n} \\ iI_{\frac{1}{2}n} & 0 \end{pmatrix},$$
$$S_\alpha = \begin{pmatrix} S'_\alpha & 0 \\ 0 & -S'_\alpha \end{pmatrix}, \quad \alpha = 1, 2, \ldots, \sigma + s + 1, \tag{11}$$

by the use of a similar transformation U with $U^{\#}U = 1$ or -1; cf. (4). Then we obtain a pre-Hurwitz algebra $\mathcal{H}_{\sigma,s-2} = \{S'_\alpha\}$ with the identity representation $\psi' : \mathcal{H}_{\sigma,s-2} \to M_{\frac{1}{2}n}(\mathbb{C})$. In this manner, we can reduce the representation to $\{\mathcal{H}_{\sigma,0}, \psi\}$ or $\{\mathcal{H}_{\sigma,1}, \psi'\}$ or $\{\mathcal{H}_{0,2}, \psi'\}$. The first representation can be reduced to $\{\mathcal{H}_{2,0}, \psi\}$ in a similar manner with a replacement of (11) by

$$S_{\sigma-1} = \begin{pmatrix} 0 & I_{\frac{1}{2}n} \\ I_{\frac{1}{2}n} & 0 \end{pmatrix}, \quad S_{\sigma} = \begin{pmatrix} 0 & iI_{\frac{1}{2}n} \\ -iI_{\frac{1}{2}n} & 0 \end{pmatrix},$$
$$S_\alpha = \begin{pmatrix} S'_\alpha & 0 \\ 0 & -S'_\alpha \end{pmatrix}, \quad \alpha = 1, 2, \ldots, \sigma. \tag{12}$$

The second representation is reduced to a reduction of $\mathcal{H}_{0,\sigma}$ by use of the duality theorem:

$$\mathcal{H}_{\sigma,1} \cong \mathcal{H}_{0,\sigma+1}, \quad \sigma \equiv 1 \pmod 2.$$

Hence we obtain $\{\mathcal{H}_{2,0}, \psi\}$ and $\{\mathcal{H}_{0,2}, \psi'\}$, as desired. By (11) or (12) we can see that these representations are irreducible if and only if $n = 2$, and this proves the assertion.

Lemma 3. *Let $\mathcal{H}_{\sigma,s}$ be a Hurwitz algebra. Then $\mathcal{H}_{\sigma,s} \otimes \mathbb{C} = M_n(\mathbb{C})$.*

We prove Lemma 3 in two steps.

Proof (first step). By the use of Lemma 2 we may assume that the representation is given by the basic construction. We prove Lemma 3 by induction. When $\sigma + s = 2$, we see that

$$\mathcal{H}_{2,0} = \{\sigma_1, \sigma_2\}, \quad \mathcal{H}_{1,1} = \{\sigma_3, i\sigma_2\}, \quad \mathcal{H}_{0,2} = \{i\sigma_1, i\sigma_2\}.$$

Since $\sigma_1\sigma_2 = i\sigma_3$, we have $\mathcal{H}_{\sigma,s} \otimes \mathbb{C} \cong M_2(\mathbb{C})$, $\sigma + s = 2$. By the fundamental relation

$$\sigma_1\sigma_2\sigma_3 = iI_2,$$

we can see that $\mathcal{H}_{\sigma,s} \cong M_2(\mathbb{C})$, $\sigma + s = 3$, where the tensoring of $\mathbb{C}$ is no longer needed. Next, by use of the assumption $\mathcal{H}_{\sigma,s} \otimes \mathbb{C} \cong M_n(\mathbb{C})$, $\sigma + s = 2m$ and $\sigma + s = 2m + 1$, $n = 2^m$, we prove that the isomorphism holds for $\mathcal{H}_{\sigma'',s''} \otimes \mathbb{C} \cong M_{2n}(\mathbb{C})$,

$\sigma'' + s'' = 2m+2$ and $2m+3$. The generators of $\mathcal{H}_{\sigma'',s''}$ are assumed to be of the form (11):

$$S_{2m+2} = \begin{pmatrix} 0 & I_n \\ -I_n & 0 \end{pmatrix}, \quad S_{2m+3} = \begin{pmatrix} 0 & iI_n \\ iI_n & 0 \end{pmatrix},$$

$$S_\alpha = \begin{pmatrix} S'_\alpha & 0 \\ 0 & -S'_\alpha \end{pmatrix}, \quad \alpha = 1,\ldots,2m+1, \quad \text{if } s \geq 2.$$

R e m a r k 10. $S_{2m+2} = \begin{pmatrix} 0 & I_n \\ -I_n & 0 \end{pmatrix}, \quad S_{2m+3} = \begin{pmatrix} 0 & iI_n \\ iI_n & 0 \end{pmatrix},$

and S_α, $\alpha = 1,\ldots,2m+1$ are given as above if $s = 0$.

P r o p o s i t i o n 1. *Let $\{S'_\alpha\}$ be generators of $\mathcal{H}_{2m+1}$. Then the following identity holds:*

$$S'_1 \ldots S'_{2m} S'_{2m+1} = cI_n \quad c \in \mathbb{C}.$$

P r o o f. The assertion can be easily checked by use of induction.

R e m a r k 11. The constant c is given by the formulae:

$$c = \begin{cases} 1 & \text{for} \quad 2m+1 \equiv 1 \quad \text{and} \quad 5 \pmod 8, \\ i & \text{for} \quad 2m+1 \equiv 3 \quad \text{and} \quad 7 \pmod 8). \end{cases}$$

P r o o f of Lemma 3 (second step). In order to complete the proof of Lemma 3 it is sufficient to show that $\mathcal{H}_{2m+3} \otimes \mathbb{C} \cong M_{2n}(\mathbb{C})$. This can be easily checked by the use of a suitable decomposition: any $X \in M_n(\mathbb{C})$ can be written as $X = X_e + X_o$, where X_e (resp. X_o) is the product of an even (resp. odd) number of copies of S_α.

R e m a r k 12. By Remark 11 we can see that

$$\mathcal{H}_{\sigma,s} \otimes \mathbb{C} \cong \mathcal{H}_{\sigma,s} \quad \text{for} \quad \sigma + s \equiv 3 \quad \text{and} \quad 7 \pmod 8).$$

With Lemma 3 we can prove our Theorem 2. As remarked before, the non-trivial part of the proof is (i) $\Longrightarrow$ (iii). P r o o f of (i) $\Longrightarrow$ (iii). Let $\phi : \mathbb{R}^p \times \mathbb{C}^n \to \mathbb{C}^n$ be an irreducible mapping. Then we have an irreducible representation $\psi_\phi : C_{\sigma,s} \to M_n(\mathbb{C})$. Hence ψ_ϕ is equivalent to the representation given in the basic construction of Section 1.4 and is isomorphic to $M_n(\mathbb{C})$, which implies (iii).

P r o o f of (iii) $\Longrightarrow$ (ii). Since $(M_n(\mathbb{C}))' \cong \mathbb{C}$, we have the assertion (ii).

P r o o f of (ii) $\Longrightarrow$ (i). Assume that (i) does not hold. Then there is a $\mathbb{C}$-subspace V of $\mathbb{C}^n$, $\{0\} \subsetneqq V \subsetneqq \mathbb{C}^n$, such that every $\mathbf{S}_\alpha$ has the following form with respect to V :

$$\mathbf{S}_\alpha = \begin{pmatrix} \mathbf{S}^{(\alpha)}_{1,1} & \mathbf{S}^{(\alpha)}_{1,2} \\ 0 & \mathbf{S}^{(\alpha)}_{2,2} \end{pmatrix}, \quad \alpha = 1,2,\ldots,p.$$

By use of changes of basis of $\mathbb{C}^n$, we may assume the metric is diagonal. Hence

$$I_{\sigma',s'} = \begin{pmatrix} I_1 & 0 \\ 0 & I_2 \end{pmatrix}.$$

By $S_\alpha^\# = S_\alpha$, $\alpha = 1, 2, \ldots, p$, we have

$$S_\alpha^\# = \begin{pmatrix} I_1 S_{1,1}^{(\alpha)*} I_1 & 0 \\ I_1 S_{1,2}^{(\alpha)*} I_2 & I_2 S_{2,2}^{(\alpha)*} I_{2,2} \end{pmatrix}.$$

Therefore $\mathbf{S}_{1,2}^{(\alpha)} = 0$, $\alpha = 1, 2, \ldots p$, and, consequently,

$$\mathbf{S}_\alpha = \begin{pmatrix} \mathbf{S}_{1,1}^{(\alpha)} & 0 \\ 0 & \mathbf{S}_{2,2}^{(\alpha)} \end{pmatrix}, \qquad \alpha = 1, 2, \ldots, p.$$

By this $\widehat{\mathcal{H}}_{\sigma,s} \otimes \mathbb{C}$ is not central, which is a contradiction, and the proof is completed.

3. Duality theorems

We give two duality theorems for an HHP which play the basic role in this paper (cf. [13]).

3.1. Preliminaries

Let $(\mathbb{C}^n(I_{\sigma',s'}), \mathbb{R}^p(I_{\sigma,s}))$ be an HHP with $s \neq 0$ and let C_α, $\alpha = 1, 2, \ldots, p$, be its Hurwitz matrices. We consider an algebra $\mathcal{A}$ which is generated by

$$C_\alpha^\# C_\beta, \qquad \alpha \leqq \beta, \ \ \alpha, \beta = 1, 2, \ldots, p.$$

We construct $\mathcal{A} \otimes \mathbb{C}$ which will be also denoted by $\mathcal{A}$. Referring to (2), we see that $\mathcal{A}$ is also generated by

$$C_\rho^\# C_\beta, \qquad \beta = 1, 2, \ldots, p$$

for a fixed ρ, $1 \leqq \rho \leqq p$. We have the involution $\sharp : \mathcal{A} \to \mathcal{A}$ with

(i) $\sharp(\xi + \eta) = \sharp\xi + \sharp\eta, \quad \xi, \eta \in \mathcal{A}$,

(ii) $\sharp(c\xi) = \bar{c}\sharp\xi, \quad c \in \mathbb{C}, \ \xi \in \mathcal{A}$,

(iii) $\sharp^2 = \mathrm{id}$.

We get

Proposition 2.

(i) *For a permutation* $\sigma = \begin{pmatrix} \alpha_1, \beta_1, \alpha_2, \beta_2, & \ldots, & \alpha_k \beta_k \\ \alpha_1', \beta_1', \alpha_2', \beta_2', & \ldots, & \alpha_k' \beta_k' \end{pmatrix}$, *we have*

$$C_{\alpha_1}^\# C_{\beta_1} \ldots C_{\alpha_k}^\# C_{\beta_k} = \operatorname{sign} \sigma \ \ C_{\alpha_i'}^\# C_{\beta_i'} \ldots C_{\alpha_k'}^\# C_{\beta_k'}. \tag{13}$$

(ii) $\{C_{\alpha_1}^\# C_{\beta_1} \ldots C_{\alpha_k}^\# C_{\beta_k}, \quad \alpha_1 < \beta_1 < \cdots < \alpha_k < \beta_k\}$ *are linearly independent over* $\mathbb{C}$.

(iii) *An element $\xi \in \mathcal{A}$ can be written uniquely as $\xi = \sum_{k=0}^{n} \xi_k$, where*

$$\begin{aligned} \xi_0 =& C_0 I_n, \\ \xi_k =& \sum_{\alpha_1<\beta_1<\cdots<\alpha_k<\beta_k} \xi_{\alpha_1\beta_1}\ldots\xi_{\alpha_k\beta_k} C_{\alpha_1}^{\#} C_{\beta_1}\ldots C_{\alpha_k}^{\#} C_{\beta_k}, \quad k \geq 1. \end{aligned}$$

(iv) *Elements of the form* (13) *are orthogonal to each other with respect to the indefinite scalar product*

$$\langle \xi, \eta \rangle := \frac{1}{n} \mathrm{Tr}(\xi^{\#} \eta), \qquad \xi, \eta \in \mathcal{A}.$$

(v) *Putting $\Phi(\xi) = \langle \xi, \xi \rangle$, we have*

$$\begin{aligned} \Phi(\xi) &= \sum \eta_{\alpha_1\beta_1}\ldots\eta_{\alpha_k\beta_k} \xi_{\alpha_1\ldots\beta_k} \overline{\xi_{\alpha_1\ldots\beta_k}} \\ \textit{for} \quad \xi &= \sum \xi_{\alpha_1\beta_1\ldots\alpha_k\beta_k} C_{\alpha_1}^{\#} C_{\beta_1}\ldots C_{\alpha_k}^{\#} C_{\beta_k}. \end{aligned} \tag{14}$$

P r o o f s can be done by use of the basic representation which will be given in (15) and may be omitted here.

3.2. Hurwitz-type duality

We proceed to the duality theorems.

T h e o r e m 3 (Hurwitz-type duality theorem). *Let $\mathcal{H}_{\sigma-1,s}$ and $\mathcal{H}_{s-1,\sigma}$ be Hurwitz algebras of an HHP. Hence $(\sigma, s) \neq$ (odd, odd). Moreover, there are $\sharp$-linear isomorphisms $\alpha^{(1)} : \mathcal{A} \to \mathcal{H}_{\sigma-1,s}$ and $\alpha^{(p)} : \mathcal{A} \to \mathcal{H}_{s-1,\sigma}$, which are called basic representations and a $\sharp$-linear isomorphism $\alpha^{(1)}_{(p)} : \mathcal{H}_{\sigma-1,s} \to \mathcal{H}_{s-1,\sigma}$, such that*

$$\begin{array}{ccc} & \mathcal{A} & \\ \alpha^{(1)} \swarrow & & \searrow \alpha^{(p)} \\ & \curvearrowright & \\ \mathcal{H}_{\sigma-1,s} & \xrightarrow{\alpha^{(1)}_{(p)}} & \mathcal{H}_{s-1,\sigma} \end{array} \tag{15}$$

P r o o f (first step). We define $\alpha^{(1)}, \alpha^{(p)}$ as follows. Let $\mathcal{H}_\rho$ be the Hurwitz algebra which is generated by $\{S_\alpha^{(\rho)}\}$, $\alpha = 1, 2, \ldots, p$, $\alpha \neq \rho$. We define $\alpha^{(\rho)} : \mathcal{A} \to \mathcal{H}_\rho$ by the conditions:

$$
(i)\ \alpha^{(\rho)}(C_\alpha^\# C_\beta) = \begin{cases} -i\ \eta_{\rho\rho} S_\beta^{(\rho)}, & \alpha=\rho,\ \beta\neq\rho, \\ i\ \eta_{\rho\rho} S_\alpha^{(\rho)}, & \alpha\neq\rho,\ \beta=\rho, \\ \eta_{\rho\rho} S_\alpha^{(\rho)} S_\beta^{(\rho)}, & \alpha\neq\rho,\ \beta\neq\rho, \\ \eta_{\rho\rho}, & \alpha=\rho,\ \beta=\rho. \end{cases}
$$

$$
\begin{aligned}
&(ii)\ \alpha^{(\rho)}(C_{\alpha_1}^\# C_{\beta_1}\dots C_{\alpha_k}^\# C_{\beta_k}) \\
&\quad = \alpha^{(\rho)}(C_{\alpha_1}^\# C_{\beta_1})\dots\alpha^{(\rho)}(C_{\alpha_k}^\# C_{\beta_k}),
\end{aligned} \tag{16}
$$

(iii) For $\xi = \sum \xi_{\alpha_1\beta_1\dots\alpha_k\beta_k} C_{\alpha_1}^\# C_{\beta_1}\dots C_{\alpha_k}^\# C_{\beta_k}$ we have

$$
\alpha^{(\rho)}(\xi) = \sum \xi_{\alpha_1\beta_1\dots\alpha_k\beta_k}\alpha^{(\rho)}\alpha^{(\rho)} : \mathcal{A}\to\mathcal{H}_{(C_{\alpha_1}^\# C_{\beta_1}\dots C_{\alpha_k}^\# C_{\beta_k})}.
$$

Then we obtain

Proposition 3 (i). $\alpha^{(\rho)}:\mathcal{A}\to\mathcal{H}_\rho$ *is a linear isomorphism.*
(ii) $\alpha^{(\rho)}$ *is a* $\natural$*-linear isomorphism, i.e.,* $\natural\alpha^{(\rho)} = \alpha^{(\rho)}\natural$.

Proof of (i). We choose a linear basis of $\mathcal{A}$, I_p and

$$
C_\rho^\# C_{\alpha_1} C_\rho^\# C_{\beta_1}\dots C_\rho^\# C_{\alpha_k} C_\rho^\# C_{\beta_k},\ \alpha_1<\beta_1<\dots<\alpha_k<\beta_k,
$$
$$
\rho\neq\alpha_j,\beta_\ell,\ j,\ell=1,\dots,k.
$$

Since

$$
\alpha^{(\rho)}(C_\rho^\# C_\beta) = -i\eta_{\rho\rho}S_\beta^{(\rho)},\ \rho\neq\beta,
$$
$$
\alpha^{(\rho)}(C_\rho^\# C_{\alpha_1}\dots C_\rho^\# C_{\beta_k}) = (-i\eta_{\rho\rho})^k S_{\alpha_1}^{(\rho)}\dots S_{\beta_k}^{(\rho)},
$$

we can see that $\alpha^{(\rho)}:\mathcal{A}\to\mathcal{H}_{\sigma-1,s}$ is a linear isomorphism.

Proof of (ii). By (16) (i), we have $\alpha^{(\rho)}\natural = \natural\alpha^{(\rho)}$ for $C_{\alpha_1}^\# C_\beta$. Since

$$
\natural(C_{\alpha_1}^\# C_{\beta_1}\dots C_{\alpha_k}^\# C_{\beta_k}) = \natural(C_{\alpha_k}^\# C_{\beta_k})\dots\natural(C_{\alpha_1}^\# C_{\beta_1}),
$$

we get the conclusion.

Proof of Theorem 3 (second step). Putting $\alpha_{(p)}^{(1)} = \alpha^{(p)}\circ\alpha^{(1)^{-1}}$, we arrive at (15), as desired.

The linear isomorphism $\alpha^{(\rho)}:\mathcal{A}\to H_\rho$ is called the *basic representation of* $\mathcal{H}_\rho$.

3.3. Space-time duality

Next we consider the duality between space and time. We know that there exists a linear isomorphism between Clifford algebras $C_{\sigma,s}$ and $C_{s,\sigma}$. Also, we can show that there exists a $\natural$-linear isomorphism between Hurwitz algebras $\mathcal{H}_{\sigma,s}$ and $\mathcal{H}_{s,\sigma}$ whenever they exist. In this part of Section 3 we shall state the sharper form of the space-time duality, which will be extended to that between the solutions of Dirac equations. We shall prove the following

Theorem 4 (the space-time duality theorem). *Let* $\sigma+s\equiv 0 \pmod 2$, $\sigma>0$, $s>0$. *We have*

(i) *If* $\sigma \equiv 0 \pmod 2$ *and* $s \equiv 0 \pmod 2$, *we get* $\mathcal{H}_{\sigma-1,s}$, $\mathcal{H}_{\sigma,s}$, *and* $\mathcal{H}_{s,\sigma}$. *Moreover:*

(a) *there exist* $\natural$*-injective morphisms*

$$\tau_{\sigma,s} : \mathcal{H}_{\sigma-1,s} \to \mathcal{H}_{\sigma,s} \quad \text{and} \quad \tau_{s,\sigma} : \mathcal{H}_{\sigma-1,s} \to \mathcal{H}_{s,\sigma};$$

(b) *there exist a* $\natural$*-isomorphism* $\gamma^{(\sigma)}_{(s)} : \mathcal{H}_{\sigma,s} \to \mathcal{H}_{s,\sigma}$ *and surjective morphisms*

$$\pi_{\sigma,s} : \mathcal{H}_{\sigma,s} \to \mathcal{H}_{\sigma-1,s}, \qquad \pi_{\sigma,s} \circ \tau_{\sigma,s} = \mathrm{id},$$

$$\pi_{s,\sigma} : \mathcal{H}_{s,\sigma} \to \mathcal{H}_{\sigma-1,s}, \qquad \pi_{s,\sigma} \circ \tau_{s,\sigma} = \mathrm{id},$$

such that

$$\begin{array}{ccc} \mathcal{H}_{\sigma,s} & \overset{\gamma^{(\sigma)}_{(s)}}{\longrightarrow} & \mathcal{H}_{s,\sigma} \\ \pi_{\sigma,s} \searrow & \curvearrowright & \swarrow \pi_{s,\sigma} \\ & \mathcal{H}_{\sigma-1,s} & \end{array} \tag{17}$$

(ii) *If* $\sigma \equiv 1 \pmod 2$ *and* $s \equiv 1 \pmod 2$, *we get* $\mathcal{H}_{\sigma,s-1}$, $\mathcal{H}_{\sigma,s}$, *and* $\mathcal{H}_{s,\sigma}$. *Moreover,* (a) *and* (b) *hold for* $\mathcal{H}_{\sigma,s-1}$, $\mathcal{H}_{\sigma,s}$ *and* $\mathcal{H}_{s,\sigma}$.

P r o o f of Theorem 4 (i-a). We define a linear morphism $\tau_{\sigma,s} : \mathcal{H}_{\sigma-1,s} \to \mathcal{H}_{\sigma,s}$ by

$$\tau_{\sigma,s}(S_{\alpha_1} \dots S_{\alpha_k}) = \tilde{S}_{\alpha_1} \dots \tilde{S}_{\alpha_k},$$

which satisfies the condition in (i). In a similar manner, we can define $\tau_{s,\sigma} : \mathcal{H}_{\sigma-1,s} \to \mathcal{H}_{s,\sigma}$.

P r o o f of (i-b). We define $\pi_{\sigma,s} : \mathcal{H}_{\sigma,s} \to \mathcal{H}_{\sigma-1,s}$ by

$$\pi_{\sigma,s}(\widetilde{S}_{\alpha_1} \dots \widetilde{S}_{\alpha_k}) = S_{\alpha_1} \dots S_{\alpha_k}, \qquad \alpha_j \neq \sigma + s,\ j = 1, 2, \dots, k,$$

$$\pi_{p,q}(\widetilde{S}_{\alpha_1} \dots \widetilde{S}_{\alpha_k}) = 0 \quad \text{otherwise.}$$

Then we have a $\natural$-linear surjective morphism satisfying $\pi_{\sigma,s} \circ \tau_{\sigma,s} = 1$. In a similar manner, we have $\pi_{s,\sigma}$. Here we define $\gamma^{(\sigma)}_{((s)} : \mathcal{H}_{\sigma,s} \to \mathcal{H}_{s,\sigma}$ by $\gamma^{(\sigma)}_{(s)}(\widetilde{S}_{\alpha_1} \dots \widetilde{S}_{\alpha_k}) = \widehat{S}_{\alpha_1} \dots \widehat{S}_{\alpha_k}$, which yields a $\natural$-linear isomorphism with the commuting diagram (17).

P r o o f of (ii) is similar and may be omitted.

References

[1] J. Cnops, *Hurwitz pairs and Clifford-valued inner products* in: Generalization of Complex Analysis and Their Applications in Physics, J. Ławrynowicz (ed.), Banach Center Publications **37,** Institute of Mathematics, Polish Academy of Sciences 1996, 195–208.

[2] R. Delanghe, F. Sommen, and V. Souček, *Clifford Algebras and Spinor Valued Functions. A Function Theory for the Dirac Operator,* Series: Mathematics and Its Applications **53,** Kluwer Academic, Dordrecht (1994), xviii + 485 pp. + diskette.

[3] I. Furuoya, S. Kanemaki, J. Ławrynowicz, and O. Suzuki, *Hermitian Hurwitz pairs,* in: Deformations of Mathematical Structures II. Hurwitz-Type Structures and Applications to Surface Physics, J. Ławrynowicz (ed.), Kluwer Academic, Dordrecht (1994), 135–154.

[4] J. Ławrynowicz and V. Dietrich, *Type-changing transformations of pseudo-euclidean Hurwitz pairs, Clifford analysis, and particle lifetimes,* in: Clifford Algebras and Their Applications in Mathematical Physics, K. Habetha and G. Yank, (eds.), Kluwer Academic, Dordrecht (1998), 217–226.

[5] J. Ławrynowicz, K. Kedzia, and O. Suzuki, *Supercomplex structures, surface soliton equations, and quasiconformal mappings,* Ann. Polon. Math. **55** (1991), 245–268.

[6] J. Ławrynowicz, R. M. Porter, E. Ramírez de Arellano, and J. Rembieliński, *On dualities generated by the generalized Hurwitz problem,* in: Deformations of Mathematical Structures II. Hurwitz-Type Structures and Applications to Surface Physics, J. Ławrynowicz (ed.), Kluwer Academic, Dordrecht (1994), 189–208.

[7] J. Ławrynowicz, E. Ramírez de Arellano, and J. Rembieliński, *The correspondence between type-reversing transformations of pseudo-euclidean Hurwitz pairs and Clifford algebras I-II,* Bull. Soc. Sci. Lettres Łódź **40,** Sér. Rech. Déform. **8** (1990), 67–97 and 99–129.

[8] J. Ławrynowicz and J. Rembieliński, *Hurwitz pairs equipped with complex structures,* in: Seminar on Deformations, Proceedings, Łódź-Warsaw (1982) **84,** J. Ławrynowicz (ed.), Lecture Notes in Math. 1165, Springer, Berlin 1985, 185–195.

[9] J. Ławrynowicz and J. Rembieliński, *Pseudo-euclidean Hurwitz pairs and generalized Fueter equations* in: Clifford Algebras and Their Applications in Mathematical Physics, Proceedings, Canterbury 1985, J.S.R. Chisholm and A.K. Common (eds.), NATO-ASI Series C: Mathematical and Physical Sciences 183, D. Reidel, Dordrecht (1986), 39–48.

[10] J. Ławrynowicz and J. Rembieliński, *Complete classification for pseudo-euclidean Hurwitz pairs including the symmetry operators,* Bull. Soc. Sci. Lettres Łódź **36,** Sér. Rech. Déform. **4**, no. 39 (1986), 1–15.

[11] J. Ławrynowicz and J. Rembieliński, *Pseudo-euclidean Hurwitz pairs and the Kałuża-Klein theories,* J. Phys. A: Math. Gen. **20** (1987), 5831–5848.

[12] J. Ławrynowicz and J. Rembieliński, *On the composition of nondegenerate quadratic forms with an arbitrary index,* Ann. Fac. Sci. Toulouse Math. (5) **10** (1989), 141–168 [due to a printing error in vol. **10** the whole article was reprinted in vol. **11** (1990), no. 1, of the same journal, 141–168].

[13] J. Ławrynowicz and O. Suzuki, *The duality theorem for the Hurwitz pairs of bidimension* $(8,5)$ *and the Penrose theory,* in: Deformations of Mathematical Structures II. Hurwitz-Type Structures and Applications to Surface Physics, J. Ławrynowicz (ed.), Kluwer Academic, Dordrecht (1994), 209–212.

[14] J. Ławrynowicz and O. Suzuki, *Hurwitz duality theorems for Fueter and Dirac equations,* Advances Appl. Clifford Algebras **7** (1997), 113–132.

[15] J. Ławrynowicz and O. Suzuki, *The twistor theory of the Hermitian Hurwitz pair* $(\mathbb{C}^4(I_{2,2}), \mathbb{R}^5(I_{2,3}))$, ibid. **8** (1998), 147–179.

[16] J. Ławrynowicz and L. Wojtczak in cooperation with S. Koshi and O. Suzuki, *Statistical mechanics of particle systems in Clifford-analytical formulation related to*

Hurwitz pairs of bidimension (8, 5), in: Deformations of Mathematical Structures II. Hurwitz-Type Structures and Applications to Surface Physics, J. Ławrynowicz (ed.), Kluwer Academic, Dordrecht (1994), 213–262.

[17] J. H. Schwarz, *Superstring theory*, Phys. Rep., **89**, no. 3 (1982), 223–322.

Julian Ławrynowicz
Institute of Mathematics,
Polish Academy of Sciences
Łódź Branch, Narutowicza 56
PL-90-136 Łódź, Poland
Chair of Solid State Physics
University of Łódź
Pomorska 149/153
PL-90-236 Łódź, Poland

Osamu Suzuki
Department of Mathematics,
College of Humanities and Sciences
Nihon University
Sakurajosui 3-25-40, Setagaya-ku
Tokyo 156, Japan

Operator Theory:
Advances and Applications, Vol. 114

On the problem of deciding whether a holomorphic vector field is complete

Jorge L. López and Jesús Muciño-Raymundo

1. Introduction.

Let M be a complex manifold provided with a holomorphic vector field X. We say that X is *complete* if its flow $\Phi : \mathbb{C} \times M \to M$ is well defined for all complex values of the time, otherwise X is incomplete.

Complete holomorphic vector fields are interesting from several points of view:

* In differential geometry, complete holomorphic vector fields describe monoparametric groups of holomorphic automorphisms. In particular if M is compact they form the Lie algebra of the Lie group of holomorphic automorphisms of the manifold, [18] p. 77. Over non compact manifolds the situation is more complicated. For $\mathbb{C}^n$, $n \geq 2$, the group of complex automorphisms is infinite dimensional, see [2], [18] p. 77, however very few automorphisms are the time–1 map of complete holomorphic vector fields [4]. On the other hand, a bounded domain has a real Lie group of complex automorphisms, but not a non–identically zero complete holomorphic vector field, a result due to H. Cartan, [1], [18] p. 78, and Section 3.

* The iteration dynamics of complex automorphisms of M is easier to describe if the iterated functions are the time–1 map of complete holomorphic vector fields, [4], [9] p. 44.

* Algebraic or semi–algebraic $\mathbb{C}$ or $\mathbb{C}^*$–actions, in affine and projective complex manifolds, produce complete holomorphic vector fields. See [19] for a list of results and problems in this area, and [8], [34].

* In singular holomorphic foliation theory, foliations coming from complete holomorphic vector fields are rare. In fact, they are singular holomorphic foliations by complex curves having the simplest intrinsic types of leaves: planes $\mathbb{C}$, cylinders $\mathbb{C}^*$, or tori $\mathbb{C}/\Lambda$, [6], [10]. In particular foliations by complex curves having hyperbolic leaves, which is the generic case on projective manifolds [12], never support complete holomorphic vector fields.

1991 *Mathematics Subject Classification.* 34A20.
Key words and phrases. holomorphic vector fields, complete flows.
Partially suported by DGAPA-UNAM and CONACYT 28492-E.

* The problem of completeness (or incompleteness) for real analytic vector fields on open manifolds, is also very interesting. For example, in the n–body problem, it is related to the problem of finding non–collision singularities [29]. See also [15] for results about completeness of real vector fields and relations with the completeness problem for partial differential equations. Hence, we can see the solution of the more rigid problem of completeness for holomorphic vector fields, as a first step to understand the real analytic case.

It is elementary that holomorphic vector fields on compact manifolds are always complete. In this case the main problem is to determine whether a non-identically zero holomorphic vector field exists, see for example [1], [5], [11], [18], [22], [23].

Moreover, the construction and/or recognizing of complete holomorphic vector fields on open manifolds remains as a very interesting problem [4], [10], [11], [19], [28], [34], even in the simplest non compact complex manifold $M = \mathbb{C}^n$, $n \geq 2$, or in the case of quasi–projective manifolds, where holomorphic vector fields always exist.

The inspiration for this paper is the recent works of J. C. Rebelo [28], F. Forsterneric [10], G. T. Buzzard and J. E. Fornæss [4].

Our aim is to survey some well known facts on complete holomorphic vector fields, giving some simple ideas on the following problem.

Given an open complex manifold M and a holomorphic vector field X, recognize in an effective way whether X is complete.

Obviously this is a very difficult task, because it is almost always impossible to compute the flow directly from X (that is, to solve explicitly the associated system of holomorphic differential equations). We split the original problem in the following more geometric subproblems:

i) To classify all complete holomorphic vector fields on Riemann surfaces.

ii) To determine where the original vector field X assumes the above one-dimensional models, in trajectories across the zeros of X.

iii) To study how the complex trajectories of X escape to infinity in M, and to compute the flow along these escapes (for this we only consider the case $M = \mathbb{C}^2$ and X polynomial).

Here is the outline of the paper.

In Section 2 we study flat structures induced by a meromorphic vector field in a Riemann surface, as our main tool for the problem. Recall that natural one to one correspondences exist between: meromorphic vector fields, meromorphic forms, and orientable meromorphic quadratic differentials. Using this idea we identify the flat metric coming from the $(\mathbb{C}, +)$–action of a meromorphic vector field with the metric of the associated quadratic differential. From the classical description of zeros and poles in quadratic differential theory, we describe normal forms for poles and zeros of meromorphic vector fields.

The classification of complete holomorphic vector fields on arbitrary Riemann surfaces is given in Section 3, solving subproblem (i). Six families of vector fields appear. We also remark that only zeros of order one or two appear.

For subproblem (ii), by a *separatrix* we understand a complex analytic curve $\mathcal{L} \subset M$ (probably with singularities) such that it is invariant under the local flow of X, having a discrete and non empty intersection with the zeros of X. Following J. C. Rebelo [28], we show that a complete vector field X with a separatrix by a zero $p \in M$, is such that the order of X at p is at most two. Basically, we follow the seminal idea of J. C. Rebelo; however, our proof uses the explicit classification of complete holomorphic vector fields on Riemann surfaces. Several simple facts follow easily, for example, if the manifold M is Stein and has a separatrix, the order of a complete vector field X at the corresponding zero is exactly one.

We show several simple examples of complete (or incomplete) vector fields in Section 5.

In Section 6, we consider basic properties of holomorphic automorphisms that are the time–1 map of some holomorphic vector field. As a first step, we show that the time–1 map of a complete holomorphic vector field, having zeros of order one and separatrices across these zeros, has cylinders $\mathbb{C}^* \subset M$ of periodic points. In particular, hyperbolic holomorphic automorphisms are never the time–1 map of complete vector fields.

In Section 7, we address subproblem (iii) for polynomial vector fields in $\mathbb{C}^2$. One basic fact is that a polynomial vector field extends to a rational vector field in the compactification given by the complex projective plane $\mathbb{C}P^2 = \mathbb{C}^2 \cup \mathbb{C}P^1_\infty$. We introduce simple elements from singular holomorphic foliation theory in $\mathbb{C}P^2$. Polynomial vector fields of degree at least two which are (almost everywhere) transverse to the line at infinity $\mathbb{C}P^1$ are incomplete. On the other hand, vector fields having the line at infinity as a leaf can be either complete or incomplete. The problem can be localized on:

i) The type of the flow at the singular points of the associated foliation in the line at infinity.

ii) The existence of non-trivial recurrences of some leaves around the line at infinty.

Here we only study (i). We generalize the concept of indetermination point for rational functions to rational vector fields in $\mathbb{C}P^2$. Then we study how the flow is of a polynomial vector field restricted to the separatrices by the singularities of its foliation in the line at infinity. Some results are:

A polynomial vector field in $\mathbb{C}^2$ having a separatrix $\mathcal{L}$ by some point p in the line at infinity, such that the vector field induces a regular point or a pole at p is incomplete in $\mathbb{C}^2$, see 7.6.

A polynomial vector field in $\mathbb{C}^2$ having a polynomial first integral, such that on every separatrix $\mathcal{L}$ by points $\{p\}$ in the line at infinity, induces zeros at $p \in \mathcal{L}$, is complete in $\mathbb{C}^2$, see 7.8.

A polynomial Hamiltonian vector field in $\mathbb{C}^2$, such that one of its leaves intersects the line at infinity in at least three points (set theoretically) is incomplete in $\mathbb{C}^2$, see 7.9.

For the convenience of the freshman reader, we include several simple examples, complete proofs and extensive bibliography.

2. Singular flat metrics from meromorphic vector fields.

2.1. Flat geometry at regular points.

To reduce the background on quadratic differential theory, we describe some simple ideas from it. Let $f = u + \sqrt{-1}v : \Omega \subset \mathbb{C} \to \mathbb{C}P^1$ be a meromorphic function in some domain Ω. We have the following associated objects:

* A meromorphic vector field

$$X = f(z)\frac{\partial}{\partial z}.$$

* A meromorphic differential form

$$\omega = \frac{dz}{f(z)}.$$

* An orientable meromorphic quadratic differential

$$\omega \otimes \omega = \frac{dz}{f(z)} \otimes \frac{dz}{f(z)} = \frac{dz^2}{f(z)^2}.$$

* A pair of smooth vector fields in Ω − {poles and zeros of f}

$$\Re e(X) = (X + \overline{X}) = u\frac{\partial}{\partial x} + v\frac{\partial}{\partial y}\ , \quad \Im m(X) = J(X - \overline{X}) = -v\frac{\partial}{\partial x} + u\frac{\partial}{\partial y}\ ,$$

here $\overline{X}$ means the conjugate vector field and $J : T\mathbb{R}^2 \to T\mathbb{R}^2$ is the usual complex structure on $\mathbb{R}^2 \cong \mathbb{C}$.

* A smooth flat Riemannian metric in Ω − {poles and zeros of f} given by

$$g_f = \begin{pmatrix} \frac{1}{u^2+v^2} & 0 \\ 0 & \frac{1}{u^2+v^2} \end{pmatrix}.$$

Some features and relations between the above objects are as follow:

The vector field X has ω as time–form, namely $w(X) \equiv 1$, and for all smooth trajectories γ from z_0 to z_1 in Ω − {poles and zeros of f}, the number

$$\int_\gamma \frac{dz}{f(z)}$$

is the complex time required to travel from z_0 to z_1 along γ under the field X.

For $z_0 \in \Omega$ − {poles and zeros of f} the holomorphic function

$$F(z) = \int_{z_0}^{z} \frac{dw}{f(w)} : B(z_0) \subset \Omega \to \mathbb{C}\ ,$$

for z in some disk $B(z_0)$ around z_0 free of poles and zeros, is called a local parameter for the quadratic differential $\omega \otimes \omega$. If $f(z_0) \neq 0, \infty$, then the differential F_* maps:

$$F_*(X) = \frac{\partial}{\partial z}\ , \quad F_*(\Re e(X)) = \frac{\partial}{\partial x}\ , \quad F_*(\Im m(X)) = \frac{\partial}{\partial y}\ .$$

The first equality says that F is a holomorphic flow box for X. The last two say that the real trajectories of $\Re e(X)$ are (by definition) the horizontal trajectories of the quadratic differential $\omega \otimes \omega$, and the real trajectories of $\Im m(X)$ are the vertical ones.

From the above it is easy to compute the commutator

$$[\Re e(X), \Im m(X)] = [u\frac{\partial}{\partial x} + v\frac{\partial}{\partial y}\ ,\ -v\frac{\partial}{\partial x} + u\frac{\partial}{\partial y}] \equiv 0\ .$$

Since the Riemannian metric g_f has $\Re e(X)$, $\Im m(X)$ as orthonormal frame, it is well known that the curvature of g_f is identically zero, [31] p. 261. Moreover, every map

$$F(z) : B(z_0) \subset (\Omega - \{\text{poles and zeros of } f\}, g_f) \to (\mathbb{C}, \delta)$$

is a local isometry, where δ is the usual flat metric.

Let us give some elementary examples of the above.

2.1 Example. Consider $c = a + \sqrt{-1}b \in \mathbb{C}^*$, and let $X = c\frac{\partial}{\partial z}$ be the complex vector field on $\Omega = \mathbb{C}$. The space $(\mathbb{C}, g_f)$ is isometric with the usual flat plane $\mathbb{R}^2$, where the isometry is given by the map $x + \sqrt{-1}y \mapsto \frac{1}{|c|}(x, y)$. The real associated vector field is

$$\Re e(c\frac{\partial}{\partial z}) = a\frac{\partial}{\partial x} + b\frac{\partial}{\partial y}\ ,$$

having as trajectories straight lines of slope $arg(c)$.

2.2 Example. For $f(z) = z$, let $\omega = dz/z$ be the meromorphic form in $\Omega = \mathbb{C}$. The metric space $(\mathbb{C} - \{0\}, g_f)$ is isometric to the cylinder $S^1_{2\pi} \times \mathbb{R}$, where the subindex 2π means the length of the closed geodesics. The associated real vector field is

$$\Re e(z\frac{\partial}{\partial z}) = x\frac{\partial}{\partial x} + y\frac{\partial}{\partial y}\ ,$$

having as trajectories straight lines through 0 in $\mathbb{C}$. The associated imaginary vector field is

$$\Im m(z\frac{\partial}{\partial z}) = -y\frac{\partial}{\partial x} + x\frac{\partial}{\partial y}\ ,$$

having as trajectories closed circles around zero; they correspond to closed geodesics in the cylinder $S^1_{2\pi} \times \mathbb{R}$.

2.3 Example. For $f(z) = 1/z$, let $\omega = zdz$ on $\Omega = \mathbb{C}$. The metric space $(\mathbb{C} - \{0\}, g_f)$ is isometric to the glue of four copies of the half flat plane $\{(x, y) \in \mathbb{R}^2 \mid y \geq 0\}$. We glue the positive x–axis of one copy with the negative part of the x–axis in another, using isometries. The real vector field is

$$\Re e(\frac{1}{z}\frac{\partial}{\partial z}) = \frac{x}{x^2 + y^2}\frac{\partial}{\partial x} - \frac{y}{x^2 + y^2}\frac{\partial}{\partial y}$$

in $\mathbb{C}$, having as trajectories hyperbolas. Is easy to see that the real flow is incomplete along the trajectories in the x and y axes.

2.4 Example. Consider $f(z) = e^z$, and let $X = e^z\frac{\partial}{\partial z}$ be the complex vector field in $\Omega = \mathbb{C}$. The space $(\mathbb{C}, g_f)$ is obtained gluing an infinite number of copies of the usual half flat plane. Here we glue the positive x–axis of one copy with the negative part of the x–axis in another. The real associated vector field is

$$\Re e(e^z \frac{\partial}{\partial z}) = e^x (\cos y \frac{\partial}{\partial x} + \sin y \frac{\partial}{\partial y}) .$$

An elementary exercise shows that the associated real vector field is incomplete on the real trajectories $\{y = k\pi\}$, for $k \in \mathbb{Z}$.

2.2. Flat geometry at poles and zeros.

We start by studying the normal forms for meromorphic vector fields at poles and zeros.

2.5 Lemma. *Let $f(z)\frac{\partial}{\partial z}$ be a meromorphic vector field in a neighborhood $B(0) \subset \mathbb{C}$ of 0, up to holomorphic change of coordinates it is as follows:*
1.- If $f(z)\frac{\partial}{\partial z}$ has a zero of order one in 0, then it is

$$\lambda z \frac{\partial}{\partial z} ,$$

for $\lambda = f'(0)$.
2.- If $f(z)\frac{\partial}{\partial z}$ has a zero of order $s \geq 2$ in 0, then it is

$$\frac{z^s}{1 + \lambda z^{s-1}} \frac{\partial}{\partial z} ,$$

for λ the residue of the associated differential form at 0.
3.- If $f(z)\frac{\partial}{\partial z}$ has a pole of order $-k \leq -1$ in 0, then it is

$$\frac{1}{z^k} \frac{\partial}{\partial z} .$$

Proof. We use the obvious idea, consider a local holomorphic change of coordinates, compute how the vector field changes in the new coordinates. For cases (1) and (2), see [3]. Moreover, also for cases (1)–(3) we can consider the associated quadratic differential $dz^2/f(z)^2$, and apply the corresponding theory, see [32] p. 29–31. □

To describe in an explicit way the geometry of the metric g_f and the singular foliation by horizontal trajectories at poles and zeros, we need some preliminary definitions.

Consider the Riemann sphere $\mathbb{C}P^1 = \mathbb{C} \cup \{\infty\}$ provided with the natural flat metric on $\mathbb{C}$, and where ∞ is a "singular point" of this metric. Also introduce in $\mathbb{C}P^1$ the singular real foliation by trajectories of $\frac{\partial}{\partial x}$ in $\mathbb{C}$, this foliation is singular at ∞. Define a half sphere as the subset $\mathcal{H} = \{z \in \mathbb{C} \mid Im(z) \geq 0\} \cup \{\infty\} \subset \mathbb{C}P^1$. A *flat hyperbolic sector* is an open neighborhood of $0 \in \mathcal{H}$ (which does not contains ∞).

A *flat elliptic sector* is an open neighborhood of $\infty \in \mathcal{H}$ (which does not contains 0).

Both types of sectors are Riemannian surfaces with boundary, and having a foliation by unitary geodesics.

2.6 Lemma. *Let $f(z)\frac{\partial}{\partial z}$ be a meromorphic vector field on a neighborhood $B(0) \subset \mathbb{C}$ of 0.*
1.- If $f(z)\frac{\partial}{\partial z}$ has a zero of order one at 0, then $(B(0) - \{0\}, g_f)$ is isometric to the end of an infinite euclidean cylinder $S^1_T \times (0, \infty)$. The real trajectories of the vector field assume one of the following models: center, source or sink.
2.- If $f(z)\frac{\partial}{\partial z}$ has a zero of order $s \geq 2$ at 0, then $(B(0) - \{0\}, g_f)$ is isometric to a suitable glue of $2s - 2$ flat elliptic sectors (see the proof for full details).
3.- If $f(z)\frac{\partial}{\partial z}$ has a pole of order $-k \leq -1$ at 0, then $(B(0) - \{0\}, g_f)$ is isometric to the glue of $2k + 2$ flat hyperbolic sectors (the metric has a point of cone angle $(2k + 2)\pi$).
4.- In each case the Poincaré–Hopf index for the real singular foliation is equal to the order of $f(z)$ at $0 \in \mathbb{C}$.

Proof. Cases (1) and (3) follow from the classical theory of quadratic differentials [32].

Let us explain (2) in more detail. When the order of the zero is $s \geq 2$, there are two isometric invariants of $(B(0) - \{0\}, g_f)$: the order $s \in \mathbb{N}$, and $\lambda \in \mathbb{C}$ the residue of $dz/f(z)$ at $0 \in \mathbb{C}$. We make the description of the metric by cut and paste methods.

Assume $\lambda = 0$, the glue of $2s - 2$ flat elliptic sectors produce the metric space $(B(0) - \{0\}, g_f)$.

Now assume $\lambda = a + \sqrt{-1}b \neq 0$ and $s = 2$. Consider the above global model $\mathbb{C}P^1 = \mathbb{C} \cup \{\infty\}$ with $\lambda = 0$. It is necessary to consider two bands

$$A = \{x + \sqrt{-1}y \in \mathbb{C} \mid a \geq x \geq 0,\ y \geq 0\}\,,\ B = \{x + \sqrt{-1}y \in \mathbb{C} \mid b \geq y \geq 0,\ x \geq 0\}\,.$$

Remove the bands from $\mathbb{C}P^1$. Now we glue the boundaries, using isometries:
In A, glue x to $x + \sqrt{-1}b$ for $x \geq a$.
In B, glue $\sqrt{-1}y$ to $a + \sqrt{-1}y$ for $y \geq b$.
Then an open neighborhood of the point coming from $\infty \in \mathbb{C}P^1$, in the new flat surface, is the local model for $[z^2/(1 + \lambda z)]\frac{\partial}{\partial z}$ having $s = 2$ and $\lambda = a + \sqrt{-1}b$.

The case $\lambda = a + \sqrt{-1}b \neq 0$ and $s \geq 3$ is now easy, following the same ideas.
□

2.3. Global description.

The following is well known for the specialist:

2.7 Lemma. *Let $\mathcal{L}$ be a Riemann surface. One to one correspondences exist between: meromorphic vector fields, meromorphic forms and orientable meromorphic quadratic differentials, given locally as:*

$$f(z)\frac{\partial}{\partial z} \leftrightarrow \frac{dz}{f(z)} \leftrightarrow \frac{dz^2}{f(z)^2}\,.$$

Proof. It is an easy computation with local charts. Let $\{z\}$ and $\{w\}$ be local holomorphic charts for $\mathcal{L}$, with transition function $w = T(z)$. Two local meromorphic vector fields $f(z)\frac{\partial}{\partial z}$ and $g(w)\frac{\partial}{\partial w}$ define the same meromorphic vector field in $\mathcal{L}$ iff

$$T'(z)f(z) = g(T(z)) = g(w) \ .$$

The associated one forms $dz/f(z)$ and $dw/g(w)$ define a differential form in $\mathcal{L}$ iff

$$(T^{-1})'(w)\frac{1}{f(z)} = \frac{1}{g(T(z))} = \frac{1}{g(w)} \ .$$

This is equivalent with the above equality by the inverse function theorem. □

As elementary application, since every Riemann surface admits meromorphic functions then it also admits meromorphic one forms and meromorphic vector fields. It is easy to see that the above correspondence (between vector fields and one forms) is well defined on manifolds modeled on a field. For its applications to real one–dimensional manifolds see [17].

Another explicit description of the associated flat structure is:

2.8 Corollary. *Let $\mathcal{L}$ be a Riemann surface provided with a meromorphic vector field X, or equivalently with a meromorphic form ω. There exists in $\mathcal{L}$ − {poles and zeros of X} a flat holomorphic atlas $\{(B(p_i), F_i)\}$ with coordinate functions given by*

$$F_i(p) = \int_{p_i}^{p} \omega : B(p_i) \subset \mathcal{L} \to \mathbb{C} \ ,$$

having as transition functions euclidean translations $F_i \circ F_j^{-1} : z \mapsto z + c_{ij}$. □

As usual the zeros and poles of global meromorphic objects obey some rules:

2.9 Corollary. *Let $\mathcal{L}$ be a compact connected Riemann surface of genus* g. *For a meromorphic vector field X on $\mathcal{L}$:*

$$\text{zeros}(X) + \text{poles}(X) = 2 - 2\text{g} = c_1(T\mathcal{L}) \ .$$

For ω a meromorphic form on $\mathcal{L}$:

$$\text{zeros}(\omega) + \text{poles}(\omega) = 2\text{g} - 2 = c_1(\mathcal{K}_{\mathcal{L}}) \ .$$

Here zeros and poles are counted with multiplicities, $T\mathcal{L}$, $\mathcal{K}_{\mathcal{L}}$ are the tangent and cotangent holomorphic line bundles respectively, and c_1 represents the Chern class, see [16] p. 139.

Proof. There are several ways, depending on the reader's background. For example, compute the Poincaré–Hopf index for the real singular foliation given by the real trajectories of $\Re e(X)$. □

In consequence, for compact Riemann surfaces only the Riemann sphere and the tori admit holomorphic vector fields.

3. Complete holomorphic vector fields on Riemann surfaces.

Now we are ready to classify complete holomorphic vector fields on arbitrary Riemann surfaces.

3.1 Lemma. *Let $\mathcal{L}$ be a connected Riemann surface, and X a non–identically zero complete holomorphic vector field in $\mathcal{L}$. Then, up to biholomorphism, X and $\mathcal{L}$ are as follow:*

1.- $\lambda z \frac{\partial}{\partial z}$ in $\mathbb{C}P^1$.
2.- $z^2 \frac{\partial}{\partial z}$ in $\mathbb{C}P^1$.
3.- $\frac{\partial}{\partial z}$ in $\mathbb{C}$.
4.- $\lambda z \frac{\partial}{\partial z}$ in $\mathbb{C}$.
5.- $\lambda z \frac{\partial}{\partial z}$ in $\mathbb{C}^*$.
6.- $\lambda \frac{\partial}{\partial z}$ in $\mathbb{C}/\Lambda$.
Here $\lambda \in \mathbb{C}^*$, and Λ is a rank two lattice.

Proof. Start by noting that if $\mathcal{L}$ is covered by the Poincaré disk Δ, then it does not have a non–identically zero complete holomorphic vector field. By contradiction, assume that a such X exists. Let $\Phi(t, p_0) : \mathbb{C} \to \mathcal{L}$ be the complex flow, for $p_0 \in \mathcal{L}$ such that $X(p_0) \neq 0$. By analytic continuation we can lift Φ to a non constant holomorphic function $\widetilde{\Phi}(t, p_0) : \mathbb{C} \to \Delta$, which is a contradiction.

It follows that $\mathcal{L}$ supporting a nontrivial complete holomorphic vector field has as universal cover $\mathbb{C}$ or $\mathbb{C}P^1$.

If $\mathcal{L}$ is compact we have two possibilities. First $\mathcal{L} = \mathbb{C}/\Lambda$ is a torus, where the holomorphic vector fields are the constant (the existence of a zero will imply the existence of poles, by Corollary 2.9), we have the case (6). Second $\mathcal{L} = \mathbb{C}P^1$, by Corollary 2.9 X has two simple zeros giving case (1), or one double zero giving case (2).

In fact, if two simple zeros appear, we can move them by a biholomorphism of $\mathbb{C}P^1 = \mathbb{C} \cup \{\infty\}$ to the points 0 and ∞. In the affine chart $\mathbb{C}$ containing 0, the resulting vector field assumes the explict form in (1). If one double zero appears, moving it to the point $0 \in \mathbb{C}P^1$, the expression (2) follows.

If $\mathcal{L}$ is noncompact then it is a copy of $\mathbb{C}$ or the cylinder $\mathbb{C}^*$. Applying suitable biholomorphisms as above, we get normal forms (3), (4) and (5). □

To be more familiar with the above complete vector fields, we recognize them as the Lie algebra of complex Lie groups of automorphisms on Riemann surfaces:

Riemann surface:	Lie algebra of complete holomorphic vector fields:	Dimension of the Lie algebra:	Lie group of holomorphic automorphisms:
$\mathbb{C}$	$(\lambda z + \mu)\frac{\partial}{\partial z}$	2	$\text{Affine}(\mathbb{C}) = \{z \mapsto az + b\}$
$\mathbb{C}^*$	$\lambda z \frac{\partial}{\partial z}$	1	$\mathbb{C}^* = \{z \mapsto az\}$
$\mathbb{C}P^1$	$\lambda(z - \mu)(z - \eta)\frac{\partial}{\partial z}$	3	$PSL(2, \mathbb{C}) = \{z \mapsto \frac{az+b}{cz+d}\}$
$\mathbb{C}/\Lambda$	$\lambda \frac{\partial}{\partial z}$	1	$\mathbb{C}/\Lambda = \{z \mapsto z + a\}$

where $\lambda, \mu, \eta, a, b, c, d \in \mathbb{C}$, $a \neq 0$ in the first and second line, and $ad - bc = 1$ in the third line.

In the negative sense we have:

3.2 Corollary. *Let X be a holomorphic vector field on a Riemann surface $\mathcal{L}$. If X has a zero of order greater than or equal to three, then the vector field is incomplete.* □

The usual theory of differential equations can not be applied to vector fields having poles or essential singularities at its singular points. However the next result will be very useful in the last section.

3.3 Corollary. *Let $\mathcal{L}$ be a Riemann surface, and $p \in \mathcal{L}$. If X is a holomorphic vector field on $\mathcal{L} - \{p\}$, which extends to p having a pole or an essential singularity, then the vector field on $\mathcal{L} - \{p\}$ is incomplete.*

Proof. By contradiction assume that X is complete over $\mathcal{L} - \{p\}$. Since p is a conformal puncture, X near p looks like the conformal punctures of the Riemann surfaces in Lemma 3.1, cases (3), (4) or (5). But in any case the local model in the punctured neighborhood of p implies that X has a zero of order 1 or 2 at p, which is a contradiction.

For the case of poles at p, a direct proof follows from the local model of a pole having order $-k$ (the glue of $2k+2$ flat hyperbolic sectors). This singular flat model is not geodesically complete. See also examples 2.3 and 2.4. □

For more examples of meromorphic vector fields and the dynamical study of their $(\mathbb{R}, +)$ and $(\mathbb{C}, +)$–actions, see [24], [25].

On the other hand we have:

3.4 Example. *Two families of complex manifolds without complete holomorphic vector fields.* It is well known that a compact complex manifold M is Kobayashi hyperbolic if and only if every holomorphic map from $\mathbb{C}$ to M is constant, see [21] p. 166. Hence in Kobayashi hyperbolic or moreover in Brody hyperbolic manifolds, the unique complete holomorphic vector field is the identically zero vector field, see [21] for explicit examples. Another large family of complex manifolds without non–identically zero complete holomorphic vector fields are the manifolds covered by bounded domains in $\mathbb{C}^n$. Use the idea in the proof of Lemma 3.1.

4. Separatrices of complete holomorphic vector fields.

Following an idea of J. C. Rebelo [28], in order to get the existence of zeros of order ≥ 3 as an obstruction to the completeness for vector fields in higher dimensional complex manifolds, it is necessary compute the flow at the separatrices, i.e., complex trajectories across the zeros of the vector field. The main problems are: the existence of separatrices, and that separatrices can be singular.

Let M be a complex manifold, provided with a holomorphic vector field X. A complex analytic curve $\mathcal{L} \subset M$ is a *separatrix* of the vector field iff:
i) $\mathcal{L}$ has only one irreducible component.

ii) $\mathcal{L}$ is tangent to the vector field X.
iii) $\mathcal{L} \cap \text{zeros}(X)$ is a discrete and nonempty set.

Since any complex analytic curve admits a finite number of irreducible components, see [20] II.5, we assume (i) only for simplicity. Condition (ii) says that $\mathcal{L}$ is union of complex orbits or equivalently invariant under the local complex flow of X.

The problem of deciding whether a holomorphic vector field has a separatrix is famous and very difficult. It was first proposed by Briot and Bouquet in 1854. In 1982, C. Camacho and P. Sad proved that a germ of a two dimensional holomorphic vector field with an isolated zero always admits a separatrix [7]. X. Gómez–Mont and I. Luengo showed in [14] examples of germs of holomorphic vector fields with isolated zeros in $\mathbb{C}^3$ without separatrix. Moreover, J. Olivares–Vásquez has constructed additional examples without separatrix in $\mathbb{C}^{3n}$ [26], [27].

In what follows of this section we assume that X has at least one separatrix $\mathcal{L}$. See the work of J. C. Rebelo [28] for further results in the absence of separatrices. As first step we want to define the order of the zero of a holomorphic vector field at a separatrix.

Consider two cases:

Let $p \in M$ be a zero of X. If $\mathcal{L}$ is smooth by p, then X restricts to $\mathcal{L}$ given a holomorphic vector field $X|_{\mathcal{L}}$. In particular the order of the zero for $X|_{\mathcal{L}}$ is well defined. Denote it by $\text{order}(X|_{\mathcal{L}}, p)$.

Assume $\mathcal{L}$ is non-smooth for some $p \in \mathcal{L} \cap \text{zeros}(X)$. It is well known that given a germ $(\mathcal{L}, p)$ of a singular complex analytic curve in M at p as above, we can always resolve p, see [33]. That is, there exists a non singular complex analytic curve $(\widetilde{\mathcal{L}}, 0)$ called the strict transform and a holomorphic map $\alpha : (\widetilde{\mathcal{L}}, 0) \to (\mathcal{L}, p)$ which is a local biholomorphic map from some punctured neighborhood of $\widetilde{\mathcal{L}} - \{0\}$ to $\mathcal{L} - \{p\}$, for each point $0 \in \alpha^{-1}(p)$. Hence, the vector field X on $\mathcal{L} \subset M$ can be lifted to a unique holomorphic vector field $\alpha^*(X) = \widetilde{X}$ on $\widetilde{\mathcal{L}}$. However resolutions are not at all unique.

4.1 Remark. *The positive number* $\text{order}(X|_{\mathcal{L}}, p)$ *is independent of the resolution.* Note that $\alpha^{-1}(p)$ is a finite set, we fix some point $0 \in \alpha^{-1}(p)$, and define the order for this point. By abuse of notation we do not make explicit this in the notation. Now the order of the zero of $\widetilde{X}$ at some $0 \in \{\alpha^{-1}(p)\}$ is equal to the Poincaré–Hopf index of the associated real vector field, by Lemma 2.6.4. Every resolution α is a local biholomorphism from some punctured neighborhood of $\widetilde{\mathcal{L}} - \{0\}$ to $\mathcal{L} - \{p\}$. The Poincaré–Hopf index is independent of the resolution, and hence the order is also independent. Also note that $\alpha^*(X)$ is holomorphic and has a zero at 0 (using for example the removable singularity Theorem). In particular the order is positive.

4.2 Definition. Let $\mathcal{L}$ be a separatrix for the vector field X at $p \in M$. The *order of X restricted to $\mathcal{L}$ at p* is the order of the vector field $\alpha^*(X)$ at 0, for some resolution $\alpha : (\widetilde{\mathcal{L}}, 0) \to (\mathcal{L}, p)$. By simplicity, in the singular or smooth case we denote the order by:

$$\mathrm{order}(X|_{\mathcal{L}}, p) \in \mathbb{N} .$$

Recall that for X any non–identically zero holomorphic vector field in M, given $p \in \mathrm{zeros}(X)$, the *order of* X *at* p is the lowest degree of the monomials that appear in the power series expansion of X around p. We denote this number as $\mathrm{order}(X,p) \in \mathbb{N}$. It is a very simple invariant of vector fields under holomorphic change of coordinates.

4.3 Proposition. *Let* M *be a complex manifold,* X *a non–identically zero holomorphic vector field, and* $p \in M$ *a zero of* X. *Assume that* $\mathcal{L} \subset M$ *is a separatrix of* X *by* p, *and* $\alpha : (\widetilde{\mathcal{L}}, 0) \to (\mathcal{L}, p)$ *is a local resolution of* $\mathcal{L}$ *at* p.
1.- Then

$$\mathrm{order}(X|_{\mathcal{L}}, p) \geq \mathrm{order}(\alpha, 0) \cdot \mathrm{order}(X, p) - \mathrm{order}(\alpha, 0) + 1,$$

where $\mathrm{order}(\alpha, 0)$ *is the order of the resolution at* p.
2.- In particular, if $\mathcal{L}$ *is smooth at* p, *then*

$$\mathrm{order}(X|_{\mathcal{L}}, p) \geq \mathrm{order}(X, p) .$$

Proof. For convenience of the freshman reader we give an elementary proof for (2). Assume without loss of generality that there exists a local holomorphic coordinate in M such that p corresponds to $0 \in \mathbb{C}^n$ and the separatrix trajectory $\mathcal{L}$ is the z_1–axis in $(\mathbb{C}^n, 0)$. Suppose that X looks like

$$X(z_1, ..., z_n) = X_1(z_1, ..., z_n)\frac{\partial}{\partial z_1} + ... + X_n(z_1, ..., z_n)\frac{\partial}{\partial z_n} ,$$

in $(\mathbb{C}^n, 0)$, for holomorphic functions $X_j : (\mathbb{C}^n, 0) \to \mathbb{C}$. The restriction of the vector field to the separatrix $\mathcal{L}$ is:

$$X|_{\mathcal{L}}(z_1) = X_1(z_1, 0, ..., 0)\frac{\partial}{\partial z_1} .$$

Since $\mathcal{L}$ intersects the zero set of X in a discrete set, the above vector field has an isolated zero at $z_1 = 0$. Note that the power series of $X_1(z_1, 0, ..., 0)$ contains at least one monomial of type $a_i z_1^i$, for $i \in \mathbb{N}$ and $a_i \in \mathbb{C}^*$. Hence $\mathrm{order}(X|_{\mathcal{L}}, 0)$, that is by definition the lowest power of the series $\Sigma a_j z_1^j$ of $X_1(z_1, 0, ..., 0)$, is greater or equal than the original $\mathrm{order}(X(z_1, ..., z_n), 0)$. This finishes part (2).

For the singular case (1), introduce local coordinates $(\mathbb{C}^n, 0)$ for M at p. Let α be a local representative of the resolution of $\mathcal{L}$ at the singular point, this means a germ of a holomorphic parametrization

$$\begin{array}{rcl} \alpha : (\mathbb{C}, 0) & \to & \mathcal{L} \subset (\mathbb{C}^n, 0) \\ t & \mapsto & (z_1(t), z_2(t), ..., z_n(t)) \end{array}$$

such that
i) $\alpha(0) = 0$ and $\frac{d\alpha}{dt}(0) = 0$,
ii) $\alpha(t)$ is a local biholomorphic mapping from a punctured neighborhood in $(\mathbb{C}, 0)$ to $\mathcal{L} - \{p\}$.

Define the order$(\alpha, 0)$ as the lowest number in $\{\text{order}(z_i(t), 0)\}$. Since $\mathcal{L}$ is singular at 0, it follows that order$(\alpha, 0) \geq 2$.

Consider an auxiliary holomorphic vector field $f(t)\frac{\partial}{\partial t}$ in $(\mathbb{C}, 0)$ which satisfies

$$\alpha_* \left(f(t)\frac{\partial}{\partial t} \right) = X(\alpha(t)) \ .$$

Hence by definition order$(X|_{\mathcal{L}}, p) = $ order$(f(t), 0)$. The above vectorial equation is equivalent with the system of equations

$$\frac{dz_i(t)}{dt} f(t) = X_i(z_1(t), z_2(t), ..., z_n(t)) \quad , \text{ for } i = 1, 2, ..., n \ .$$

Since X and $\frac{d\alpha}{dt}$ are $\mathbb{C}$–linearly dependent along $\mathcal{L}$, a holomorphic function $f(t)$ as above exists.

Assume that for some index i, the function $z_i(t)$ realizes the order of α at 0, and moreover X_i is non–identically zero. In fact, if $X_i(z_1(t), ..., z_n(t)) \equiv 0$, where $t \in (\mathbb{C}, 0)$, then $dz_i(t)/dt \equiv 0$. This is impossible, since the order of $z_i(t)$ is finite and positive.

We have that

$$\text{order}(\frac{dz_i(t)}{dt}, 0) + \text{order}(f(t), 0) = \text{order}(X_i(z_1(t), z_2(t), ..., z_n(t)), 0) \ ,$$

which implies

$$\begin{aligned} \text{order}(f(t), 0) &= \text{order}(X_i(z_1(t), z_2(t), ..., z_n(t)), 0) - \text{order}(\tfrac{dz_i(t)}{dt}, 0) \\ &\geq \text{order}(\alpha(t), 0) \cdot \text{order}(X, 0) - \text{order}(\tfrac{d\alpha(t)}{dt}, 0) \ . \end{aligned}$$

This implies

$$\text{order}(X|_{\mathcal{L}}, p) \geq \text{order}(\alpha, 0) \cdot \text{order}(X, p) - \text{order}(\frac{d\alpha(t)}{dt}, 0) \ ,$$

and the result follows. □

4.4 Example. Consider in $\mathbb{C}^2$, the polynomial vector field

$$X(z_1, z_2) = z_1^n \frac{\partial}{\partial z_1} + z_2^m \frac{\partial}{\partial z_2} \ ,$$

where n, m are natural numbers. At $(0,0) \in \mathbb{C}^2$, the axes $\{z_1 = 0\}$ and $\{z_2 = 0\}$ are smooth separatrices for the field X. If $m > n$, then exist separatrices $\mathcal{L}$, $\mathcal{J}$ such that order$(X|_{\mathcal{L}}, 0) = $ order$(X, 0) = n$ and $m = $ order$(X|_{\mathcal{J}}, 0) > $ order$(X, 0) = n$, hence part (2) in the Proposition is the best possible.

4.5 Example. Consider the holomorphic function $H : \mathbb{C}^2 \to \mathbb{C}$ given by $H(z_1, z_2) = z_1^n - z_2^m$, where $(n, m) = 1$. Its Hamiltonian vector field is

$$X(z_1, z_2) = m z_2^{m-1} \frac{\partial}{\partial z_1} + n z_1^{n-1} \frac{\partial}{\partial z_2} \ .$$

The origin is a zero of X having as singular separatrix $\mathcal{L} = \{z_1^n - z_2^m = 0\}$. The usual resolution is given by the Puiseux parametrization

$$\begin{array}{rcl} \alpha : (\mathbb{C},0) & \to & \mathcal{L} \subset (\mathbb{C}^2,0), \\ t & \mapsto & (t^m, t^n). \end{array}$$

Since

$$\begin{aligned} \alpha_* f(t)\frac{\partial}{\partial t} &= mt^{m-1} f(t)\frac{\partial}{\partial z_1} + nt^{n-1} f(t)\frac{\partial}{\partial z_2} \\ &= mt^{(m-1)n}\frac{\partial}{\partial z_1} + nt^{(n-1)m}\frac{\partial}{\partial z_2} = X(\alpha(t)) , \end{aligned}$$

it follows that

$$\alpha^* X = t^{(n-1)(m-1)}\frac{\partial}{\partial t} \quad \text{in } (\mathbb{C},0) .$$

4.6 Theorem. *Let M be a complex manifold, X a complete non–identically zero holomorphic vector field, and $p \in M$ a zero of X. Assume that $\mathcal{L} \subset M$ is a separatrix of X by p. Then*
1.- $\operatorname{order}(X|_{\mathcal{L}}, p)$ *is 1 or 2.*
2.- $\operatorname{order}(X,p)$ *is 1 or 2.*
3.- If $\operatorname{order}(X|_{\mathcal{L}}, p) = 1$, *then the separatrix trajectory contains an embedded copy of* $\mathbb{C}^* \subset M$.
4.- If $\operatorname{order}(X|_{\mathcal{L}}, p) = 2$, *then the separatrix trajectory contains an embedded copy of* $\mathbb{C} \subset M$.

Proof. Since X is complete, it follows that its flow must be complete restricted to the resolution of the separatrix $\widetilde{\mathcal{L}}$. This Riemann surface belongs to the list in Section 3, so (1) follows.

For (2), using the same notation as in 4.3, completeness hypothesis gives

$$2 \geq \operatorname{order}(X|_{\mathcal{L}}, p) \geq \operatorname{order}(\alpha,0)\cdot\operatorname{order}(X,p) - \operatorname{order}(\alpha,0) + 1 \geq 1,$$

which implies

$$1 \geq \operatorname{order}(\alpha,0)(\operatorname{order}(X,p)-1) \geq 0,$$

so the assertion (2) follows. Also by Lemma 3.1, (3) and (4) follow by simple inspection. Note that the copies of $\mathbb{C}$, $\mathbb{C}^*$ are embedded, since X is nonzero there. □

Note that Examples 4.4 and 4.5 are complete if and only if $n = 1 = m$.

5. Some examples.

5.1 Example. *Linear vector fields.* Let $A = (a_{ij})$ be a linear function in $\mathbb{C}^{n+1}$, having $n+1$ different eigenvectors. Consider the associated complete holomorphic vector field in $\mathbb{C}^{n+1}$, given by

$$X(z_0, \ldots, z_n) = (\sum_i a_{i1} z_i)\frac{\partial}{\partial z_0} + \ldots + (\sum_i a_{in+1} z_i)\frac{\partial}{\partial z_n} .$$

From each eigenvector we have an smooth separatrix across the zero at $0 \in \mathbb{C}^{n+1}$. Under the usual projection $\pi : \mathbb{C}^{n+1} - \{0\} \to \mathbb{C}P^n$ the vector field X defines a holomorphic vector field in complex projective space. Each eigenvector produces a zero, assume that they are $\{[0, ..., 1, ..., 0]\}$. The projective lines $\mathbb{C}P^1 \subset \mathbb{C}P^n$ given by the 2–planes "$z_i z_j$" in $\mathbb{C}^{n+1}$ are smooth separatrices, as in case (1) of the Lemma 3.1. See [6] for a dynamical description in $\mathbb{C}^{n+1}$ and $\mathbb{C}P^n$.

5.2 Example. *Vector fields from overshears.* In $\mathbb{C}^n$ for $n \geq 2$, let

$$X(z_1, ..., z_n) = (f(z_2, ..., z_n)z_1 + g(z_2, ..., z_n))\frac{\partial}{\partial z_1}$$

be a holomorphic vector field, where $f, g : \mathbb{C}^{n-1} \to \mathbb{C}$ are entire functions. The vector field X has zeros at the hypersurface $\{f(z_2, ..., z_n)z_1 + g(z_2, ..., z_n) = 0\}$. The lines $\{z_2 = c_2, \ ..., \ z_n = c_n\}$ intersecting the hypersurface are smooth separatrices for the vector field. By simple integration, the flow is given by overshears, see [2], in $\mathbb{C}^n$ of the form

$$\Phi(t, (z_1, \ldots, z_n))$$
$$= \left(\frac{[f(z_2, \ldots, z_n)z_1 + g(z_2, \ldots, z_n)]e^{f(z_2, \ldots, z_n)t} - g(z_2, \ldots, z_n)}{f(z_2, \ldots, z_n)}, z_2, \ldots, z_n \right),$$

when $f(z_2, ..., z_n) \neq 0$. Also by Lemma 3.1, it follows that X is complete.

5.3 Example. *Product manifolds.* Let M be any complex manifold and $\mathcal{L}$ a Riemann surface as in Lemma 3.1. The product $M \times \mathcal{L}$ has a complete holomorphic vector field. If $\mathcal{L} = \mathbb{C}$ this family of examples are known as cylinder–like manifolds, see [8].

5.4 Corollary. *Let M be a complex manifold, X a complete holomorphic vector field (non–identically zero). Assume that $\mathcal{L} \subset M$ is a separatrix of X. The intersection $\mathcal{L} \cap \{zeros\ of\ X\}$ has one or two points.*

Proof. Apply Lemma 3.1 to the resolution of the separatrix. □

In manifolds without rational curves we can say a little bit more:

5.5 Corollary. *Let M be a complex manifold, assume that every holomorphic map from $\mathbb{C}P^1$ to M is constant, for example for $M = \mathbb{C}^n$ or M a Stein manifold. Let X be a complete non–identically zero holomorphic vector field and $\mathcal{L} \subset M$ a separatrix of X. Then*

1.- order$(X, p) = 1$, *where p is a zero of X.*

2.- The intersection $\mathcal{L} \cap \{$zeros of $X\}$ *is one point.* □

For example, holomorphic vector fields in $\mathbb{C}^n$ having a complex saddle conection (i.e., a complex analytic curve having at least two zeros in its closure) are incomplete.

As simple applications we have:

5.6 Example. *The sum of two complete vector fields does not need to be complete.* Consider two complete holomorphic vector fields in $\mathbb{C}^2$:

$$z_1 z_2 \frac{\partial}{\partial z_1} \ , \ z_2 z_1 \frac{\partial}{\partial z_2} \ .$$

The sum

$$X(z_1, z_2) = z_1 z_2 (\frac{\partial}{\partial z_1} + \frac{\partial}{\partial z_2}) \ ,$$

has as zero set $\{z_1 z_2 = 0\}$, and the lines $\{z_1 - z_2 = c\}$ are separatrices. By 5.5 part (2), it follows that X is incomplete.

There are very interesting examples of complete holomorphic vector fields without zeros in $\mathbb{C}^n$ but non holomorphically equivalent to the trivial $\frac{\partial}{\partial z_1}$, see [34].

Obviously, holomorphic vector fields having trajectories or separatrices with fundamental group different from $\{e\}$, $\mathbb{Z}$, or $\mathbb{Z} \oplus \mathbb{Z}$ are incomplete. For example:

5.7 Corollary. *Let (M, ω) be a complex holomorphic symplectic surface, $H : M \to \mathbb{C}$ a nonconstant holomorphic function. If some level set curve $\{H^{-1}(c)\} \subset M$ has fundamental group nonisomorphic with $\{e\}$, $\mathbb{Z}$, or $\mathbb{Z} \oplus \mathbb{Z}$, then the complex Hamiltonian vector field X_H is incomplete.* □

A consequence of the classical genus formula for algebraic curves in $\mathbb{C}P^2$, see [16] p. 220, is that generically polynomial functions H in $\mathbb{C}^2$, of degree at least four, produce incomplete Hamiltonian vector fields. See also [10] Section 7, and our Proposition 7.9.

6. Periodic points of time–1 maps.

The inspiration for this Section is the work of G. T. Buzzard and J. E. Fornæss [4]. In fact, from the explicit knowledge of the complex trajectories of complete vector fields in Section 3 we can study periodic points for time-1 maps.

Let us recall that $p \in M$ is of *minimal period* $n \geq 2$ for some holomorphic automorphisms $\Phi : M \to M$, if $\Phi^{(n)}(p) = p$, but $\Phi^{(m)}(p) \neq p$ for $m = 1, 2, \ldots, n-1$.

6.1 Proposition. *Let M be a complex manifold, X a non–identically zero complete holomorphic vector field, Φ its flow and $\Phi_1 : M \to M$ its time–1 map. Let $p \in M$ be a periodic point of Φ_1 having minimal period $n \geq 2$.*
1.- If the Φ–orbit of p is non compact, then it is an embedded cylinder $\mathbb{C}^* \subset M$ of periodic points.
2.- If the Φ–orbit of p is compact, then it is an embedded torus $\mathbb{C}/\Lambda \subset M$ of periodic points.

Proof. Consider the real vector field $\Re e(X)$, as in Section 2. The trajectory of p under $\Re e(X)$ is a circle $S^1 \subset M$. Every point on this circle has the same period as p under Φ_1. Recall that the above circle is a closed geodesic in $(\mathcal{L}, g_X)$ for the flat metric g_X as in Section 2. Use the classification in Lemma 3.1 to show that S^1 is in $\mathcal{L}$ biholomorphic to $\mathbb{C}^*$ or $\mathbb{C}/\Lambda$. □

As one consequence of the existence of cylinders of periodic points we have the following:

6.2 Corollary. *Let $\Phi_1 : M \to M$ be the time–1 map of X a complete non–identically zero holomorphic vector field, with a periodic point $p \in M$ of minimal period $n \geq 2$. Then the differential $D\Phi_n(p)$ has 1 as eigenvalue.*

Proof. Since p is not a fixed point, $X(p) \neq 0$ and the complex trajectory of p under Φ is an embedded copy of $\mathbb{C}^*$ or $\mathbb{C}/\Lambda \subset M$. The time–$n$ map has p as fixed point and is the identity map restricted to the complex Φ trajectory. Hence the eigenvalue corresponding to eigenvector tangent to the Φ complex trajectory is 1. □

Separatrix trajectories for X also give origin to periodic points in the flow:

6.3 Corollary. *Assume that the complete vector field X has a separatrix trajectory $\mathcal{L}$ by p, such that $order(X|_{\mathcal{L}}, p) = 1$. Given a number $n \geq 2$, there exists a complex time $T(n) \in \mathbb{C}^*$ such that $\Phi_{T(n)} : M \to M$, the time–$T(n)$ map, has a cylinder $\mathbb{C}^*$ of periodic points of minimal period n.*

Proof. Since $\mathrm{order}(X|_{\mathcal{L}}, p) = 1$, by Lemma 3.1 the normalization of $\mathcal{L}$ has an embedded copy of $\mathbb{C}^*$. Assume that the vector field X looks like $\lambda z \frac{\partial}{\partial z}$ in this copy of $\mathbb{C}^*$, for λ a non zero complex number (recall that by definition of separatrix p is a isolated zero in $X|_{\mathcal{L}}$). Making the choice $T(n) = (2\pi\sqrt{-1}/n\lambda) \in \mathbb{C}^*$, the time–$T(n)$ map of X is the same as the time–1 map of $T(n)X$. Note that

$$T(n)X = \frac{2\pi\sqrt{-1}}{n} z \frac{\partial}{\partial z} \quad \text{on } \mathbb{C}^* \subset \mathcal{L},$$

and hence its flow has only periodic points of primitive period n. See also [25] for applications of the idea of rotating the vector field X by some complex number. □

If in the above Corollary M is a complex Stein surface and X has an isolated zero, then by the Camacho–Sad theorem [7], a separatrix of X by p always exists. It follows that many time–$T(n)$ maps give origin to cylinders of periodic points.

7. Indetermination points for rational vector fields.

In this part we consider polynomial vector fields in $\mathbb{C}^2$, giving some simple ideas on the flow behavior at infinity (for simplicity we restrict our attention to dimension two).

It is elementary that every polynomial vector field in $\mathbb{C}^2$ extends in a unique way to a rational vector field on $\mathbb{C}P^2$.

Recall some notation. Let $\mathbb{C}^3 = \{(z_0, z_1, z_2)\}$ be the complex space, the complex projective plane is $\mathbb{C}P^2 = \mathbb{C}^3 - \{0\}/\mathbb{C}^*$. We consider on the affine chart

$\mathbb{C}^2 = \{z_0 = 1\} \subset \mathbb{C}P^2$ the polynomial vector field P. The coordinate charts are as usual $\{\phi_i : U_i \subset \mathbb{C}P^2 \to \mathbb{C}^2 \mid i = 0, 1, 2\}$, where $U_i = \{z_i \neq 0\}$, and

$$\begin{array}{rcl} \phi_0 : U_0 & \to & \mathbb{C}^2 \\ (z_0, z_1, z_2) & \mapsto & (\frac{z_1}{z_0}, \frac{z_2}{z_0}) \ , \end{array}$$

etc.

The changes of coordinates applied to P give rational vector fields on the other charts

$$\{(\phi_j \circ \phi_0^{-1})_* P \ \ in \ \ \mathbb{C}^2 = \{z_j = 1\} \mid for \ j = 1, 2\} \ .$$

The *associated singular holomorphic foliation* $\mathcal{F}(P)$ is defined by the above vector fields. To do this, we remove the poles at $\mathbb{C}P^1_\infty = \{z_0 = 0\}$ from each vector field $(\phi_j \circ \phi_0^{-1})_* P$ (multiplying by a suitable z_0^k), obtaining holomorphic vector fields on each coordinate chart U_i, the associated foliations glue to give $\mathcal{F}(P)$ on $\mathbb{C}P^2$. See [13] for the explicit computation.

The behavior of polynomial vector fields at infinity is well understood, let us recall also from [13].

7.1 Proposition. *Let*

$$P(z_1, z_2) = P_1(z_1, z_2)\frac{\partial}{\partial z_1} + P_2(z_1, z_2)\frac{\partial}{\partial z_2}$$

be a polynomial vector field in $\mathbb{C}^2 = \{z_0 = 1\} \subset \mathbb{C}P^2$ *of degree* m *(i.e.* m *is the highest degree of the polynomials* P_j*).*
1.- Then P *has a pole of order* $1-m$ *at the line at infinity* $\mathbb{C}P^1_\infty = \{z_0 = 0\} \subset \mathbb{C}P^2$*, unless the terms of degree* m *of* P *have the form*

$$g(z_1, z_2)(z_1\frac{\partial}{\partial z_1} + z_2\frac{\partial}{\partial z_2}) \ ,$$

for g *a polynomial of degree* $m-1$*, in which case it has a pole of order* $2-m$ *at the line at infinity.*
2.- Then the singular holomorphic foliation $\mathcal{F}(P)$ *has* $\mathbb{C}P^1_\infty$ *as a leaf if and only if the terms of degree* m *of* P *can not be expressed as*

$$g(z_1, z_2)(z_1\frac{\partial}{\partial z_1} + z_2\frac{\partial}{\partial z_2}) \ ,$$

for g *a polynomial of degree* $m-1$*.*
3.- If $\mathbb{C}P^1_\infty$ *is a leaf of* $\mathcal{F}(P)$*, then* $\mathcal{F}(P)$ *has a finite number of singularities in* $\mathbb{C}P^1_\infty$*.* □

Strictly speaking, $\mathbb{C}P^1_\infty$ *is a leaf of* $\mathcal{F}(P)$, means that $\mathbb{C}P^1_\infty - \{$ *singular points of* $\mathcal{F}(P)\}$ is a leaf in the usual sense.

The application to our problem is as follows. Given a polynomial vector field P on $\mathbb{C}^2$ of degree $m \geq 2$ (the case $m = 1$ always defines complete vector fields). A *separatrix of the singular foliation* $\mathcal{F}(P)$ is a complex analytic curve that is a finite union of non-singular leaves of $\mathcal{F}(P)$ and having singularities of $\mathcal{F}(P)$ in its closure. Roughly speaking, we consider each separatrix $\{\mathcal{L}\}$ of the singular foliation

$\mathcal{F}(P)$ intersecting the line of poles $\mathbb{C}P^1_\infty$ in a discrete set, and we understand how the restricted vector field is on $\mathcal{L}$.

As a first problem, note that some $\mathcal{L}$ can be singular at points on the line at infinity. This is solved as in Section 4 by considering its resolution.

Following 7.1 we have two possibilities for the behavior at infinity of singular holomorphic foliations in $\mathbb{C}P^2$ coming from polynomial vector fields. The first is:

7.2 Corollary. *Let P be a polynomial vector field in $\mathbb{C}^2$ of degree $m \geq 2$. If $\mathbb{C}P^1_\infty$ is not a leaf of the singular holomorphic foliation defined by P in $\mathbb{C}P^2$, then P is incomplete on the trajectories across $\mathbb{C}P^1_\infty - \{singular\ points\ of\ \mathcal{F}(P)\}$.*

Proof. Assume the degree of P is at least three. Let $\mathcal{L}$ be a leaf of the foliation $\mathcal{F}(P)$ in $\mathbb{C}P^2$. Since the singularities of $\mathcal{F}(P)$ are isolated, it is easy to choose a leaf such that $p = \mathcal{L} \cap \mathbb{C}P^1_\infty$ is a nonsingular point of $\mathcal{F}(P)$. The vector field X defines on $\mathcal{L}$ near p a meromorphic vector field with a pole at p (since $\mathbb{C}P^1_\infty$ is not a leaf). From Corollary 3.3 it follows that the flow of X along $\mathcal{L}$ is incomplete in $\mathbb{C}^2$.

For P of degree two, an explicit computation in the other affine charts shows that the vector field is transverse and holomorphic at the line at infinity. Hence it is incomplete in $\mathbb{C}^2$. □

7.3 Example. Consider the vector field

$$P(z_1, z_2) = g(z_1, z_2)(z_1\frac{\partial}{\partial z_1} + z_2\frac{\partial}{\partial z_2}) \quad \text{at } z_0 = 1,$$

where g is a polynomial of degree $m - 1 \geq 1$, and $\{g = 0\}$ as an algebraic curve does not contain complex lines through $(z_1, z_2) = (0, 0)$. The foliation $\mathcal{F}(P)$ in $\mathbb{C}P^2$ is a pencil of lines through $[1, 0, 0]$. The associated rational vector field has a pole of order $2 - m$ in the line at infinity. Every complex line $\mathcal{L}$ through $[1, 0, 0]$ is such that

$$\text{zeros}\,(P|_{\mathcal{L}}) + \text{poles}\,(P|_{\mathcal{L}}) = (1 + m - 1) + (2 - m) = 2\ ,$$

according to Corollary 2.9. Following 7.2 the vector field is incomplete in $\mathbb{C}^2$. For the particular subcase

$$P(z_1, z_2) = (az_1 + bz_2)(z_1\frac{\partial}{\partial z_1} + z_2\frac{\partial}{\partial z_2}) \quad \text{at } z_0 = 1.$$

The affine degree is two; however the associated rational vector field in $\mathbb{C}^3$ is the linear:

$$-(az_1 + bz_2)\frac{\partial}{\partial z_0}\ ,$$

giving a holomorphic vector field in $\mathbb{C}P^2$. The line $\mathbb{C}P^1_\infty$ is not invariant under its flow, hence this vector field is incomplete in the affine plane $\mathbb{C}^2 = \{z_0 = 1\}$.

As second possibility; the foliation defined by P has $\mathbb{C}P^1_\infty$ as a leaf, we know that $\mathbb{C}P^1_\infty$ is always a pole for P.

Roughly speaking the flow of P will be incomplete along the leaves of $\mathcal{F}(P)$ that are separatrices for the singular points of $\mathcal{F}(P)$ in $\mathbb{C}P^1_\infty$. The next example shows that the above can be false.

7.4 Example. Let

$$P(z_1, z_2) = g(z_2)\frac{\partial}{\partial z_1}$$

be a polynomial vector field, where g is a polynomial of degree $m \geq 2$. The associated singular holomorphic foliation in $\mathbb{C}P^2 = \{[z_0, z_1, z_2]\}$ has a unique singularity in $[0,1,0]$, and the leaves are given by the pencil of lines by this point. Every leaf of $\mathcal{F}(P)$ is a separatrix across $[0,1,0]$. However the vector field is holomorphic and complete in $\mathbb{C}^2$.

What kind of point is $[0,1,0]$ for the associated rational vector field on $\mathbb{C}P^2$?

On Riemann surfaces the types of singularities of a complex analytic vector field are zeros, poles or essential singularities. However on complex manifolds of dimension two or more, there can exist points where the zeros of a rational vector field intersect its poles. This is the phenomenon of indetermination points, well known in several complex variables for functions and maps, see [16] p. 490–491. For simplicity the next definition is formulated locally in $(\mathbb{C}^2, 0)$, hence can be applied to meromorphic vector fields in complex surfaces:

7.5 Definition. Let $A, B, C, D : (\mathbb{C}^2, 0) \to \mathbb{C}$ be holomorphic functions. The point $0 \in \mathbb{C}^2$ is an *indetermination point* of the meromorphic vector field

$$X(z_1, z_2) = \frac{A(z_1, z_2)}{B(z_1, z_2)}\frac{\partial}{\partial z_1} + \frac{C(z_1, z_2)}{D(z_1, z_2)}\frac{\partial}{\partial z_2},$$

iff $\{A = 0\} \cap \{B = 0\} = 0$ and/or $\{C = 0\} \cap \{D = 0\} = 0$.

Moreover, the singular holomorphic foliation $\mathcal{F}(X)$ associated to X can be described by the holomorphic one form

$$\mathcal{F}(X) = \{\ C(z_1, z_2)B(z_1, z_2)dz_1 - A(z_1, z_2)D(z_1, z_2)dz_2 = 0\ \}.$$

However for our problem it is necessary to understand the original vector field at the separatrices at the origin for this foliation $\mathcal{F}(X)$.

It is possible to show that for a polynomial vector field P in $\mathbb{C}^2$, the indetermination points of its associated rational vector field in $\mathbb{C}P^2$ always define (probably removable) singularities of the singular holomorphic foliation $\mathcal{F}(P)$ in $\mathbb{C}P^2$.

7.6 Theorem. *Let P be a polynomial vector field on $\mathbb{C}^2$ of degree $m \geq 2$, having $\mathbb{C}P^1_\infty$ as a leaf of its associated foliation $\mathcal{F}(P)$. Let $\mathcal{L}$ be a separatrix of $\mathcal{F}(P)$ through a singularity $p \in \mathbb{C}P^1_\infty$ of $\mathcal{F}(P)$. Assume that $P|_{\mathcal{L}}$ at p has:*
1.- A regular point.
2.- A zero of order greater or equal to three.
3.- A pole.
4.- An essential singularity.
Then P is incomplete (in $\mathbb{C}^2$).

An easy exercise shows that possibility (4) never appears, our proof is independent of this.

Proof. Consider $\alpha : (\mathbb{C}, 0) \to (\mathcal{L}, p)$ a local representative of the resolution of $\mathcal{L}$ at p. It follows that $\alpha^* P$ is a holomorphic vector field in a punctured neighborhood of 0 in $(\mathbb{C}, 0)$. Following the classical theory of complex analytic functions in one variable, $\alpha^* P$ has at 0 a regular point, a zero, a pole or an essential singularity. Case (1) follows from the classical theory of ordinary differential equations, since $p \in \mathbb{C}P^1_\infty$ the flow leaves the affine $\mathbb{C}^2 = \{z_0 = 1\}$ in finite time. Moreover by Corollaries 3.2, 3.3, results (2)–(4) follow. □

7.7 Example. Consider in $\mathbb{C}^2$ the polynomial vector field

$$P(z_1, z_2) = z_1^2 \frac{\partial}{\partial z_1} + z_2^2 \frac{\partial}{\partial z_2} \quad \text{at } z_0 = 1 \, .$$

This defines an incomplete vector field in $\mathbb{C}^2$. The associated rational vector fields in the other affine charts of $\mathbb{C}P^2$ are:

$$-\frac{\partial}{\partial z_0} + \frac{z_2(z_2 - 1)}{z_0} \frac{\partial}{\partial z_2} \quad \text{at } z_1 = 1,$$

$$-\frac{\partial}{\partial z_0} + \frac{z_1(z_1 - 1)}{z_0} \frac{\partial}{\partial z_1} \quad \text{at } z_2 = 1.$$

It follows that $[1, 0, 0]$ is a zero, and $\{[0, 1, 0], \ [0, 1, 1], \ [0, 0, 1]\}$ are indetermination points, for P as a rational vector field. Moreover, the vector field has a rational first integral

$$\frac{z_0 z_1 - z_0 z_2}{z_1 z_2} : \mathbb{C}P^2 \to \mathbb{C}P^1 \, .$$

Hence, the topology of the associated foliation is given by a pencil of quadrics in $\mathbb{C}P^2$, having as base locus $\{[1, 0, 0], [0, 1, 0], [0, 0, 1]\}$. Note that $[0, 1, 1]$ is a critical point for the rational first integral.

It is easy to compute that each separatrix of $\mathcal{F}(P)$ is isomorphic to $\mathbb{C}P^1$ having a holomorphic vector field $\alpha^*(P)$ with a double zero coming from $[1, 0, 0]$, and that $\mathcal{L} \cap \mathbb{C}P^1_\infty$ is a regular point for $\alpha^*(P)$.

The affirmative result is as follows:

7.8 Theorem. *Let P be a polynomial vector field in $\mathbb{C}^2$ of degree $m \geq 2$, having $\mathbb{C}P^1_\infty$ as a leaf of its associated foliation $\mathcal{F}(P)$, and a polynomial first integral. The following assertions are equivalent:*
1.- P is complete (in $\mathbb{C}^2$).
2.- For every separatrix $\mathcal{L}$ of $\mathcal{F}(P)$ through the singularities of $\mathcal{F}(P)$ in $\mathbb{C}P^1_\infty$, the restrictions $P|_{\mathcal{L}}$ have zeros.

Proof. Assume (1), it follows by Theorem 7.6 that any other possibility for $P|_{\mathcal{L}}$ in a separatrix through a point at infinity, must imply that X is incomplete.

For the converse, let $\mathcal{L} \subset \mathbb{C}P^2$ be a leaf of $\mathcal{F}(P)$ irreducible as a complex analytic curve. Since P has a polynomial first integral, the closure $\overline{\mathcal{L}}$ is an algebraic curve in $\mathbb{C}P^2$.

Note that $\overline{\mathcal{L}} \cap \mathbb{C}P^1_\infty$ is non empty. P induces a complete holomorphic vector field on the resolution of $\overline{\mathcal{L}}$ if and only if assertion (2) is true. The result follows. □

The existence of a polynomial first integral is a strong hypothesis. Recall in this direction the nice result of G. Darboux; a polynomial vector field having enough algebraic curves as solutions has in fact a polynomial first integral, [30] p. 440.

7.4 Example (revised). For the vector field

$$P(z_1, z_2) = g(z_2)\frac{\partial}{\partial z_1}$$

the expression at the indetermination point $[0,1,0]$ is given as:

$$-z_0 g(\frac{z_2}{z_0})(z_0\frac{\partial}{\partial z_0} + z_2\frac{\partial}{\partial z_2}) \quad \text{at} \quad z_1 = 1.$$

It follows that the vector field restricted to each separatrix by $(z_0, z_2) = (0,0)$, given by a parametrization

$$\alpha : t \mapsto (t, \lambda t) \ , \text{ for } \lambda \in \mathbb{C} \ , \ t \in (\mathbb{C}, 0) \ ,$$

is such that

$$\alpha^* X = -t^2 g(\lambda)\frac{\partial}{\partial t},$$

which is holomorphic at $\alpha(0) = [0,1,0]$. Hence the indetermination point is a double zero for each complex line through it, except the line at infinity $\mathbb{C}P^1_\infty = \{z_0 = 1\}$, or the lines that are zeros of g.

The simplest examples of polynomial vector fields with polynomial first integral are Hamiltonian vector fields. We give a very simple result for the incompleteness of Hamiltonian vector fields in $\mathbb{C}^2$, compare with [10], 7.1.

7.9 Proposition. *Let H be a polynomial in $\mathbb{C}^2$. If a projectivized curve $\{H = \lambda_0/\mu_0\} \subset \mathbb{C}P^2$, for $\lambda_0/\mu_0 \in \mathbb{C}$, intersects $\mathbb{C}P^1_\infty$ in at least three different points (set theoretically), then its Hamiltonian vector field X_H is incomplete in $\mathbb{C}^2$, for all the complex leaves $\{H = \lambda/\mu\}$.*

Proof. Consider H as a rational function, this means

$$H = \frac{\widetilde{H}(z_0, z_1, z_2)}{z_0^m} : \mathbb{C}P^2 \to \mathbb{C}P^1 \ ,$$

where $\widetilde{H}$ is the homogenized associated polynomial, and m is the degree of H. The indetermination points of this function are by definition

$$\{\widetilde{H} = 0\} \cap \{z_0^m = 0\} \subset \mathbb{C}P^2 \ .$$

Every curve $\{H = \lambda/\mu\}$, for $\lambda/\mu \in \mathbb{C}$, can be written in $\mathbb{C}P^2$ as $\{\mu\widetilde{H} - \lambda z_0^m = 0\}$. Hence all the curves intersect the indetermination points of the function. By hypothesis, the indetermination set of $\widetilde{H}/z_0^m$ has at least three different points. They give origin to the same number of indetermination points for the Hamiltonian vector field X_H, as a rational vector field in $\mathbb{C}P^2$.
Case 1. Assume $\{H = \lambda/\mu\}$ is an irreducible curve in $\mathbb{C}^2$. The resolution for every $\{H = \lambda/\mu\}$ has at least three points coming from $\mathbb{C}P^1_\infty$. If the three points are zeros for the Hamiltonian vector field, we have a Riemann surface with three zeros, and by Lemma 3.1, the vector field is not complete. In any other case (regular points, poles, essential singularities), the result follows from Theorem 7.6.
Case 2. Assume $\{H = \lambda/\mu\}$ is a reducible curve of degree three. Suppose it has two irreducible components, say $\mathcal{L}$, $\mathcal{J} \subset \mathbb{C}P^2$. By the hypothesis the intersection $\{H = \lambda/\mu\} \cap \mathbb{C}P^1_\infty$ is given by three different points, hence one of the components is a quadric, for example say $\mathcal{L}$. Then $\mathcal{L}$ has three special points:
Two points from $\mathcal{L} \cap \mathbb{C}P^1_\infty$.
One additional point from $\mathcal{L} \cap \mathcal{J}$ that is a point in the affine $\mathbb{C}^2 = \{z_0 = 1\}$, giving origin to a zero in the restricted vector field.
Hence P restricted to $\mathcal{L}$ is an incomplete vector field in $\mathbb{C}^2$.

All other cases: $\{H = \lambda/\mu\}$ with three components of degree one, or $\{H = \lambda/\mu\}$ of higher degree, follow from the same type of argument. □

From the classical Bezout's Theorem in $\mathbb{C}P^2$, it follows that for H of degree $m \geq 3$, the hypothesis in the Proposition is generically true. We leave to the interested reader the problem of finding necessary and sufficient conditions.

7.10 Corollary. *Let P be a polynomial vector field on $\mathbb{C}^2$ having degree $m \geq 2$. If the associated foliation $\mathcal{F}(P)$ in $\mathbb{C}P^2$ has a leaf intersecting $\mathbb{C}P^1_\infty \cup \{zeros(P)$ in $\mathbb{C}^2\}$ in at least three points (set theoretically), then P is incomplete in $\mathbb{C}^2$.*

Proof. Follows the same idea as for the leaves of the Hamiltonian in 7.9. □

Note that the hypothesis of P having a polynomial first integral or P Hamiltonian, avoid complex recurrences on the leaves of $\mathcal{F}(P)$ near the leaf at infinity. The presence of these recurrences also give origin to incompleteness. It is probably the most difficult step in the characterization of complete polynomial vector fields.

Acknowledgments. We express our gratitude to the editors and to professor R. Michael Porter for their assistance in the preparation of this paper.

References

[1] D. Akhiezer, *Lie Group Actions in Complex Analysis*, Vieweg (1995).

[2] E. Andersen, L. Lempert, *On the group of holomorphic automorphisms of* $\mathbb{C}^n$, Invent. Math. 110 (1992) 371–388.

[3] L. Brickman, E. S. Thomas, *Conformal equivalence of analytic flows*, Journal of Differential Equations 25 (1977) 310–324.

[4] G. T. Buzzard, J. E. Fornæss, *Complete holomorphic vector fields and time-1 maps*, Indiana University Mathematics Journal 44 (1995) 1175–1182.

[5] J. Carrell, A. Howard, C. Kosniowski, *Holomorphic vector fields on complex surfaces*, Math. Ann. 204 (1973) 303–309.

[6] C. Camacho, N. Kuiper, J. Palis, *The topology of holomorphic flows with singularity*, Pub. Math. IHES 48 (1978) 5–38.

[7] C. Camacho, P. Sad, *Invariant varieties through singularities of holomorphic vector fields*, Annals of Math. 115 (1982) 579–595.

[8] K. Fieseler, *On complex affine surfaces with* $\mathbb{C}^+$*-action*, Comment. Math. Helvetici 69 (1994) 5–27.

[9] J. E. Fornæss, *Dynamics in several complex variables*. Regional Conference Series in Mathematics A. M. S., 87 (1996).

[10] F. Forstneric, *Actions of* $(\mathbb{R}, +)$ *and* $(\mathbb{C}, +)$ *on complex manifolds*, Math. Z. 223 (1996) 123–153.

[11] E. Ghys, J.-C. Rebelo, *Singularités des flots holomorphes II*. Ann. Inst. Fourier, Grenoble 47, 4 (1997) 1117–1174.

[12] A. A. Glutsyuk, *Hyperbolicity of leaves of a generic one-dimensional holomorphic foliation on a nonsingular projective algebraic manifold*, Proceedings of the Steklov Institute of Mathematics, 213 (1996) 90–111.

[13] X. Gómez-Mont, *On families of rational vector fields*, Coloquio de Sistemas Dinámicos, J. Seade and G. Sienra (eds.) Aportaciones Matemáticas 1 (1985) 36–65.

[14] X. Gómez-Mont, I. Luengo, *Germs of holomorphic vector fields in* $\mathbb{C}^3$ *without a separatrix*, Inv. Math 109 (1990).

[15] A. Goriely, C. Hyde, *Necessary and sufficient conditions for finite time blow-up in ordinary differential equations*, Preprint (1997).

[16] Ph. Griffiths, J. Harris, *Principles of Algebraic Geometry*, Wiley Interscience (1978).

[17] N. Hitchin, *Vector fields in the circle*, 200 *Years after Lagrange*, M. Francaviglia (ed.), Elsevier Science Publishers (1991) 359–378.

[18] S. Kobayashi, *Transformation Groups in Differential Geometry*, Springer–Verlag (1972).

[19] H. Kraft, *Challenging problems on affine n-space*, Séminaire Bourbaki, Astérisque 237 (1996) 295-317.

[20] S. Lojasiewicz, *Introduction to Complex Analytic Geometry*, Birkhaüser (1991).

[21] S. Lang, *Hyperbolic and Diophantine analysis*. Bulletin (New Series) of the A. M. S. 14, 2 (1986) 159–205.

[22] D. Lieberman, *Holomorphic vector fields on projective varieties*, Proceedings of Symposia in Pure Mathematics 30 (1997) 273–276.

[23] Y. Matsushima, *Holomorphic Vector Fields on Compact Kähler Manifolds*, Regional Conference Series on Mathematics 7, A. M. S. (1971).

[24] J. Muciño–Raymundo, *Complex structures adapted to smooth vector fields*, Preprint (1997).

[25] J. Muciño–Raymundo, C. Valero–Valdés, *Bifurcations of meromorphic vector fields on the Riemann sphere*. Ergodic Theory and Dynamical Systems 15 (1995) 1211–1222.

[26] J. Olivares–Vázquez, *On vector fields in* $\mathbb{C}^3$ *without separatarix*. Revista Matemática de la Universidad Complutense de Madrid. vol. 5 núm. 1 (1992) 13–34.

[27] J. Olivares–Vázquez, *On the problem of existence of germs of holomorphic vector fields in* $\mathbb{C}^m$*, without separatrix, (*$m \geq 3$*)*, Ecuaciones Diferenciales Singularidades, J. Mozo (ed.), Universidad de Valladolid (1997) 317–351.

[28] J. C. Rebelo, *Singularités des flots holomorphes*, Ann. Inst. Fourier, Grenoble 46, 2 (1996) 411-428..

[29] D. G. Saari, Z. Xia, *Off to infinity in finite time*, Notices of the AMS, 42, 5 (1995) 538-546.

[30] D. Schlomiuk, *Algebraic and geometric aspects of the theory of polynomial vector fields*, Bifurcations and periodic orbits of vector fields, D. Schlomiuk (ed.), Kluwer (1993) 429–467.

[31] M. Spivak, *A Comprehensive Introduction to Differential Geometry II*, Publish or Perish (1979).

[32] K. Strebel, *Quadratic Differentials*, Springer–Verlag (1984).

[33] B. Teissier, *Introduction to curve singularities*, Singularity Theory, D. T. Le, K. Saito, B. Teissier (eds.) World Scientific (1995) 866–893.

[34] J. Winkelmann, *On free holomorphic* $\mathbb{C}$*-actions on* $\mathbb{C}^n$ *and homogeneous Stein manifolds*, Math. Ann. 286, (1990) 593–612.

Jorge L. López
Instituto de Física y Matemáticas,
Universidad Michoacana,
Morelia, 58060,
Michoacán, México
E-mail address: jorge@itzel.ifm.umich.mx

Jesús Miciño Raymundo
Instituto de Matemáticas UNAM,
Unidad Morelia,
Nicolás Romero 150,
Col. Centro, Morelia 58000,
Michoacán, Mexico
E-mail address: jmucino@zeus.ccu.umich.mx

Operator Theory:
Advances and Applications, Vol. 114

Variations on a theorem of Severi

D. Napoletani, I. Sabadini, and D. C. Struppa

Abstract. We consider several questions related to the removability of singularities for regular functions of quaternionic variables. In particular we use an old idea of Severi to prove that a function $f : \mathbb{H} \times \mathbb{R} \to \mathbb{H}$ regular in the first variable and real analytic in the second variable, cannot have compact singularities; we show how this result must be modified in the case of functions defined on biquaternions. Finally, we construct a class of regular functions on $\mathbb{H}^2$ for which we can remove singularities such as $K \times \mathbb{R}^3$ with K compact in $\mathbb{R}^5$.

1. It sometimes happens that a theorem, despite its interest or relevance, remains at the margins of the literature for a long time, almost as a reminder of the destiny of human things.The purpose of this paper is to discuss one such case, related to a generalization due to Severi [10] of the well known Hartogs' theorem on the removability of compact singularities for holomorphic functions in $\mathbb{C}^n$, $(n \geq 2)$; more precisely, Severi's result shows that compact singularities can be removed for functions defined on $\mathbb{C} \times \mathbb{R}$ and which are holomorphic in the first variable and real analytic in the second one. In this paper, we will discuss the case in which holomorphic functions on $\mathbb{C}$ and $\mathbb{C}^n$ are replaced by regular functions in one or more quaternionic variables ($\mathbb{H}$ and $\mathbb{H}^n$), and we will show that Severi's idea can be applied to this setting to provide some new results on the removability of compact and tubular singularities for such functions.

In this introduction we will state the original Hartogs' theorem and we will give a sketch of its most general proof, which is due to Ehrenpreis [4].

Theorem 1.1. *Let Ω be an open set in $\mathbb{C}^n$, $n \geq 2$, and let K be a compact subset of Ω such that $\Omega \backslash K$ is connected. Then every function f holomorphic on $\Omega \backslash K$ extends uniquely to a function g holomorphic on all of Ω.*

This theorem, first formulated and proved by Hartogs in 1906, has originated several interesting generalizations (see [11] for a rather detailed history of Hartogs' theorem and its variations), and has been given several different and interesting proofs. The fundamental enlightening, however, was reached when Ehrenpreis, in 1960, showed how the phenomenon of removability of compact singularities had nothing to do with holomorphic functions per se, but was rather to be ascribed to general properties of the Cauchy-Riemann system, and was therefore generalizable to a large class of systems. We give here the statement and a sketch of the proof of this result for its intrinsic interest as well as because it is the basis for our own theorems. Indeed, the algebraic techniques we will employ can be considered a refined extension of the fundamental ideas of Ehrenpreis. As standard notation, if P is a polynomial in $\mathbb{C}^n$, the symbol $P(D)$ will denote the linear constant coefficient partial differential operator obtained by replacing the variables in P by the symbol $D = (-i\frac{\partial}{\partial x_1}, \dots, -i\frac{\partial}{\partial x_n})$, where $x = (x_1, \dots, x_n)$ is the variable in $\mathbb{R}^n$.

Theorem 1.2. *Let $P_1, \dots, P_r$ be r polynomials ($r \geq 2$) in $\mathbb{R}^n$ with no common factors; let K be a compact set in $\mathbb{R}^n$ such that $\mathbb{R}^n \backslash K$ is connected, and let f be a hyperfunction which, on $\mathbb{R}^n \backslash K$, satisfies the system of differential equations*

$$P_1(D)f = \dots = P_r(D)f = 0. \tag{1}$$

Then f extends to a hyperfunction $\tilde{f}$ which satisfies the system (1) *everywhere on $\mathbb{R}^n$.*

Proof. We only sketch the proof here, because of its intrinsic interest, and because we will need it later on to generalize our results. Consider the hyperfunction f; because of the flabbiness of the sheaf of hyperfunctions (i.e. because of the fact that every hyperfunction on an open set can be extended to a hyperfunction defined everywhere), we can extend f to a hyperfunction g defined on $\mathbb{R}^n$. Such a hyperfunction, in principle, will not be a solution of the system (1), but if we define $g_k = P_k(D)f$ we see immediately that the hyperfunctions g_k are supported in K (since they coincide with f outside of K, and f is a solution of (1) outside of K). Now, the commutativity of constant coefficient operators shows that for every pair of different indices k and j the following cocycle relationship holds between K-supported hyperfunctions:

$$P_k(D)g_j = P_j(D)g_k.$$

If we now take the Fourier transform of this relation, and we call G_j the Fourier transform of g_j, we can define a holomorphic function H by setting

$$H(z) = \frac{G_j(z)}{P_j(z)} = \frac{G_k(z)}{P_k(z)}.$$

Since H is holomorphic outside the variety defined by the common zeroes of the polynomials P_j, and because these polynomials have no common factors, then H is in fact an entire function (removability of codimension 2 singularities). However, H is the quotient of the Fourier transform of a K-supported hyperfunction by a polynomial, and therefore (by using a variation of the Ehrenpreis-Malgrange lemma, see [11]), we see that H is in fact the Fourier transform of a hyperfunction h whose support is still K. Now it is an immediate computation to show that $\tilde{f} = g - h$ is the required extension and the theorem is completely proved.

■

Remark 1.2.. In algebraic terms this proof shows that the first extension module associated to the system of differential equations vanishes; this can also be rephrased in terms of the vanishing of the first cohomology group with compact support of the sheaf of solutions of the system (see [5]). It is immediate to see that the classical Hartogs' theorem is an immediate consequence of Theorem 1.1, once the differential operators $P_1(D), \ldots, P_r(D)$ are taken to be the Cauchy-Riemann operators and $2r = n \geq 4$. It is interesting to note that Ehrenpreis' argument actually works even for non–scalar systems of differential operators; this is described in some detail in Ehrenpreis' book [5] and in Palamodov's one [9]; these ideas were used in [1] and subsequent papers such as [2] to prove compact singularity removability theorems for regular functions of quaternionic variables (in this case, the Cauchy-Riemann system has to be replaced by the so called Cauchy-Fueter system). Our interest in this paper, however, stems from the different kind of extension which Severi proved in 1932; in order to provide a new proof of Hartogs' theorem, Severi proved that compact singularities can be removed for functions defined on $\mathbb{C} \times \mathbb{R}$ which are holomorphic in the first variable and real analytic in the second one. In this paper, we will use Severi's ideas to prove a similar result for functions defined on $\mathbb{H} \times \mathbb{R}$ and $\mathbb{BH} \times \mathbb{R}$; we then revert to Ehrenpreis' method, more algebraic in nature, to answer (at least partially) a problem posed by Palamodov on the existence of a certain kind of tubular singularities for regular functions of two quaternionic variables. We will then discuss several corollaries and applications of this approach.

Acknowledgements. The third author would like to thank Professor Shapiro for the invitation to the "1996 Simposio Internacional de Analisis Complejo y Temas Afines" in Cuernavaca where these and related results were first presented. The authors also wish to thank Professor Loustaunau for helping them with some algebraic computations.

2. As we mentioned before, Severi developed in [10] a new method for studying Hartogs' phenomena; he called it the "real to complex" method (*passaggio dal reale al complesso*); using this technique he was able to obtain not only Hartogs' theorem (at least in the case of two complex variables), but also an interesting and not trivial extension, namely the existence of a Hartogs phenomenon for functions $f : \mathbb{C} \times \mathbb{R} \to \mathbb{C}$ such that $f(z, x)$ holomorphic in z and is real analytic in x. We will see in what follows that the techniques used by Severi can be easily used to generalize our own theorem [1] on the removability of singularities for regular functions of several quaternionic variables.

We begin by giving the basic notation and terminology. We refer the reader to [12] for further details. $\mathbb{H}$ will denote the associative algebra of quaternions with respect to the basis $\{1, \mathbf{i}, \mathbf{j}, \mathbf{k}\}$ and a quaternion q will be written as $q = x_0 + \mathbf{i}x_1 + \mathbf{j}x_2 + \mathbf{k}x_3$. Given an open set U in $\mathbb{H}$, a real differentiable function $f : U \subseteq \mathbb{H} \longrightarrow \mathbb{H}$ is said to be *left regular* on U if

$$\frac{\partial f}{\partial \bar{q}} = \frac{\partial f}{\partial x_0} + \mathbf{i}\frac{\partial f}{\partial x_1} + \mathbf{j}\frac{\partial f}{\partial x_2} + \mathbf{k}\frac{\partial f}{\partial x_3} = 0.$$

Let us consider now $\mathbb{H}^n$ whose coordinates are $(q_1, \dots, q_n)$. A real differentiable function $f : U \subseteq \mathbb{H}^n \to \mathbb{H}$ is said to be *left regular* if

$$\frac{\partial f}{\partial \bar{q}_1} = \dots = \frac{\partial f}{\partial \bar{q}_n} = 0. \tag{2}$$

It is possible to give analogous definitions for the right regularity, but since the two theories are equivalent, we will consider only left regularity and from now on we will omit the adjective "left".

We have the following theorem:

Theorem 2.1. *Let U be an open set in $\mathbb{H} \times \mathbb{R}$ and let K be a compact set in U. Let $f(q,x)$ be regular in q and real analytic in x on $U \backslash K$. Then f extends uniquely to a function $\tilde{f}$ which is regular and real analytic in all of U.*

Proof. Given K, we can find another compact set K' with real analytic boundary such that $K \subset K' \subset U$, and such that every hyperplane $x =$ const that intersects K', does it along a closed manifold without singular points. It is obviously possible to find a hyperplane $x = a_1$ which cuts K' without intersecting K and such that K is all on one side of the given hyperplane, and another hyperplane $x = a_2$ that cuts K' and that lies in the half space that does not contain K. Define now, for every t, the function

$$\tilde{f}(q,t) = \frac{1}{8\pi^2} \int_\gamma \frac{(p-q)^{-1}}{|p-q|^2} Dpf(p,t),$$

where

$$Dp = dx_1 \wedge dx_2 \wedge dx_3 - \mathbf{i} dx_0 \wedge dx_2 \wedge dx_3 + \mathbf{j} dx_0 \wedge dx_1 \wedge dx_3 - \mathbf{k} dx_0 \wedge dx_1 \wedge dx_2$$

and $\gamma = \gamma_t = \partial K' \cap \{x = t\}$. If the slice $K' \cap \{x = t\}$ does not contain any singularity for f, then $\tilde{f}$ coincides with f; if on the other hand the slice contains singular points for f, $\tilde{f}$ still defines a function which is regular with respect to q and outside $\partial K'$. Therefore $\tilde{f}(q,t)$ is regular with respect to q and real analytic with respect to t inside K'. But since f and $\tilde{f}$ coincide in the portion of K' between $x = a_1$ and $x = a_2$, which has nonempty interior, and where f is regular and real analytic, we have that $\tilde{f}$ is the required extension. ∎

A more accurate analysis of the proof shows that the result can be improved as follows:

Corollary 2.1. *Let U be an open set in $\mathbb{H} \times \mathbb{R}$ and let K be a set in U such that there exist two real constants a_1 and a_2 for which $K \cap \{a_1 \le x \le a_2\} = \emptyset$ and the sections $K \cap \{x = const\}$ are relatively compact with respect to x. Let $f(q,x)$ be regular in q and real analytic in x on $U \backslash K$. Then f extends uniquely to a function $\tilde{f}$ which is regular and real analytic in all of U.*

Corollary 2.2. *Let U be an open set in $\mathbb{H} \times \mathbb{R}^n$ and let K be a compact in U. Let $f(q,x)$ be regular in q and real analytic in x on $U \backslash K$. Then f extends uniquely to a function $\tilde{f}$ which is regular and real analytic in all of U.*

Remark 2.1. It is obviously impossible to use Theorem 2.1 or Corollary 2.1 to deduce Hartogs' theorem for two quaternionic variables as a corollary. In the complex setting, as we mentioned above, Severi was able to use his result to derive Hartogs' theorem, using the fact that any function defined on $\mathbb{C} \times \mathbb{R}$, holomorphic in $\mathbb{C}$ and real analytic on $\mathbb{R}$ extends uniquely to a holomorphic function of two variables; in the case of regular functions of several quaternionic variables the extension of a function $f : \mathbb{H} \times \mathbb{R} \to \mathbb{H}$ to a regular function on $\mathbb{H}^2$ is not necessarily unique. However, since every real analytic function on $\mathbb{R}^3$ extends uniquely to a regular function we can use this argument to prove the removability of compact singularities for regular functions on $\mathbb{H}^2$ as a consequence of Corollary 2.2.
The discussion above is related to the following question posed by Palamodov [8]: *Is it possible to remove singularities of regular functions of two quaternionic variables of the type $K \times \mathbb{R}^{8-q}$ where K has real dimension q and is relatively compact in $\mathbb{R}^q$?*
For $q = 6, 7, 8$ the answer follows easily from more general results in [1], where an algebraic approach is used. We can use these techniques to prove a partial answer to the previous question for $q = 5$.

Let us begin by writing the CauchyFueter system (2) for two quaternionic variables in matrix form; if $f : \mathbb{H}^2 \longrightarrow \mathbb{H}$ is written as $f = f_0 + \mathbf{i}f_1 + \mathbf{j}f_2 + \mathbf{k}f_3$, with variables in $\mathbb{H}^2$ defined by $(q_1, q_2) = (x_0 + \mathbf{i}x_1 + \mathbf{j}x_2 + \mathbf{k}x_3, y_0 + \mathbf{i}y_1 + \mathbf{j}y_2 + \mathbf{k}y_3)$, the Cauchy–Fueter system can be written as

$$\begin{bmatrix} \frac{\partial}{\partial x_0} & -\frac{\partial}{\partial x_1} & -\frac{\partial}{\partial x_2} & -\frac{\partial}{\partial x_3} \\ \frac{\partial}{\partial x_1} & \frac{\partial}{\partial x_0} & -\frac{\partial}{\partial x_3} & \frac{\partial}{\partial x_2} \\ \frac{\partial}{\partial x_2} & \frac{\partial}{\partial x_3} & \frac{\partial}{\partial x_0} & -\frac{\partial}{\partial x_1} \\ \frac{\partial}{\partial x_3} & -\frac{\partial}{\partial x_2} & \frac{\partial}{\partial x_1} & \frac{\partial}{\partial x_0} \\ \frac{\partial}{\partial y_0} & -\frac{\partial}{\partial y_1} & -\frac{\partial}{\partial y_2} & -\frac{\partial}{\partial y_3} \\ \frac{\partial}{\partial y_1} & \frac{\partial}{\partial y_0} & -\frac{\partial}{\partial y_3} & \frac{\partial}{\partial y_2} \\ \frac{\partial}{\partial y_2} & \frac{\partial}{\partial y_3} & \frac{\partial}{\partial y_0} & -\frac{\partial}{\partial y_1} \\ \frac{\partial}{\partial y_3} & -\frac{\partial}{\partial y_2} & \frac{\partial}{\partial y_1} & \frac{\partial}{\partial y_0} \end{bmatrix} \begin{bmatrix} f_0 \\ f_1 \\ f_2 \\ f_3 \end{bmatrix} = 0$$

where now each f_i is real function of 8 variables (see [1] for details). Now the fact that this matrix is of maximal rank 4 and that all its 4×4 minors are relatively prime allows us to deduce (see again [1]) that $\mathrm{Ext}^1(M, R) = 0$ where R is the ring $\mathbb{C}[x_0, \ldots, x_3, y_0, \ldots, y_3]$ and M is the module associated to our matrix via the Fourier transform; using CoCoA[1] to perform the computations of the resolution of the module (in the case of theorem 1.2 this was done directly in the proof) it is possible to deduce also that $\mathrm{Ext}^2(M, R) = 0$,but $\mathrm{Ext}^3(M, R) \neq 0$. This (using general results from Chapter VIII, sect. 13 of [9]) is sufficient to answer Palamodov's question for $q = 6, 7, 8$.

[1]CoCoA is a special purpose computer system for performing computations in commutative algebra. it is the ongoing product of a research team in Computer Algebra at the University of Genova (Italy). Their program can be obtained by sending an e-mail to `cocoa@dima.unige.it`

We now define a proper subclass $\mathcal{S}$ of the class $\mathcal{R}$ of regular functions on $\mathbb{H}^2$ defined by a system whose module M_1 satisfies $\mathrm{Ext}^3(M_1, R) = 0$. In this way we will be able to show that, for such a subclass, we can give a positive answer to Palamodov's question.

To describe the subclass $\mathcal{S}$, we rewrite q_2 as

$$q_2 = y_0 + \mathbf{i}y_1 + \mathbf{k}(y_3 + \mathbf{i}y_2) = z_1 + \mathbf{k}z_2$$

and $f = F + G\mathbf{j}$ where $F = f_0 + \mathbf{i}f_1$, $G = f_2 + \mathbf{i}f_3$. We now define $\mathcal{S}$ to be the set of functions $F + \mathbf{j}G : \mathbb{H}^2 \longrightarrow \mathbb{H}$ which are regular in q_1 and such that both F and G are holomorphic in (z_1, z_2). In matrix terms, this amounts to asking that f be a solution of

$$\begin{bmatrix} \frac{\partial}{\partial x_0} & -\frac{\partial}{\partial x_1} & -\frac{\partial}{\partial x_2} & -\frac{\partial}{\partial x_3} \\ \frac{\partial}{\partial x_1} & \frac{\partial}{\partial x_0} & -\frac{\partial}{\partial x_3} & \frac{\partial}{\partial x_2} \\ \frac{\partial}{\partial x_2} & \frac{\partial}{\partial x_3} & \frac{\partial}{\partial x_0} & -\frac{\partial}{\partial x_1} \\ \frac{\partial}{\partial x_3} & -\frac{\partial}{\partial x_2} & \frac{\partial}{\partial x_1} & \frac{\partial}{\partial x_0} \\ \frac{\partial}{\partial y_0} & -\frac{\partial}{\partial y_1} & 0 & 0 \\ \frac{\partial}{\partial y_1} & \frac{\partial}{\partial y_0} & 0 & 0 \\ 0 & 0 & \frac{\partial}{\partial y_0} & -\frac{\partial}{\partial y_1} \\ 0 & 0 & \frac{\partial}{\partial y_1} & \frac{\partial}{\partial y_0} \\ \frac{\partial}{\partial y_3} & -\frac{\partial}{\partial y_2} & 0 & 0 \\ \frac{\partial}{\partial y_2} & \frac{\partial}{\partial y_3} & 0 & 0 \\ 0 & 0 & \frac{\partial}{\partial y_3} & -\frac{\partial}{\partial y_2} \\ 0 & 0 & \frac{\partial}{\partial y_2} & \frac{\partial}{\partial y_3} \end{bmatrix} \begin{bmatrix} f_0 \\ f_1 \\ f_2 \\ f_3 \end{bmatrix} = 0 \qquad (3)$$

and therefore it is immediate to notice that $\mathcal{S} \subseteq \mathcal{R}$.

We have the following

Theorem 2.2. *Let L be a 5-dimensional linear set in $\mathbb{H}^2 = \mathbb{R}^8$ and let $L^\perp$ be its orthogonal complement. Let K be a compact set contained in L and let $f \in \mathcal{S}(\mathbb{H}^2 \backslash (K \times L^\perp))$. Then f can be uniquely extended to $\tilde{f} \in \mathcal{S}(\mathbb{H}^2)$.*

Proof. Let M_1 the module associated to the matrix (3). A CoCoA calculation shows that $\mathrm{Ext}^i(M_1, R) = 0$ for $i = 1, 2, 3$. The thesis follows from general results in Chapter VIII, Sect. 13 of [9]. ∎

Another immediate corollary of [9] and the vanishing of the $\mathrm{Ext}^i(M_1, R)$ for $i = 1, 2, 3$ is the following result, which provides an interesting improvement on the removability of non compact singularities.

Corollary 2.3. *Let K be a compact set in $\mathbb{H}^2$ such that $\mathbb{H}^2 \backslash K$ is connected. Let Σ_i, $i = 1, 2, 3$ be closed half spaces in $\mathbb{H}^2$ and $\Sigma = \cup_{i=1}^3 \Sigma_i$. Then if $f \in \mathcal{S}(\mathbb{H}^2 \backslash (K \cup \Sigma)$ then f can be extended to $\tilde{f} \in \mathcal{S}(\mathbb{H}^2 \backslash \Sigma)$.*

Let us point out that, in general, an analogous statement can be proved for functions of two quaternionic variables but involving only two half spaces. Corollary 2.3 shows that, at least for the subclass $\mathcal{S}$, it is possible to improve the result.

Remark 2.2. Since $\mathcal{S} \subseteq \mathcal{R}$, the theorem above gives a positive answer to Palamodov's question, at least for the subclass $\mathcal{S}$.

Remark 2.3. From the algebraic point of view it is interesting to point out that the resolution of the module M_1 (available from the authors upon request) involves an alternance of quadratic and linear maps; up to now, in all the (regularity) systems which we have studied, we always found that the first syzygies were quadratic, and that all the other syzygies were linear; this fact was pointed out by Loustaunau [7], and we hope to be able to come back to its analytic significance in a future joint paper.

We will denote by $\tilde{S}$ the set of functions $f = (F, G) : \mathbb{H} \times \mathbb{C} \times \mathbb{R} \to \mathbb{H}$ which are regular with respect to the quaternionic variable, real analytic with respect to the real variable and such that F and G are holomorphic with respect to the complex variable.

Corollary 2.4. *Let $f \in \tilde{S}(\mathbb{H} \times \mathbb{C} \times \mathbb{R} \backslash (K \times \mathbb{R}))$ where K is a compact $K \subset \mathbb{H} \times \mathbb{C}$, then it is possible to extend f to $\tilde{f} \in \tilde{S}(\mathbb{H} \times \mathbb{C} \times \mathbb{R})$.*

Proof. Just note that f can be extended by complexifying the real variable, and that the extended function satisfies the system (3). The thesis follows immediately from the fact that $\mathrm{Ext}^1(M_1, R) = 0$.

■

Remark 2.4. We point out that it is possible to prove Corollary 2.4 in a spirit more similar to that of Severi.

Let us denote by q, z and x the quaternionic, complex and real variable respectively. If we complexify the real variable x to w, the function f is represented by two hyperfunctions F and G depending on a complex parameter w. The sheaf of hyperfunctions having a holomorphic parameter is flabby with respect to the six real components of q and z (see [6]), so we can extend F and G to two hyperfunctions $\tilde{F}$ and $\tilde{G}$ defined on $\mathbb{H} \times \mathbb{C} \times \mathbb{C}$. Given such extensions, it is possible to use Ehrenpreis' ideas (as in Section 1) to prove that $\tilde{f} = (\tilde{F}, \tilde{G})$ can be modified (essentially by proving a vanishing of an Ext–module) to an extension of f satisfying the same regularity conditions and still holomorphic with respect to w. The required extension is now obtained by restricting it to $\mathbb{H} \times \mathbb{C} \times \mathbb{R}$. Finally, we provide one last example of a Severi like theorem in a different setting that sheds a new light on this kind of problem. When we deal with function satisfying regularity conditions (we refer with this term to holomorphy in $\mathbb{C}$ or $\mathbb{C}^n$, regularity in $\mathbb{H}$ or $\mathbb{H}^n$, $n > 1$, but also monogeneity in Clifford Algebras, see for example [1]), we have that for functions depending on one variable removability of compact singularities does not occur. Also, as we have seen before, the Severi theorem is an almost direct consequence of the existence of a Cauchy formula (see Theorem 2.1). Our next result shows however that the algebraic structure of $\mathbb{H}$ and $\mathbb{C}$ (and in particular the fact that they are division algebras) plays a crucial role in the argument.

To this purpose let $\mathbb{BH}$ be the complex algebra defined on the basis $\{1, \mathbf{e_1}, \mathbf{e_2}, \mathbf{e_3}\}$ where $\mathbf{e_1}, \mathbf{e_2}, \mathbf{e_3}$ satisfy the Pauli relations $\mathbf{e_l}^2 = 1$ and $\mathbf{e_j}\mathbf{e_l} = -i\mathbf{e_k}$ where (j, l, k) is a cyclic permutation of $(1, 2, 3)$ and $i = \sqrt{-1} \in \mathbb{C}$. A biquaternion will be denoted by $Z = z_0 + z_1\mathbf{e_1} + z_2\mathbf{e_2} + z_3\mathbf{e_3}$, where $z_j = x_j + iy_j \in \mathbb{C}$. A real differentiable function $f : \mathbb{BH} \to \mathbb{BH}$ will be said to be *left regular* if it satisfies

$$\frac{\partial f}{\partial z_0} - \sum_{j=1}^{3} \mathbf{e_j} \frac{\partial f}{\partial z_j} = 0;$$

the reader is referred to [3] for the study of such functions and their relations with Maxwell's equations. It is known that, in general, the Cauchy formula holds without limitations in any real Clifford algebra. However in complexified Clifford algebras, which are not division algebras, the Cauchy kernel $G(Z-Z_P)$ at a point P is not necessarily defined everywhere and, in our case, it is defined only outside the translated light cone, i.e. in $\mathbb{BH}\setminus\{N(Z-Z_P)=0\}$, where $N(Z) = z_0^2 - z_1^2 - z_2^2 - z_3^2$ denotes the norm of the biquaternion Z. Let us indicate by $\mathbb{C}N_P$ the set $\{Z \in \mathbb{BH} \ : \ N(Z-Z_P) = 0\}$ and by S^3 the 3–sphere contained in $\mathbb{H}$

$$S^3 = \{Z_P + Z \ : \ Z = x_0 + \mathbf{i}y_1 + \mathbf{j}y_2 + \mathbf{k}y_3 \in \mathbb{H}, \ x_0^2 + \sum_{i=0}^{3} y_i^2 = 1\}.$$

A domain $\Omega \subset \mathbb{BH}$ is said to be *null-convex* if for all Z, $Z' \in \Omega$ such that $N(Z - Z') = 0$, the whole segment ZZ' belongs to Ω.

Theorem 2.3 [3]. *Let $\Omega \subset \mathbb{BH}$ be a null–convex domain and let f be regular in Ω. If $P \in \Omega$, then*

$$f(Z_P) = \frac{1}{2\pi^2} \int_\Sigma G(Z - Z_P)\ DZ\ f(Z)$$

where $\Sigma \subset \Omega$ is any cycle homologous to the 3–sphere S^3 in Ω and DZ is a suitable 3–form.

Using this Cauchy formula it is possible to prove the following

Theorem 2.4. *Let U be an open set in $\mathbb{BH} \times \mathbb{R}$ such that $U \cap \mathbb{BH}$ is null convex and let K be a compact in U. Let $f(Z,x)$ be regular in Z and real analytic in x on $U\setminus K$. Then f extends uniquely to a function $\tilde{f}$ which is regular and real analytic in all of U.*

Proof. Just follow the proof of Theorem 2.1 by applying Theorem 2.3 rather than the usual Cauchy formula.

∎

References

[1] W. W. Adams, C. A. Berenstein, P. Loustaunau, I. Sabadini and D. C. Struppa, *Regular Functions of Several Quaternionic Variables and the Cauchy-Fueter Complex*, to appear in Journal of Geometric Analysis, (1997).

[2] W. W. Adams, P. Loustaunau, V. P. Palamodov and D. C. Struppa, *Hartogs' Phenomenon for Polyregular Functions and Projective Dimension of Related Modules over a Polynomial Ring*, Ann. Inst. Fourier, **47** (1997), 623–640.

[3] F. Colombo, P. Loustaunau, I. Sabadini and D. C. Struppa, *Regular functions of biquaternionic variables and Maxwell's equations*, to appear in Jour. Geom. Phys. (1998).

[4] L. Ehrenpreis, *A new proof and an extension of Hartogs' Theorem*, Bull. A.M.S. **67** (1961), 507–509.

[5] L. Ehrenpreis, Fourier Analysis in Several Complex Variables, New York, 1970.

[6] A. Kaneko, *Introduction to Hyperfunctions*, Tokyo, 1989.

[7] P. Loustaunau, *Private communication* , February 1997.

[8] V. P. Palamodov, Letter to the authors, July 1995.

[9] V. P. Palamodov, Linear Differential Operators with Constant Coefficients, Springer Verlag, New York, (1970).

[10] F. Severi, *Una proprietá fondamentale di campi di olomorfismo di una variabile reale e di una variabile complessa*, Rend. Acc. Lincei **15**, (1932), 487–490.

[11] D. C. Struppa, *The first eighty years of Hartogs' theorem*, Sem. dell'Univ. di Bologna 1997, (1987).

[12] A. Sudbery, *Quaternionic analysis*, Math. Proc. Camb. Phil. Soc. **85**, (1979), 199–225.

D. Napoletani,
D. C. Struppa
Department of Mathematical Sciences,
George Mason University,
Fairfax (VA), USA.

E-mail address: `Dnapoletani@gmu.edu`
E-mail address: `dstruppa@turtle.gmu.edu`

I. Sabadini,
Dipartimento di Matematica,
Universitá degli Studi di Milano,
Milano, Italy.

Operator Theory:
Advances and Applications, Vol. 114

Bergman-Toeplitz and pseudodifferential operators

V. Rabinovich and N. Vasilevski

Abstract. We show that the problem of studying the Bergman-Toeplitz operators with functional symbols on a plane domain is, in a certain sense, the problem of studying the pseudodifferential operators on a half-line.

The symbols of the pseudodifferential operators we obtain are slowly varying in the additive sense at $+\infty$ and in the multiplicative sense at 0 in the variable x. With respect to the dual variable the operators are a mixture of additive Wiener-Hopf operators and multiplicative Mellin convolutions.

The essential spectra of operators from the Toeplitz operator algebra are in general massive (have positive plain Lebesgue measure).

1. Introduction

The aim of this paper is to show that the problem of studying the Bergman-Toeplitz operators with functional symbols on a plane domain is, in a certain sense, the problem of studying the pseudodifferential operators on a half-line.

Let $\Pi = \mathbb{R} + i\mathbb{R}_+$ be the upper half-plane in $\mathbb{C}$. Introduce the space $L_2(\Pi)$ with the usual Lebesgue plane measure $d\mu(z) = dxdy$, $z = x + iy$, and its Bergman subspace $\mathcal{A}^2(\Pi)$, consisting of all functions analytic in Π. Denote by B_Π the orthogonal Bergman projection of $L_2(\Pi)$ onto $\mathcal{A}^2(\Pi)$.

Let $\mathcal{A}(\Pi)$ be a C^*-subalgebra of $L_\infty(\Pi)$. Given $a \in \mathcal{A}(\Pi)$, denote by T_a the Bergman-Toeplitz, or brieftly Toeplitz, operator

$$T_a : u \in \mathcal{A}^2(\Pi) \longmapsto B_\Pi au \in \mathcal{A}^2(\Pi),$$

acting on the space $\mathcal{A}^2(\Pi)$, and denote by $\mathcal{T}(\mathcal{A}(\Pi))$ the Toeplitz operator algebra, i.e., the algebra generated by all operators T_a with $a \in \mathcal{A}(\Pi)$.

The Bergman-Toeplitz operators, as well as the algebras generated by them, have been studied very intensively, see, for example, [1], [2], [11], [15], [20], [21].

The main property of the algebras $\mathcal{A}$ considered in most of the above papers is that the commutator $[aI, B] = aB - BaI$ is compact for all functions $a \in \mathcal{A}$, or equivalently, the so-called semi-commutator $[T_a, T_b) = T_a \cdot T_b - T_{a\cdot b}$ is compact for

---

The first author was partially supported by RFFI, Grant: RFFI-98-01-01-023.
On leave from the Rostov State University, Rostov-on-Don, Russia.
The second author was partially supported by CONACYT Project 3114P-E9607, México.

all $a, b \in \mathcal{A}$. The very important question of describing of the maximal C^*-algebra satisfying this property was solved in [20] (see also [21]). One of the main features of algebras with compact semi-commutator is that the corresponding Toeplitz operator algebras $\mathcal{T}(\mathcal{A})$ admit a commutative symbolic calculus; i.e., the Fredholm symbol (Calkin) algebra $\mathrm{Sym}\,\mathcal{T}(\mathcal{A}) = \mathcal{T}(\mathcal{A})/\mathcal{K}$, where $\mathcal{K}$ is the ideal of compact operators, is commutative. That implies that the commutator $[T_a, T_b]$, for all $a, b \in \mathcal{A}$, is compact as well.

Note that under the compact semi-commutator condition the Fredholm symbol algebra $\mathrm{Sym}\,\mathcal{R}(\mathcal{A}, B) = \mathcal{R}(\mathcal{A}, B)/\mathcal{K}$ of the algebra $\mathcal{R}(\mathcal{A}, B)$, generated by the Bergman projection B and the multiplication operators aI, $a \in \mathcal{A}$, is commutative as well. At the same time, under only the assumption of compactness of the commutator, the symbol algebra $\mathrm{Sym}\,\mathcal{T}(\mathcal{A}) = \mathcal{T}(\mathcal{A})/\mathcal{K}$ still remains *commutative*, while the symbol algebra $\mathrm{Sym}\,\mathcal{R}(\mathcal{A}, B) = \mathcal{R}(\mathcal{A}, B)/\mathcal{K}$ of the algebra $\mathcal{R}(\mathcal{A}, B)$ is *non-commutative*, and may have a quite complicated structure.

Algebras with compact commutator and non compact semi-commutator (except for the algebra of piecewise continuous functions, see, for example, [15]) are practically unknown. A large class of such algebras has been constructed recently in [18]. Moreover, it is shown there that for each finite set $\Lambda = \langle n_0, n_1, \ldots, n_m \rangle$, where $1 = n_0 < n_1 < \cdots < n_m \leq \infty$, and $n_k \in \mathbb{N} \cup \{\infty\}$, there exist (and are constructed) algebras $\mathcal{A}_\Lambda$ such that the symbol algebras $\mathrm{Sym}\,\mathcal{T}(\mathcal{A}_\Lambda)$ of the algebras $\mathcal{T}(\mathcal{A}_\Lambda)$ are *commutative*, while the symbol algebras $\mathrm{Sym}\,\mathcal{R}(\mathcal{A}_\Lambda, B)$ of the algebras $\mathcal{R}(\mathcal{A}_\Lambda, B)$ have *irreducible representations exactly of dimensions* $n_0, n_1, \ldots, n_m$.

The Toeplitz operator algebra studied in this paper provides a new example of an algebra with compact commutator and non compact semi-commutator. Note that we do not even know at the moment whether the operator $T_{a \cdot b}$ belongs to our Toeplitz operator algebra $\mathcal{T}$ for every T_a, $T_b \in \mathcal{T}$.

One of the interesting features of the algebra $\mathcal{T}$ is that the essential spectra of operators from this algebra are in general massive (have positive plane Lebesgue measure). In particular, the essential spectrum of an operator $T_a \cdot T_b$ can be massive, while the essential spectra of T_a and T_b are never massive.

The principal point in the investigation is the use of the unitary operator $R : L_2(\Pi) \to L_2(\mathbb{R}_+)$ introduced in [17], which reduces Toeplitz operators T_a acting on $L_2(\Pi)$ to the unitary equivalent operators RT_aR^*, acting on $L_2(\mathbb{R}_+)$. The operator R is an exact analog of the Bargmann transform [3], mapping the Fock space $F^2(\mathbb{C}^n)$ onto $L_2(\mathbb{R}^n)$. The passing from Toeplitz operators T_a to the operators RT_aR^* is nothing but an analog of the Berezin reducing [4], [5] of operators with anti-Wick symbols ($\equiv$ Toeplitz operators) on the Fock space to Weyl pseudodifferential operators on $L_2(\mathbb{R}^n)$. This is how pseudodifferential operators appear in our context of Bergman-Toeplitz operators.

The class of pseudodifferential operators $\mathcal{R}$, we have obtained is quite interesting in itself, and extends the class of operators studied by H. Cordes [6]. The symbols of the operators are slowly varying in the additive sense at $+\infty$ and in the multiplicative sense at 0 in the variable x. With respect to the dual variable the operators are a mixture of additive Wiener-Hopf operators and multiplicative Mellin convolutions.

In Section 2 we introduce following [17] the unitary operator $R : L_2(\Pi) \to L_2(\mathbb{R}_+)$, and prepare the unitary equivalent images for future generators T_a, $a = a(y) \in L_\infty(\mathbb{R}_+)$, and T_b, $b = b(x) \in C(\overline{\mathbb{R}})$ of the Toeplitz operator algebra $\mathcal{T}$.

In Sections 3 and 4 we collect the results of [12] describing an algebra of pseudodifferential operators on $L_2(\mathbb{R})$ with slowly varying symbols, and their reformulations for Mellin pseudodifferential operators.

The Section 5 is devoted to a systematic description of an algebra $\mathcal{R}$ of pseudodifferential operators, which is adequate for studying our Toeplitz operators. In Section 6 we apply these results to describe the Fredholm symbol algebra of the Toeplitz operator algebra.

2. Bergman–Toeplitz operators on the upper half-plane

Let $\Pi = \mathbb{R}+i\mathbb{R}_+$ be the upper half-plane in $\mathbb{C}$, introduce the space $L_2(\Pi)$, with the usual Lebesgue plane measure $d\mu(z) = dxdy$, $z = x + iy$, and its subspace $\mathcal{A}^2(\Pi)$, the Bergman space, consisting of all functions analytic in Π. It is well known that the orthogonal projection B_Π of $L_2(\Pi)$ onto $\mathcal{A}^2(\Pi)$, the so-called Bergman projection, is given by

$$(B_\Pi u)(z) = \int_\Pi K_\Pi(z,\overline{\zeta})u(\zeta)\,d\mu(\zeta),$$

where the Bergman kernel function $K_\Pi(z,\overline{\zeta})$ of the upper half-plane is given by

$$K_\Pi(z,\overline{\zeta}) = -\frac{1}{\pi}\cdot\frac{1}{(z-\zeta)^2}.$$

Introduce the unitary operators

$$U_1 = F \otimes I :\ L_2(\Pi) = L_2(\mathbb{R}) \otimes L_2(\mathbb{R}_+) \longrightarrow L_2(\mathbb{R}) \otimes L_2(\mathbb{R}_+),$$

where $F : L_2(\mathbb{R}) \to L_2(\mathbb{R})$ is the Fourier transform

$$(Ff)(x) = \frac{1}{\sqrt{2\pi}}\int_\mathbb{R} e^{-ix\xi} f(\xi)\,d\xi,$$

and

$$U_2 : L_2(\Pi) = L_2(\mathbb{R}) \otimes L_2(\mathbb{R}_+) \longrightarrow L_2(\mathbb{R}) \otimes L_2(\mathbb{R}_+),$$

which is given by the rule

$$U_2 : u(x,y) \longmapsto \frac{1}{\sqrt{2|x|}}\,u(x,\frac{y}{2|x|}).$$

Then the inverse operator $U_2^{-1} = U_2^* : L_2(\mathbb{R}) \otimes L_2(\mathbb{R}_+) \longrightarrow L_2(\mathbb{R}) \otimes L_2(\mathbb{R}_+)$ is given by

$$U_2^{-1} : u(x,y) \longmapsto \sqrt{2|x|}\,u(x,2|x|\cdot y).$$

Let $\ell_0(y) = e^{-y/2}$; we have $\ell_0(y) \in L_2(\mathbb{R}_+)$ and $\|\ell_0(y)\| = 1$. Introduce the isometric imbedding

$$R_0 : L_2(\mathbb{R}_+) \longrightarrow L_2(\mathbb{R}) \otimes L_2(\mathbb{R}_+)$$

by the rule

$$(R_0 f)(x,y) = \chi_+(x)\, f(x)\, \ell_0(y).$$

The adjoint operator $R_0^* : L_2(\Pi) \to L_2(\mathbb{R}_+)$ is given by

$$(R_0^* u)(x) = \chi_+(x) \int_{\mathbb{R}_+} u(x,\eta)\, \ell_0(\eta)\, d\eta,$$

Theorem 2.1. *The operator $R : \mathcal{A}^2(\Pi) \longrightarrow L_2(\mathbb{R}_+)$, where*

$$(Ru)(x) = \sqrt{x}\, \frac{1}{\sqrt{\pi}} \int_{\Pi} u(w)\, e^{-i\overline{w}\cdot x}\, d\mu(w),$$

is an isometric isomorphism, and admits the following decomposition $R = R_0^* U_2$ $(F \otimes I)$.

Corollary 2.2. *The inverse isomorphism*

$$R^* = U^* R_0 : L_2(\mathbb{R}_+) \longrightarrow \mathcal{A}^2(\Pi)$$

is given by

$$(R^* f)(z) = \frac{1}{\sqrt{\pi}} \int_{\mathbb{R}_+} \sqrt{\xi}\, f(\xi)\, e^{iz\cdot\xi}\, dx,$$

and admits the decomposition $R^* = (F^{-1} \otimes I) U_2^{-1} R_0$.

Remark 2.3. We have

$$\begin{aligned} R R^* = I \quad &: \quad L_2(\mathbb{R}_+) \longrightarrow L_2(\mathbb{R}_+), \\ R^* R = B_\Pi \quad &: \quad L_2(\Pi) \longrightarrow \mathcal{A}^2(\Pi). \end{aligned}$$

For each function $a = a(x,y) \in L_\infty(\Pi)$ introduce the Toeplitz operator as follows:

$$T_a : u \in \mathcal{A}^2(\Pi) \longmapsto B_\Pi a u \in \mathcal{A}^2(\Pi).$$

Proposition 2.4. *If a function $a = a(y) \in L_\infty(\mathbb{R}_+)$ does not depend on x, then the Toeplitz operator T_a is unitary equivalent to the multiplication operator*

$$R T_a R^* = \gamma_a(x) I,$$

where

$$\gamma_a(x) = \int_{\mathbb{R}_+} a\Big(\frac{\eta}{2x}\Big)\, \ell_0^2(\eta)\, d\eta, \quad x \in \mathbb{R}_+. \tag{2.1}$$

Proof. In fact

$$\begin{aligned} RT_aR^* &= RB_\Pi aB_\Pi R^* = R(R^*R)a(R^*R)R^* \\ &= (RR^*)RaR^*(RR^*) = RaR^* \\ &= R_0^*U_2(F\otimes I)a(y)(F^{-1}\otimes I)U_2^{-1}R_0 \\ &= R_0^*U_2a(y)U_2^{-1}R_0 \\ &= R_0^*a(\frac{y}{2|x|})R_0. \end{aligned}$$

Calculate now

$$(R_0^*a(\frac{y}{2|x|})R_0f)(x) = \int_{\mathbb{R}_+} a(\frac{\eta}{2|x|})\, f(x)\, \ell_0^2(\eta)\, d\eta = \gamma_a(x)\cdot f(x),$$

where

$$\gamma_a(x) = \int_{\mathbb{R}_+} a(\frac{\eta}{2x})\, \ell_0^2(\eta)\, d\eta, \quad x\in\mathbb{R}_+.$$

□

Proposition 2.5. *If a function $b = b(x) \in L_\infty(\mathbb{R})$ does not depend on y, then the Toeplitz operator T_b is unitary equivalent to the following integral operator,*

$$(RT_bR^*\, f)(x) = \frac{1}{\sqrt{2\pi}}\int_{\mathbb{R}_+} \frac{2\sqrt{xt}}{x+t}\, k(t-x)\, f(t)\, dt,$$

where k is the Fourier transform of the function $b(-x)$.

Proof. Calculate

$$\begin{aligned} RT_bR^*f &= RB_\Pi bB_\Pi R^*f = R(R^*R)b(R^*R)R^*f \\ &= (RR^*)RbR^*(RR^*)f = RbR^*f \\ &= R_0^*U_2(F\otimes I)b(x)(F^{-1}\otimes I)U_2^{-1}R_0f \\ &= R_0^*U_2(F\otimes I)b(x)(F^{-1}\otimes I)U_2^{-1}(\chi_+(x)f(x)\ell_0(y)) \\ &= R_0^*U_2(F\otimes I)b(x)(F^{-1}\otimes I)(\sqrt{2x}\chi_+(x)f(x)e^{-xy}) \\ &= R_0^*U_2(\frac{1}{\sqrt{2\pi}}\int_{\mathbb{R}} \sqrt{2t}\, k(t-x)\chi_+(t)f(t)e^{-ty}\, dt) \\ &= R_0^*(\frac{1}{\sqrt{2\pi}}\int_{\mathbb{R}_+} \sqrt{\frac{t}{|x|}}\, k(t-x)f(t)e^{-\frac{ty}{2|x|}}\, dt) \\ &= \frac{1}{\sqrt{2\pi}}\int_{\mathbb{R}_+} \sqrt{\frac{t}{|x|}}\, k(t-x)f(t)\, dt \int_{\mathbb{R}_+} e^{-\frac{(x+t)y}{2x}}\, dy \\ &= \frac{1}{\sqrt{2\pi}}\int_{\mathbb{R}_+} \frac{2\sqrt{xt}}{x+t}\, k(t-x)\, f(t)\, dt. \end{aligned}$$

□

3. Pseudodifferential operators on $L_2(\mathbb{R})$ with slowly varying symbols

Denote by $S(\mathbb{R}^2)$ the class of all smooth functions $a(x,\xi)$, $(x,\xi) \in \mathbb{R}^2$, such that for all natural numbers α and β there exists a constant $c_{\alpha,\beta} = c_{\alpha,\beta}(a) > 0$ such that

$$|\partial_x^\beta \partial_\xi^\alpha a(x,\xi)| \leq c_{\alpha,\beta}.$$

To each such function $a(x,\xi)$, referred to as a *symbol*, we assign a pseudodifferential operator (ΨDO) $A = \operatorname{Op}(a)$ as follows

$$Au = \operatorname{Op}(a)\, u = (2\pi)^{-1} \int_{\mathbb{R}} d\xi \int_{\mathbb{R}} a(x,\xi)\, e^{i(x-y)\xi}\, u(y)\, dy, \tag{3.1}$$

where $u \in C_0^\infty(\mathbb{R})$. Denote by OPS the class of all ΨDO's with symbols $a(x,\xi) \in S$.

By the Calderon–Vaillancourt theorem (see, for instance [13], Chapter 13), the operator $A = \operatorname{Op}(a) \in OPS$ is bounded on the space $L_2(\mathbb{R})$ and

$$\|A\| \leq C \max_{\alpha \leq 2,\, \beta \leq 1} \sup_{(x,\xi)} |\partial_x^\beta \partial_\xi^\alpha a(x,\xi)|.$$

We say that a smooth function $a(x,y,\xi)$ belongs to $S(\mathbb{R}^3)$ if for all natural numbers α, β, and γ there exists a constant $c_{\alpha,\beta,\gamma} > 0$ such that

$$|\partial_x^\beta \partial_y^\gamma \partial_\xi^\alpha a(x,y,\xi)| \leq c_{\alpha,\beta,\gamma}.$$

For each function $a(x,y,\xi)$, which will be called a *double symbol*, assign the ΨDO $A = \operatorname{Op}(a)$ by the formula (3.1), replacing $a(x,\xi)$ by $a(x,y,\xi)$. It is well-known that the operator $A = \operatorname{Op}(a)$ with a double symbol $a(x,y,\xi)$ is, in fact, a ΨDO with the usual symbol $\sigma_A(x,\xi) \in S(\mathbb{R}^2)$, which is given by

$$\sigma_A(x,\xi) = (2\pi)^{-1} \int_{\mathbb{R}^2} a(x, x+y, \xi+\eta)\, e^{-iy\eta}\, dy d\eta,$$

where the integral is understood as an oscillatory one.

The following statements hold:

- if $A \in OPS$, then $A^* \in OPS$ as well, and

$$\sigma_{A^*}(x,\xi) = (2\pi)^{-1} \int_{\mathbb{R}^2} \overline{a(x+y,\xi+\eta)}\, e^{iy\eta}\, dy d\eta;$$

- if $A, B \in OPS$, then $AB \in OPS$ as well, and

$$\sigma_{AB}(x,\xi) = (2\pi)^{-1} \int_{\mathbb{R}^2} a(x,\xi+\eta)\, b(x+y,\xi)\, e^{iy\eta}\, dy d\eta;$$

where the integrals are understood as oscillatories.

Thus OPS is a (non-closed in $\mathcal{B}(L_2(\mathbb{R}))$ algebra with the involution $A \mapsto A^*$. Moreover OPS is inverse closed, i.e., for each invertible $A \in OPS$, we have $A^{-1} \in OPS$.

We will say that

(i) a symbol $a(x,\xi) \in S(\mathbb{R}^2)$ varies slowly as $x \to \infty$, if

$$\lim_{x\to\infty} \sup_{\xi\in\mathbb{R}} |\partial_x a(x,\xi)| = 0;$$

(ii) a double symbol $a(x,y,\xi) \in S(\mathbb{R}^3)$ varies slowly as $x \to \infty$, if

$$\lim_{x\to\infty} \sup_{\xi\in\mathbb{R},\, y\in K} |\partial_x^\beta \partial_y^\gamma a(x, x+y, \xi)| = 0,$$

where $\beta + \gamma = 1$, and K is a compact set in $\mathbb{R}$;

(iii) a symbol $a(x,\xi) \in S(\mathbb{R}^2)$ varies slowly as $\xi \to \infty$, if

$$\lim_{\xi\to\infty} \sup_{x\in\mathbb{R}} |\partial_\xi a(x,\xi)| = 0;$$

(iv) a double symbol $a(x,y,\xi) \in S(\mathbb{R}^3)$ varies slowly as $\xi \to \infty$, if

$$\lim_{\xi\to\infty} \sup_{(x,y)\in\mathbb{R}^2} |\partial_\xi a(x,y,\xi)| = 0.$$

Denote by $OP\widetilde{S}$ the class of all ΨDO's having slowly varying symbols with respect to both variables x and ξ. Let $J_{\pm\infty}$, and J_x, for each $x \in \mathbb{R}$, be the class of slowly varying symbols, such that

$$\lim_{x\to\pm\infty} \sup_{\xi\in\mathbb{R}} |a(x,\xi)| = 0,$$

and

$$\lim_{\xi\to\infty} |a(x,\xi)| = 0, \quad x \in \mathbb{R},$$

respectively. The corresponding classes of ΨDO's denote by $OPJ_{\pm\infty}$, and OPJ_x.

Let $\overline{\mathbb{R}} = \mathbb{R} \cup \{+\infty\} \cup \{-\infty\} = [-\infty, +\infty]$ be a two-point compactification of $\mathbb{R}$. Then

$$OPJ = \bigcap_{x\in\overline{\mathbb{R}}} OPJ_x \subset \mathcal{K},$$

where $\mathcal{K}$ is the ideal of all compact on $L_2(\mathbb{R})$ operators.

Proposition 3.1. *The following statements hold*

(i) *if* $A = \mathrm{Op}\,(a)$, $B = \mathrm{Op}\,(b) \in OP\widetilde{S}$, *then* $AB \in OP\widetilde{S}$, *and*

$$AB = \mathrm{Op}\,(ab) + T, \quad T \in OPJ;$$

(ii) *if* $A = \mathrm{Op}\,(a)) \in OP\widetilde{S}$, *then* $A^* \in OP\widetilde{S}$, *and*

$$A^* = \mathrm{Op}\,(\overline{a}) + T, \quad T \in OPJ;$$

(iii) *for each slowly varying double symbol* $a(x,y,\xi)$ *the operator* $A = \mathrm{Op}\,(a)$ *belongs to* $OP\widetilde{S}$, *and*

$$A = \mathrm{Op}\,(a(x,x,\xi)) + T, \quad T \in OPJ.$$

Thus the algebra $OP\widetilde{S}$ is a non-closed in $\mathcal{B}(L_2(\mathbb{R})$ algebra with involution, and the sets OPJ_x, $x \in \overline{\mathbb{R}}$, and OPJ are two-sided ideals in $OP\widetilde{S}$.

Denote by $\mathcal{A}$ the C^*-algebra, generated by all the operators from $OP\widetilde{S}$, and by $J(x)$, $x \in \overline{\mathbb{R}}$, and J the two-sided ideals in $\mathcal{A}$, generated by OPJ_x, $x \in \overline{\mathbb{R}}$, and OPJ, respectively.

We have

$$J = \bigcap_{x \in \overline{\mathbb{R}}} J(x) = \mathcal{K}. \tag{3.2}$$

We recall now a general local principle [16], based on the Dauns-Hofmann construction (see, for references, [7], [10]).

Let $\mathcal{R}$ be a C^*–algebra, and let $J_T = \{J(t) : t \in T\}$ be a system of closed two–sided ideals of $\mathcal{R}$ indexed by points of some set T. For each $t \in T$ introduce the quotient algebra

$$\mathcal{R}(t) = \mathcal{R}/J(t).$$

We denote by $a(t)$ the image of $a \in \mathcal{R}$ in $\mathcal{R}(t)$. Let

$$E = \bigsqcup_{t \in T} \mathcal{R}(t)$$

be the disjointed union of the algebras $\mathcal{R}(t)$. Denote by

$$p: \quad E \longrightarrow T$$

the following projection mapping

$$p: \quad a(t) \longmapsto t.$$

Of course, $p^{-1}(t) = \mathcal{R}(t)$.

Introduce the topologies on E and T which make the triple $\xi = (p, E, T)$ into a C^*–bundle. Each element $a \in \mathcal{R}$ generates a section $\widetilde{a} : T \to E$ of ξ by the rule

$$\widetilde{a}: \quad t \longmapsto a(t).$$

Denote by $\widetilde{\mathcal{R}}$ the set of all these sections.

For each $\varepsilon > 0$ and arbitrary $\widetilde{a} \in \widetilde{\mathcal{R}}$ introduce the set

$$U(\widetilde{a}, \varepsilon) = \{x \in E : \|x - \widetilde{a}(p(x))\| < \varepsilon\},$$

here $\|\cdot\|$ means the norm in the C^*–algebra $\mathcal{R}(p(x))$. Provide the set E with the topology whose prebase consists of with all the sets $U(\widetilde{a}, \varepsilon)$.

Remark 3.2. The topology on each fiber $\xi(t) = \mathcal{R}(t)$ generated by the quotient-norm

$$\|a(t)\| = \inf_{z \in J(t)} \|a + z\|$$

coincides with the topology on $\xi(t)$ generated by the restriction of the topology in E to $\xi(t)$.

Provide the base T of the bundle $\xi = (p, E, T)$ with the quotient topology: the strongest topology for which the projection $p : E \to T$ is continuous. This topology is usually called the $*$–bundle topology and has the following properties (see for details [7], [10]):

1. the mapping $p : E \to T$ is open,
2. this topology coincides with the weakest topology for which all the mappings $\widetilde{a} : T \to E$ are continuous,
3. the prebase of this topology can be defined by all the sets

$$V(\widetilde{a}, \varepsilon) = \{t \in T : \|\widetilde{a}(t)\| < \varepsilon\}.$$

The described C^*-bundle $\xi = (p, E, T)$ is called the *canonical C^*-bundle* defined by the C^*-algebra $\mathcal{R}$ and by the system of ideals J_T.

Let now $\xi = (p, E, T)$ be a C^*–bundle. As usual by $\Gamma^b(\xi)$ we denote the C^*–algebra of all its bounded continuous sections σ with respect to component-wise operations and the norm

$$\|\sigma\| = \sup_{t \in T} \|\sigma(t)\|.$$

Theorem 3.3. *Let $\mathcal{R}$ be a C^*–algebra, $J_T = \{J(t) : t \in T\}$ be a system of its closed two–sided ideals; let $\xi = (p, E, T)$ be the canonical C^*–bundle defined by $\mathcal{R}$ and J_T, and let $\Gamma^b(\xi)$ be C^*–algebra of bounded continuous sections of ξ. Then the mapping*

$$\widetilde{\pi} : a \in \mathcal{R} \longmapsto \widetilde{a} \in \Gamma^b(\xi)$$

is a morphism of C^*–algebras $\mathcal{R}$ and $\Gamma^b(\xi)$ under which

(i) $\ker \widetilde{\pi} = \bigcap_{t \in T} J(t)$,

(ii) $\operatorname{Im} \widetilde{\pi} = \widetilde{\mathcal{R}}$.

Let now $\mathcal{R}$ be an operator C^*-algebra which contains the ideal $\mathcal{K}$ of the compact operators, and let $\widehat{\mathcal{R}} = \mathcal{R}/\mathcal{K}$ be the quotient algebra. Denote by $\widehat{a}$ the image of an element $a \in \mathcal{R}$ under the natural projection

$$\pi : \mathcal{R} \longrightarrow \widehat{\mathcal{R}}. \tag{3.3}$$

Then the algebras $\widehat{\mathcal{R}}$ and $\widetilde{\mathcal{R}}$ are isomorphic if and only if

$$\bigcap_{t \in T} J(t) = \mathcal{K},$$

and in this case the mapping $\widetilde{\pi} : \mathcal{R} \to \widetilde{\mathcal{R}}$ coincides with the natural projection (3.3).

In fact, we have the system of so-called local algebras $\mathcal{R}(t) = \mathcal{R}/J(t)$ parameterized by points $t \in T$. Elements a_1 and a_2 from the algebra $\mathcal{R}$ are locally equivalent at the point $t \in T : a_1 \overset{t}{\sim} a_2$, if and only if, $a_1 - a_2 \in J(t)$. The natural projection

$$\pi_t : \mathcal{R} \to \mathcal{R}(t)$$

identifies the locally equivalent at the point t elements of the algebra $\mathcal{R}$. The element $a(t) = \pi_t(a)$ is called the local representative of $a \in \mathcal{R}$ in the local algebra $\mathcal{R}(t)$. Moreover,

(i) $\|\widehat{a}\| = \sup_t \|a(t)\|$,

(ii) the algebra $\widehat{\mathcal{R}}$ admits a canonical description as the algebra of continuous sections of the canonically defined C^*–bundle over T.

Return now to the algebra $\mathcal{A}$ and the system of ideals $J_{\overline{\mathbb{R}}} = \{J(x) : x \in \overline{\mathbb{R}}\}$, with the property (3.2). To describe the local algebras $\mathcal{A}(x)$, $x \in \overline{\mathbb{R}}$, and the homomorphisms

$$\pi_x : \mathcal{A} \longrightarrow \mathcal{A}(x)$$

we introduce a number of algebras.

Let $C_b(\mathbb{R}^2)$ be the algebra of all continuous bounded in $\mathbb{R}^2$ functions, and

$$\begin{aligned} C_0^{\pm}(\mathbb{R}^2) &= \{a(x,\xi) \in C_b(\mathbb{R}^2) : \lim_{x\to\pm\infty} \sup_{\xi\in\mathbb{R}} |a(x,\xi)| = 0\}, \\ C_0(\mathbb{R}) &= \{a(\xi) \in C_b(\mathbb{R}) : \lim_{\xi\to\infty} |a(\xi)| = 0\}, \\ C_0(\mathbb{R}^2) &= \{a(x,\xi) \in C_b(\mathbb{R}^2) : \lim_{(x,\xi)\to\infty} |a(x,\xi)| = 0\}. \end{aligned}$$

Lemma 3.4. *Let $x_0 \in \mathbb{R} \subset \overline{\mathbb{R}}$, then the local algebra $\mathcal{A}(x_0)$ is isometrically imbedded into the quotient algebra $C_b(\mathbb{R})/C_0(\mathbb{R})$. Under the identification of $\mathcal{A}(x_0)$ with its image, the homomorphism*

$$\pi_{x_0} : \mathcal{A} \longrightarrow \mathcal{A}(x_0) \subset C_b(\mathbb{R})/C_0(\mathbb{R})$$

is generated by the following mapping of the generators of the algebra $\mathcal{A}$

$$\pi_{x_0} : A = \mathrm{Op}\,(a(x,\xi)) \longmapsto a(x_0,\xi) + C_0(\mathbb{R}),$$

and

$$\|A(x_0)\| = \|\pi_{x_0}(A)\| = \limsup_{\xi\to\infty} |a(x_0,\xi)|.$$

Lemma 3.5. *Let $x_0 = \pm\infty \in \overline{\mathbb{R}}$, then the local algebra $\mathcal{A}(x_0)$ is isometrically imbedded into the quotient algebra $C_b(\mathbb{R}^2)/C_0^{\pm}(\mathbb{R}^2)$. Under the identification of $\mathcal{A}(x_0)$ with its image, the homomorphism*

$$\pi_{x_0} : \mathcal{A} \longrightarrow \mathcal{A}(x_0) \subset C_b(\mathbb{R}^2)/C_0^{\pm}(\mathbb{R}^2)$$

is generated by the following mapping of the generators of the algebra $\mathcal{A}$,

$$\pi_{x_0} : A = \mathrm{Op}\,(a(x,\xi)) \longmapsto a(x,\xi) + C_0^{\pm}(\mathbb{R}^2),$$

and

$$\|A(x_0)\| = \|\pi_{x_0}(A)\| = \limsup_{x\to\pm\infty} \sup_{\xi\in\mathbb{R}} |a(x,\xi)|.$$

Theorem 3.6. *The (Fredholm) symbol algebra* $\operatorname{Sym}\mathcal{A} = \mathcal{A}/\mathcal{K}$ *of the algebra* $\mathcal{A}$ *is isometrically imbedded into the quotient algebra* $C_b(\mathbb{R}^2)/C_0(\mathbb{R}^2)$. *Under the identification of* $\operatorname{Sym}\mathcal{A}$ *with its image, the symbol homomorphism*

$$\operatorname{sym} : \mathcal{A} \longrightarrow \operatorname{Sym}\mathcal{A} \subset C_b(\mathbb{R}^2)/C_0(\mathbb{R}^2)$$

is generated by the following mapping of the generators of the algebra $\mathcal{A}$,

$$\operatorname{sym} : A = \operatorname{Op}(a(x,\xi)) \longmapsto a(x,\xi) + C_0(\mathbb{R}^2),$$

and

$$\|\operatorname{sym} A\| = \limsup_{(x,\xi)\to\infty} |a(x,\xi)|.$$

Corollary 3.7. *An operator* $A = \operatorname{Op}(a)$ *from the algebra* $\mathcal{A}$ *is a Fredholm operator if and only if its symbol* $\operatorname{sym} A$ *is invertible in the algebra* $\operatorname{Sym}\mathcal{A} \subset C_b(\mathbb{R}^2)/C_0(\mathbb{R}^2)$, *i.e., if and only if*

$$\liminf_{(x,\xi)\to\infty} |a(x,\xi)| > 0.$$

The index of the Fredholm operator A *is given by*

$$\operatorname{Ind} A = -\frac{1}{2\pi}[\arg a(x,\xi)]_{\Gamma_R},$$

where Γ_R *is the (positively oriented) boundary of the square* $\Pi_R = \{(x,\xi) : |x| < R,\ |\xi| < R\}$, *and* R *is such that*

$$\inf_{(x,\xi)\in\mathbb{R}^2\setminus\Pi_R} |a(x,\xi)| > 0.$$

4. Mellin pseudodifferential operators on $\mathbb{R}_+$

Consider the class of Mellin pseudodifferential operators on $\mathbb{R}_+$ which are obtained from pseudodifferential operators on $\mathbb{R}$ of the class OPS by means of a change of variables

$$\mathbb{R} \ni x = \ln r,\ r \in \mathbb{R}_+.$$

To be more precise, introduce the unitary operator $V : L_2(\mathbb{R}) \longrightarrow L_2(\mathbb{R}_+, \frac{dr}{r})$ by the rule

$$(Vu)(r) = u(\ln r),$$

then the class of Mellin pseudodifferential operators on $\mathbb{R}_+$ which we are interested in is just $V\,OPS\,V^{-1}$. Recall that the measure dr/r coincides with the Haar measure on the multiplicative group $\mathbb{R}_+$.

We will say that a function $a(r,\lambda)$ defined on $\mathbb{R}_+ \times \mathbb{R}$ is in $\mathcal{E}$ if $a(r,\lambda)$ belongs to $C^\infty(\mathbb{R}_+ \times \mathbb{R})$ and satisfies the following estimates

$$|(r\partial_r)^k(\partial_\lambda)^j a(r,\lambda)| < \infty,$$

for all $k, j \in \mathbb{N} \cup 0$, and that $a(r,\lambda)$ $(\in \mathcal{E})$ varies slowly at the point 0 if

$$\lim_{r\to+0}\sup_{\lambda\in\mathbb{R}} |\,(r\partial_r)a(r,\lambda)| = 0.$$

Denote by $\widetilde{\mathcal{E}}$ the class of functions in $\mathcal{E}$ which vary slowly at the origin, and by $\mathcal{F}$ the class of functions as in $\mathcal{E}$ such that

$$\lim_{r\to+0}\sup_{\lambda\in\mathbb{R}} |a(r,\lambda)| = 0.$$

To each function $a(r,\lambda) \in \mathcal{E}$ assign a Mellin ΨDO $A = \mathrm{Op}_M(a)$ as follows,

$$(\mathrm{Op}_M(a)u)(r) = a(r,D_r)u = (2\pi)^{-1}\int_{\mathbb{R}} d\lambda \int_{\mathbb{R}_+} a(r,\lambda)\,(r\rho^{-1})^{i\lambda}u(\rho)\rho^{-1}d\rho, \quad (4.1)$$

where $u \in C_0^\infty(\mathbb{R}_+)$. Denote by $OP\mathcal{E}$ the class of all such operators, and by $OP\mathcal{F}$ the class of Mellin ΨDO with symbols from $\mathcal{F}$.

Denote by $\mathcal{E}_d$ the class of functions $a(r,\rho,\lambda)$ $(\in C^\infty(\mathbb{R}_+\times\mathbb{R}_+\times\mathbb{R}))$ such that

$$\sup_{\mathbb{R}_+^2\times\mathbb{R}} |(r\partial_r)^k(\rho\partial_\rho)^l\partial_\lambda^j a(r,\rho,\lambda)| < \infty,$$

for all $k, j, l \in \mathbb{N}\cup\{0\}$. Then the Mellin ΨDO with a double symbol $a(r,\rho,\lambda)$ is defined by (4.1), where $a(r,\lambda)$ is replaced by a double symbol. Let $OP\mathcal{E}_d$ stands for the class of such operators.

We will say that a double symbol $a(r,\rho,\lambda) \in \mathcal{E}_d$ *varies slowly at the point* 0 if

$$\lim_{r\to+0}\sup_{\rho\in K,\,\lambda\in\mathbb{R}} |(r\partial_r)^k(\rho\partial_\rho)^l a(r,r\rho,\lambda)| = 0,$$

for all k, l with $k + l = 1$, and for every compact $K \subset \mathbb{R}_+$.
Denote by $\widetilde{\mathcal{E}}_d$ the class of slowly varying at the point 0 double symbols and by $OP\widetilde{\mathcal{E}}_d$ the corresponding class of Mellin ΨDO.

We summarize the properties of the Mellin pseudodifferential operators:

Proposition 4.1. (i) *An operator $A \in OP\mathcal{E}$ is bounded in $L_2(\mathbb{R}_+, \frac{dr}{r})$ and*

$$\|A\|_{L_2(\mathbb{R}_+,\frac{dr}{r})} \le C \max_{\alpha\le1,\,\beta\le2}\ \sup_{\mathbb{R}_+\times\mathbb{R}} \left|(r\partial_r)^\beta\partial_\lambda^\alpha a_{ij}(r,\lambda)\right|.$$

(ii) *If $A \in OP\mathcal{E}$ is invertible in $L_2(\mathbb{R}_+, d\mu)$, then $A^{-1} \in OP\mathcal{E}$.*

(iii) *Let $A = \mathrm{Op}_M(a) \in OP\widetilde{\mathcal{E}}$ and $B = \mathrm{Op}_M(b) \in OP\widetilde{\mathcal{E}}$. Then $AB \in OP\widetilde{\mathcal{E}}$ and*

$$AB = \mathrm{Op}_M(ab) + T, \quad T \in OP\mathcal{F}.$$

(iv) *Let $A = \mathrm{Op}_M(a) \in OP\widetilde{\mathcal{E}}(N))$, then $A^* \in OP\widetilde{\mathcal{E}}$ and*

$$A = \mathrm{Op}_M(\overline{a}) + T, \quad T \in OP\mathcal{F}.$$

(v) *Let A be a Mellin pseudodifferential operator with double symbol $a(r,\rho,\lambda) \in \widetilde{\mathcal{E}}_d$. Then $A \in OP\widetilde{\mathcal{E}}$ and*

$$A = \mathrm{Op}_M(a(r,r,\lambda)) + T, \quad T \in OP\mathcal{F}.$$

Let $\varphi(x) \in C_b^\infty(\ R)$ be such that $0 \le \varphi(x) \le 1$, $\varphi(x) = 1$ for $x \ge 2$, $\varphi(x) = 0$ for $x \le 1$, and $\varphi_R(x) = \varphi(x/R), R > 0$. We will call $\varphi(x)$ a cut-off function of the point $+\infty$. Introduce a cut-off function of the point 0 as follows:

$$\varphi_{0,R}(t) = \varphi_R(-\ln t).$$

Note that $\varphi_{0,R}(t) = 1$, if $0 < t \le \exp(-2R)$, and $\varphi_{0,R}(t) = 0$, if $t \ge \exp(-R)$. Moreover, $\varphi_{0,R}(t)$ satisfies the estimates

$$\left(t\frac{d}{dt}\right)^k \varphi_{0,R}(t) = O\left(\frac{1}{R^k}\right), \quad R \to \infty.$$

Proposition 4.2. *Let $\varphi_{0,R}$ be a cut-off function of the point 0, and let $A \in OP\mathcal{F}$. Then*

$$\lim_{R\to\infty} \|\varphi_{0,R}(t)A\|_{B(L_2(\mathbb{R}_+,d\mu))} = \lim_{R\to\infty} \|A\varphi_{0,R}(t)\|_{B(L_2(\mathbb{R}_+,d\mu))} = 0$$

Hence $OP\widetilde{\mathcal{E}}$ is a non-closed in $B(L_2(\mathbb{R}_+, \frac{dr}{r}))$ algebra with involution, and $\mathcal{F}$ is a two-sided ideal in $OP\widetilde{\mathcal{E}}$. Denote by $\mathcal{W}$ the C^*-algebra generated by all operators from $OP\widetilde{\mathcal{E}}$ and by $\mathcal{W}_0$ its two-sided ideal generated by $OP\mathcal{F}$.

Let $\mathcal{W}(0) = \mathcal{W}/\mathcal{W}_0$ be the local algebra at the point $0 \in \overline{\mathbb{R}}_+$ and let

$$\pi_0 : \mathcal{W} \longrightarrow \mathcal{W}(0)$$

be the canonical homomorphism. The set

$$C_0(\mathbb{R}_+ \times \mathbb{R}) = \{a(r,\lambda) \in C_b(\mathbb{R}_+ \times \mathbb{R}) : \lim_{r\to 0} \sup_{\lambda\in\mathbb{R}} |a(r,\lambda)| = 0\}$$

is obviously an ideal in the algebra $C_b(\mathbb{R}_+ \times \mathbb{R})$.

Corollary 4.3. *The (local) algebra $\mathcal{W}(0)$ is isometrically imbedded into the quotient algebra $C_b(\mathbb{R}_+ \times \mathbb{R})/C_0(\mathbb{R}_+ \times \mathbb{R})$. Under the identification of $\mathcal{W}(0)$ with its image, the homomorphism*

$$\pi_0 : \mathcal{W} \longrightarrow \mathcal{W}(0) \subset C_b(\mathbb{R}_+ \times \mathbb{R})/C_0(\mathbb{R}_+ \times \mathbb{R})$$

is generated by the following (local symbolic) mapping of the generators of the algebra $\mathcal{W}$

$$\pi_0 : A = \mathrm{Op}_M\,(a(r,\lambda)) \longmapsto a(r,\lambda) + C_0(\mathbb{R}_+ \times \mathbb{R}),$$

and

$$\|\pi_0(A)\| = \limsup_{r\to 0} \sup_{\lambda\in\mathbb{R}} |a(r,\lambda)|.$$

An operator $A \in \mathcal{W}$ is locally invertible at the point 0 if and only if the coset $A+\mathcal{W}_0$ is invertible in $\mathcal{W}(0)$, that is $\pi_0(A)$ is invertible in $C_b(\mathbb{R}_+ \times \mathbb{R})/C_0(\mathbb{R}_+ \times \mathbb{R})$, or

$$\liminf_{r\to 0} \sup_{\lambda\in\mathbb{R}} |a(r,\lambda)| > 0,$$

where $a(r,\lambda) \in \pi_0(A)$.

5. On a C^*-algebra of integral operators on $\mathbb{R}_+$

Consider the class of integral operators of the form

$$(Au)(x) = \frac{1}{\sqrt{2\pi}} \int_{\mathbb{R}_+} b(x,y)k(x-y)u(y)dy, \quad x \in \mathbb{R}_+ \tag{5.1}$$

where

(i) the Fourier transform $\widehat{k}(\xi)$ of the function $k(x)$ satisfies the following estimates

$$|\widehat{k}^{(j)}(\xi)| \le C_j \langle \xi \rangle^{-j}, \quad j = 0, 1, ..., \quad \langle \xi \rangle = \sqrt{1+\xi^2} \tag{5.2}$$

(ii) there exist limits

$$\widehat{k}_\pm = \lim_{\xi \to \pm\infty} \widehat{k}(\xi),$$

and

$$\widehat{k_0}(\xi) = \widehat{k}(\xi) - \widehat{k}_+ \chi_+(\xi) - \widehat{k}_- \chi_-(\xi) \tag{5.3}$$

is the Fourier transform of a function $k_0(x) \in L_1(\mathbb{R})$,

(iii) $b(x,y) \in C^\infty(\mathbb{R}_+ \times \mathbb{R}_+)$, and $b(tx, ty) = b(x,y)$, for all $t > 0$,

(iv)

$$\int_{\mathbb{R}_+} \frac{|b(t,1)|}{\sqrt{t}(1+t)} dt < \infty. \tag{5.4}$$

Rearranging constants in the kernel of the integral operator (5.1), consider new functions

$$b'(x,y) = \frac{b(x,y)}{b(1,1)},$$
$$k'(x-y) = b(1,1) \cdot k(x-y).$$

Thus we will always assume that the function $b(x,y)$ satisfies the extra condition:

$$b(1,1) = 1.$$

Introduce an important characteristic of the operator A, the Mellin transform of the function $-i\sqrt{\frac{2}{\pi}\frac{b(t,1)}{(1-t)}}$,

$$q_A(\lambda) = M\left((-i)\sqrt{\frac{2}{\pi}\frac{b(t,1)}{1-t}}\right)(\lambda) = \frac{1}{i\pi}\int_{\mathbb{R}_+} t^{1/2-i\lambda}\frac{b(t,1)}{1-t}\frac{dt}{t}, \quad \lambda \in \mathbb{R}, \tag{5.5}$$

where the Mellin transform $M : L_2(\mathbb{R}_+) \longrightarrow L_2(\mathbb{R})$ is given by

$$(Mu)(\lambda) = \frac{1}{\sqrt{2\pi}} \int_{\mathbb{R}_+} t^{1/2-i\lambda} u(t) \frac{dt}{t}.$$

We will need the following known result (see, for example, [19], p. 102).

Lemma 5.1. *Let $f(x)$ be a measurable function on $\mathbb{R}$ satisfying the conditions:*

(i)

$$\int_{\mathbb{R}} \frac{|f(x)|}{1+|x|} dx < \infty,$$

(ii) *$f(x)$ is continuously differentiable in a neighborhood of the origin.*
Then

$$\lim_{\lambda \to \pm\infty} \frac{1}{i\pi} \int_{\mathbb{R}} e^{i\lambda x} f(x) \frac{dx}{x} = \pm f(0).$$

Proposition 5.2. *Under the above conditions we have*

$$\lim_{\lambda \to \pm\infty} q_A(\lambda) = \pm 1.$$

Proof. Let $t = e^x$, then

$$q_A(\lambda) = \frac{1}{i\pi} \int_{\mathbb{R}} e^{(1/2 - i\lambda)x} \frac{b(e^x, 1)}{1 - e^x} dx$$

Set

$$f(x) = \frac{e^{x/2} b(e^x, 1)}{1 - e^x},$$

and

$$f(x) = f(x)\varphi(x) + f(x)(1 - \varphi(x)) = f_1(x) + f_2(x),$$

where $\varphi(x)$ is a cut-off function of the point 0. The function $f_2(x)$ is absolutely integrable by condition (5.4). This yields that

$$\lim_{\lambda \to \pm\infty} \int_{\mathbb{R}} e^{-i\lambda x} f_2(x) dx = 0.$$

Let now

$$f_1(x) = \frac{f_1(x)x}{x} = \frac{f_3(x)}{x}.$$

The function $f_3(x)$ satisfies the conditions of Lemma 5.1, hence

$$\lim_{\lambda \to \pm\infty} q_A(\lambda) = \lim_{\lambda \to \pm\infty} \frac{1}{i\pi} \int_{\mathbb{R}} e^{-i\lambda x} f_3(x) \frac{dx}{x} = \mp f_3(0).$$

Finally,

$$f_3(0) = \lim_{x \to 0} \frac{e^{x/2} b(e^x, 1) x \varphi(x)}{1 - e^x} = -b(1, 1) = -1.$$

□

Proposition 5.3. *Let*

$$\sup_{\lambda \in \mathbb{R}} |q_A(\lambda)| < \infty. \tag{5.6}$$

Then the operator A is bounded on $L_2(\mathbb{R}_+)$.

Proof. Let $\varphi_{0,R}$ be a cut-off function of the point 0, and $\varphi_{2,R}$ be a cut-off function of the point $+\infty$. Set $\varphi_{1,R} = 1-\varphi_{0,R}-\varphi_{2,R}$. Let $\psi_{j,R}$ be such that $\varphi_{j,R}\psi_{j,R} = \psi_{j,R}$, and $\psi_{j,R}$, $j = 0, 2$, are cut-off functions of the points 0, $+\infty$, respectively. We have

$$A = \sum_{j=0}^{2} \varphi_{j,R}A = \sum_{j=0}^{2} \varphi_{j,R}A\psi_{j,R} + \sum_{j=0}^{2} \varphi_{j,R}A(1-\psi_{j,R}) \tag{5.7}$$

The estimate (5.2) and integrating by parts in the Fourier integral provide the following estimates for $k(x)$,

$$\left|\left(\frac{d}{dx}\right)^m k(x)\right| \leq C_{mn}\,|x|^{m-2n} \tag{5.8}$$

for arbitrary m, n such that $2n-m > 1$. The condition $\operatorname{supp}\varphi_{j,R}\cap\operatorname{supp}(1-\psi_{j,R}) = \emptyset$ and the estimate (5.8) provide the boundedness of the operator

$$T = \sum_{j=0}^{2} \varphi_{j,R}A(1-\psi_{j,R}).$$

The operator $\sum_{j=1}^{2}\varphi_{j,R}A\psi_{j,R}$ is bounded as an operator of the class OPS. Let us consider the operator $\varphi_{0,R}A\varphi_{0,R}$. It follows from (5.3) that

$$\frac{1}{\sqrt{2\pi}}k(x) = \frac{1}{2}\left[(\widehat{k}_+ + \widehat{k}_-)\delta(x) + (\widehat{k}_+ - \widehat{k}_-)\frac{1}{i\pi}\mathcal{P}(\frac{1}{x})\right] + \frac{1}{\sqrt{2\pi}}k_0(x) \tag{5.9}$$

where $\delta(x)$ is the Dirac distribution, and $\mathcal{P}(\frac{1}{x})$ is the distribution defining the principal value integral, $k_0(x) \in L_1(\mathbb{R})$. The condition $k_0 \in L_1(\mathbb{R}_+)$ implies that the operator

$$\varphi_{0,R}K_0\psi_{0,R}u(x) = \frac{1}{\sqrt{2\pi}}\int_{\mathbb{R}_+} \varphi_{0,R}(x)b(x,y)k_0(x-y)\psi_{0,R}(y)u(y)dy, \quad x \in \mathbb{R}_+$$

is bounded. Hence we should prove the boundedness of the operator $\varphi_{0,R}Q\psi_{0,R}$ where

$$Qu(x) = \frac{1}{i\pi}\int_{\mathbb{R}_+}\frac{b(x,y)}{y-x}u(y)dy = \frac{1}{\sqrt{2\pi}}\int_{\mathbb{R}_+}(-i)\sqrt{\frac{2}{\pi}}\frac{b(x/y,1)}{1-x/y}u(y)\frac{dy}{y}, \quad x \in \mathbb{R}_+ \tag{5.10}$$

The operator (5.10) is a Mellin convolution operator. The condition (5.6) provides the boundedness of Q in $L_2(\mathbb{R}_+)$ and at the same time the boundedness of $\varphi_{0,R}Q\psi_{0,R}$. □

Denote by $SV^\infty(\mathbb{R}_+)$ the class of smooth functions $a(x)$ on $\mathbb{R}_+$ which vary slowly at infinity in the additive sense and vary slowly at the point 0 in the multiplicative sense, i.e., the functions $a(x) \in C^\infty(\mathbb{R}_+)$, such that for each $j \in \mathbb{N}$ and $\varepsilon > 0$

$$\sup_{x\in[\varepsilon,+\infty)} |a^{(j)}(x)| \leq c_j, \qquad \sup_{x\in(0,\varepsilon]}\left|\left(x\frac{d}{dx}\right)^j a(x)\right| \leq p_j,$$

and

$$\lim_{x\to\infty} a'(x) = 0,$$

$$\lim_{x\to 0} xa'(x) = 0.$$

Let SV be the closure of SV^∞ in $C_b(\mathbb{R}_+)$. Denote by $\mathcal{R}$ the C^*-algebra, generated by integral operators of the form (5.1) with the property (5.6), and by the multiplication operators $a(x)I$, with $a \in SV^\infty$.

Proposition 5.4. *Let A be an operator of the form (5.1) and such that the condition (5.6) holds, and let $a(x) \in SV^\infty(\mathbb{R}_+)$. Then the commutator $[a(x)I, A] = a(x)A - Aa(x)I$ is compact on $L_2(\mathbb{R}_+)$.*

Proof. We take advantage of the representation of A in the form (5.7). This implies that

$$[a, A] = \sum_{j=0}^{2} \varphi_{j,R}[a, A]\psi_{j,R} + \sum_{j=0}^{2} \varphi_{j,R}[a, A](1 - \psi_{j,R})$$

It is evident that the operator $\sum_{j=0}^{2} \varphi_{j,R}[a, A](1 - \psi_{j,R})$ is compact for arbitrary R. Moreover, the operator $\sum_{j=1}^{2} \varphi_{j,R}[a, A]\psi_{j,R}$ is compact for arbitrary R, because a and A are pseudodifferential operators in the class $OP\widetilde{S}$ outside a neighborhood of the point 0.

Let us consider the operator $\varphi_{0,R}[a, A]\varphi_{0,R}$. Use the representation (5.9). The operator $\varphi_{0,R}K_0\psi_{0,R}$ is not only bounded, but also is compact. The operator $\varphi_{0,R}[a, K_0]\psi_{0,R}$ is a compact operator for arbitrary R as well. Let us prove that $\varphi_{0,R}[a, Q]\psi_{0,R}$ has the norm tending to 0 if $R \to \infty$. Indeed, the operator Q in a neighborhood of the point 0 is a Mellin pseudodifferential operator with symbol $q_A(\lambda) \in OP\widetilde{\mathcal{E}}$. Moreover $a(x) \in SV^\infty$, which implies $a \in OP\widetilde{\mathcal{E}}$ also. From Proposition 4.1 and Proposition 4.2 it follows that

$$\lim_{R\to\infty} \|\varphi_{0,R}[a, Q]\psi_{0,R}\| = 0$$

Thus there exists a sequence of compact operators tending to $[a, A]$ with respect to the norm. Hence $[a, A]$ is compact on $L_2(\mathbb{R}_+)$. □

Corollary 5.5. *Let $a(x) \in C(\overline{\mathbb{R}}_+)$, then $[a(x)I, A] \in \mathcal{K}(L_2(\mathbb{R}_+))$.*

Proof. This is evident for $a(x) \in C_b^\infty \cap C(\overline{\mathbb{R}}_+) \subset SV^\infty$; then use standard approximation arguments. □

To describe the (Fredholm) symbol algebra $\operatorname{Sym}\mathcal{R} = \mathcal{R}/\mathcal{K} = \widehat{\mathcal{R}}$ of the algebra $\mathcal{R}$ we will use the standard local principle (see, for example, [8], [14]). The algebra $C(\overline{\mathbb{R}}_+)$ is obviously a central commutative subalgebra of $\widehat{\mathcal{R}} = \operatorname{Sym}\mathcal{R}$. For each $x_0 \in \overline{\mathbb{R}}_+$ denote by J_{x_0} the maximal ideal of $C(\overline{\mathbb{R}}_+)$ corresponding to x_0, and by

$J(x_0)$ the closed two-sided ideal of the algebra $\widehat{\mathcal{R}}$, generated by J_{x_0}. Then the local algebra at the point x_0 is defined as $\mathcal{R}(x_0) = \widehat{\mathcal{R}}/J(x_0)$, and the natural projection

$$\pi_{x_0} : \mathcal{R} \longrightarrow \operatorname{Sym}\mathcal{R} \longrightarrow \mathcal{R}(x_0)$$

identifies locally equivalent at the point x_0 elements of the algebra $\mathcal{R}$.

Note that this local principle is a special case of the general one, considered for the following system of ideals

$$J'(t) = \pi^{-1}(J(t)), \quad t \in T = \overline{\mathbb{R}}_+.$$

Let $x_0 \in (0, +\infty) \subset \overline{\mathbb{R}}_+$. The multiplication operator $a(x)I$, where $a \in SV$, is locally equivalent at the point x_0 to the scalar operator $a(x_0)I$. Now let $\varphi_{x_0,R}$ be a cut-off function at the point $x_0 \in \mathbb{R}_+$, and let $\widetilde{\varphi}_{x_0,R}$ be a cut-off function such that $\varphi_{x_0,R}\widetilde{\varphi}_{x_0,R} = \widetilde{\varphi}_{x_0,R}$. Then the operator A is locally equivalent at the point x_0 to the operator $\varphi_{x_0,R}A\widetilde{\varphi}_{x_0,R}$, which is a pseudodifferential operator in $OP\tilde{S}$ with the double symbol

$$\frac{1}{\sqrt{2\pi}}\varphi_{x_0,R}(x)\widetilde{\varphi}_{x_0,R}(y)b(x,y)\hat{k}(\xi).$$

As follows from Proposition 3.4 (iii),

$$\varphi_{x_0,R}A\widetilde{\varphi}_{x_0,R} = \frac{1}{\sqrt{2\pi}}\operatorname{Op}(\widetilde{\varphi}_{x_0,R}(x)b(x,x)\hat{k}(\xi)) + K$$

where K is a compact operator. It is well known that modulo a compact operator the operator $\operatorname{Op}(\widetilde{\varphi}_{x_0,R}(x)b(x,x)\hat{k}(\xi))$ coincides with $\operatorname{Op}(\widetilde{\varphi}_{x_0,R}(x)b(x,x)(\hat{k}_+\chi_+(\xi) + \hat{k}_-\chi_-(\xi))$, where $\chi_\pm(\xi)$ is the characteristic function of $\mathbb{R}_\pm$. Finally,

$$\begin{aligned}
&\operatorname{Op}(\widetilde{\chi}_{x_0,R}(x)b(x,x)(\hat{k}_+\chi_+(\xi) + \hat{k}_-\chi_-(\xi)) \\
&\qquad = \widetilde{\chi}_{x_0,R}(x)b(1,1)(\hat{k}_+F^{-1}\chi_+(\xi)F + F^{-1}\hat{k}_-\chi_-(\xi)F), \\
&\qquad = \widetilde{\chi}_{x_0,R}(x)(\hat{k}_+F^{-1}\chi_+(\xi)F + \hat{k}_-F^{-1}\chi_-(\xi)F),
\end{aligned}$$

and the last operator is locally equivalent at the point x_0 to the operator

$$\hat{k}_+F^{-1}\chi_+(\xi)F + \hat{k}_-F^{-1}\chi_-(\xi)F.$$

Thus we have the following lemma.

Lemma 5.6. *Given $x_0 \in (0, +\infty)$, the local algebra $\mathcal{R}(x_0) \cong \mathbb{C}^2$ is two-dimensional. The (local symbol) homomorphism*

$$\pi_{x_0} : \mathcal{R} \longrightarrow \mathcal{R}(x_0) \cong \mathbb{C}^2$$

is generated by the following mapping of the generators of the algebra $\mathcal{R}$

$$\begin{aligned}
\pi_{x_0} : a(x)I \quad &\longmapsto \quad (a(x_0),\, a(x_0)), \\
\pi_{x_0} : \; A \quad &\longmapsto \quad (\hat{k}_+,\, \hat{k}_-),
\end{aligned}$$

where $a(x) \in SV$, and A is of the form (5.1).

Let now $x_0 = +\infty \in \overline{\mathbb{R}}_+$. Two multiplication operators $a_1(x)I$ and $a_2(x)I$, where $a_1, a_2 \in SV$, are locally equivalent at the point x_0 if and only if

$$\lim_{x \to +\infty} (a_1(x) - a_2(x)) = 0.$$

Let $\varphi_{x_0,R}$ be a cut-off function at the point $x_0 = +\infty$, and let $\widetilde{\varphi}_{x_0,R}$ be a cut-off function such that $\varphi_{x_0,R}\widetilde{\varphi}_{x_0,R} = \widetilde{\varphi}_{x_0,R}$. Again, the operator A is locally equivalent at the point x_0 to the operator $\varphi_{x_0,R}A\widetilde{\varphi}_{x_0,R}$, which is a pseudodifferential operator in $OP\widetilde{S}$ with a double symbol. Thus

$$\begin{aligned}\varphi_{x_0,R}A\widetilde{\varphi}_{x_0,R} &= \mathrm{Op}\,(\varphi_{x_0,R}(x)b(x,y)\widetilde{\varphi}_{x_0,R}(y)\hat{k}(\xi)) \\ &= \mathrm{Op}\,(\widetilde{\varphi}_{x_0,R}(x)b(x,x)\hat{k}(\xi)) + K \\ &= \widetilde{\varphi}_{x_0,R}(x)F^{-1}\hat{k}(\xi)F + K,\end{aligned}$$

where K is a compact operator. Now the operator $\widetilde{\varphi}_{x_0,R}(x)F^{-1}\hat{k}(\xi)F$ is locally equivalent at the point $x_0 = +\infty$ to the operator

$$F^{-1}\hat{k}(\xi)F.$$

To formulate our result introduce the ideal

$$C_0^{+\infty}(\mathbb{R}_+) = \{a(x) \in SV : \lim_{x \to +\infty} a(x) = 0\}$$

of the algebra SV.

Lemma 5.7. *For $x_0 = +\infty$ the local algebra $\mathcal{R}(+\infty) \cong SV/C_0^{+\infty}(\mathbb{R}_+) \otimes C(\overline{\mathbb{R}})$ is commutative. The (local symbol) homomorphism*

$$\pi_{+\infty} : \mathcal{R} \longrightarrow \mathcal{R}(+\infty) \cong SV/C_0^{+\infty}(\mathbb{R}_+) \otimes C(\overline{\mathbb{R}})$$

is generated by the following mapping of the generators of the algebra $\mathcal{R}$

$$\begin{aligned}\pi_{+\infty} : a(x)I &\longmapsto a(x) + C_0^{+\infty}(\mathbb{R}_+), \\ \pi_{+\infty} : A &\longmapsto \hat{k}(\xi),\end{aligned}$$

where $a(x) \in SV$, and A is of the form (5.1).

Let finally $x_0 = 0 \in \overline{\mathbb{R}}_+$. Two multiplication operators $a_1(x)I$ and $a_2(x)I$, where $a_1, a_2 \in SV$, are locally equivalent at the point $x_0 = 0$ if and only if

$$\lim_{x \to 0} (a_1(x) - a_2(x)) = 0.$$

Let $\varphi_{x_0,R}$ be a cut-off function at the point $x_0 = 0$, and let $\widetilde{\varphi}_{x_0,R}$ be a cut-off function such that $\varphi_{x_0,R}\widetilde{\varphi}_{x_0,R} = \widetilde{\varphi}_{x_0,R}$. Again, the operator A is locally equivalent at the point x_0 to the operator $\varphi_{x_0,R}A\widetilde{\varphi}_{x_0,R}$.

Introduce the operators

$$\begin{aligned} A_0u(x) &= \frac{1}{2}\left[(\widehat{k}_+ + \widehat{k}_-)u(x) + (\widehat{k}_+ - \widehat{k}_-)\frac{1}{i\pi}\int_{\mathbb{R}_+}\frac{b(x,y)}{y-x}u(y)dy\right], \\ &= \frac{1}{2}\left[(\widehat{k}_+ + \widehat{k}_-)u(x) + (\widehat{k}_+ - \widehat{k}_-)\frac{1}{i\pi}\int_{\mathbb{R}_+}\frac{b(x/y,1)}{1-x/y}u(y)\frac{dy}{y}\right], \\ K_0u(x) &= \frac{1}{\sqrt{2\pi}}\int_{\mathbb{R}_+} b(x,y)k_0(x-y)u(y)dy. \end{aligned}$$

Then by (5.9) we have

$$\varphi_{x_0,R}A\widetilde{\varphi}_{x_0,R} = \varphi_{x_0,R}A_0\widetilde{\varphi}_{x_0,R} + \varphi_{x_0,R}K_0\widetilde{\varphi}_{x_0,R},$$

where the last summand is obviously compact. Finally, the operator $\varphi_{x_0,R}A_0\widetilde{\varphi}_{x_0,R}$ is locally equivalent at the point $x_0 = 0$ to the Mellin convolution operator

$$A_0 = M^{-1}\frac{1}{2}[(\widehat{k}_+ + \widehat{k}_-) + (\widehat{k}_+ - \widehat{k}_-)q_A(\lambda)]M,$$

where M and M^{-1} are direct and inverse Mellin transforms, and the function $q_A(\lambda)$ is given by (5.5).

Introduce the ideal

$$C_0^0(\mathbb{R}_+) = \{a(x) \in SV : \lim_{x\to 0} a(x) = 0\}$$

in the algebra SV.

Lemma 5.8. *For $x_0 = 0$ the local algebra $\mathcal{R}(0) \cong SV/C_0^0(\mathbb{R}_+) \otimes C(\overline{\mathbb{R}})$ is commutative. The (local symbol) homomorphism*

$$\pi_0 : \mathcal{R} \longrightarrow \mathcal{R}(0) \cong SV/C_0^0(\mathbb{R}_+) \otimes C(\overline{\mathbb{R}})$$

is generated by the following mapping of the generators of the algebra $\mathcal{R}$

$$\begin{aligned} \pi_0 : a(x)I &\longmapsto a(x) + C_0^0(\mathbb{R}_+), \\ \pi_0 :\ A &\longmapsto \frac{1}{2}[(\widehat{k}_+ + \widehat{k}_-) + (\widehat{k}_+ - \widehat{k}_-)q_A(\lambda)], \end{aligned}$$

where $a(x) \in SV$, and A is of the form (5.1).

Remark 5.9. As follows from Proposition 5.2,

$$\lim_{\lambda\to\pm\infty}\frac{1}{2}[(\widehat{k}_+ + \widehat{k}_-) + (\widehat{k}_+ - \widehat{k}_-)q_A(\lambda)] = \widehat{k}_\pm.$$

To describe the Fredholm symbol algebra $\operatorname{Sym}\mathcal{R} = \mathcal{R}/\mathcal{K} = \widehat{\mathcal{R}}$ we introduce some notation. Denote by $\mathfrak{S}$ the C^*-algebra of all vector-valued functions σ continuous on $\overline{\mathbb{R}}_+ = [0, +\infty]$, where $\sigma(x) \in \mathcal{R}(x)$, for each $x \in [0, +\infty]$, with point-wise operations and

$$\|\sigma\| = \sup_{x\in\overline{\mathbb{R}}_+}\|\sigma(x)\|.$$

For each $x \in (0,+\infty)$ we have $\sigma(x) = (\sigma^+(x), \sigma^-(x)) \in \mathbb{C}^2 = \mathcal{R}(x)$, and a vector-function $\sigma|_{(0,+\infty)}$ is continuous, if and only if the complex-valued functions $\sigma^+ = \sigma^+(x)$ and $\sigma^- = \sigma^-(x)$ are continuous in $(0,+\infty)$.

Now $\mathcal{R}(+\infty) = SV/C_0^{+\infty}(\mathbb{R}_+) \otimes C(\overline{\mathbb{R}}) = (SV \otimes C(\overline{\mathbb{R}}))/C_0^{\infty}(\mathbb{R}_+ \times \overline{\mathbb{R}})$, where

$$C_0^{\infty}(\mathbb{R}_+ \times \overline{\mathbb{R}}) = \{a(x,\xi) \in SV \otimes C(\overline{\mathbb{R}}) : \lim_{x\to+\infty} \sup_{\xi\in\mathbb{R}} |a(x,\xi)| = 0\}.$$

Given $\sigma(+\infty) \in \mathcal{R}(+\infty)$, denote by $\sigma(x,\xi)$ a function from $SV \otimes C(\overline{\mathbb{R}})$, whose image in $\mathcal{R}(+\infty)$ coincides with $\sigma(+\infty)$. Then a vector-valued function $\sigma \in \mathfrak{S}$ is continuous at the point $+\infty$ if and only if

$$\begin{aligned} \sigma(x,+\infty) - \sigma^+(x) &\in C_0^{+\infty}(\mathbb{R}_+), \\ \sigma(x,-\infty) - \sigma^-(x) &\in C_0^{+\infty}(\mathbb{R}_+). \end{aligned}$$

Finally, $\mathcal{R}(0) = SV/C_0^0(\mathbb{R}_+) \otimes C(\overline{\mathbb{R}}) = (SV \otimes C(\overline{\mathbb{R}}))/C_0^0(\mathbb{R}_+ \times \overline{\mathbb{R}})$, where

$$C_0^0(\mathbb{R}_+ \times \overline{\mathbb{R}}) = \{a(x,\xi) \in SV \otimes C(\overline{\mathbb{R}}) : \lim_{x\to 0} \sup_{\xi\in\mathbb{R}} |a(x,\xi)| = 0\}.$$

Given $\sigma(0) \in \mathcal{R}(0)$, denote by $\sigma(x,\lambda)$ a function from $SV \otimes C(\overline{\mathbb{R}})$ whose image in $\mathcal{R}(0)$ coincides with $\sigma(0)$. Then a vector-valued function $\sigma \in \mathfrak{S}$ is continuous at the point 0 if and only if

$$\begin{aligned} \sigma(x,+\infty) - \sigma^+(x) &\in C_0^{+\infty}(\mathbb{R}_+), \\ \sigma(x,-\infty) - \sigma^-(x) &\in C_0^{+\infty}(\mathbb{R}_+). \end{aligned}$$

Lemmas 5.6, 5.7, and 5.8, together with the local principle, lead to the following theorem.

Theorem 5.10. *The Fredholm symbol algebra* $\operatorname{Sym}\mathcal{R} = \mathcal{R}/\mathcal{K} = \widehat{\mathcal{R}}$ *of the algebra* $\mathcal{R}$ *is isomorphic and isometric to the algebra* $\mathfrak{S}$. *Under their identification the symbol homomorphism*

$$\operatorname{sym} : \mathcal{R} \longrightarrow \operatorname{Sym}\mathcal{R} = \mathfrak{S}$$

is generated by the following mapping of the generators of the algebra $\mathcal{R}$:

$$\operatorname{sym} : a(x)I \longmapsto \begin{cases} a + C_0^0(\mathbb{R}_+), & x = 0 \\ (a(x),\, a(x)), & x \in (0,+\infty)\,, \\ a + C_0^{+\infty}(\mathbb{R}_+), & x = +\infty \end{cases}$$

$$\operatorname{sym} : A \longmapsto \begin{cases} \frac{1}{2}[(\widehat{k}_+ + \widehat{k}_-) + (\widehat{k}_+ - \widehat{k}_-)q_A(\lambda)], & x = 0 \\ (\hat{k}_+,\, \hat{k}_-), & x \in (0,+\infty)\,, \\ \hat{k}(\xi), & x = +\infty \end{cases}$$

where $a(x) \in SV$, *and* A *is of the form (5.1).*

Corollary 5.11. *An operator* B *from the algebra* $\mathcal{R}$ *is Fredholm if and only if its symbol* $\operatorname{sym} B$ *is invertible in the algebra* $\operatorname{Sym}\mathcal{R} = \mathfrak{S}$.

Let $B \in \mathcal{R}$ be a Fredholm operator, and $\operatorname{sym} B$ be its (Fredholm) symbol. Denote by $\operatorname{sym} B(x,\lambda)$ and $\operatorname{sym} B(x,\xi)$ the functions from $SV \otimes C(\overline{\mathbb{R}})$ whose images in $\mathcal{R}(0)$ and $\mathcal{R}(+\infty)$ coincide with $\operatorname{sym} B(0)$ and $\operatorname{sym} B(+\infty)$, respectively.

Theorem 5.12. *The index of a Fredholm operator $B \in \mathcal{R}$ is given by*

$$\begin{aligned} \operatorname{Ind} B &= -\frac{1}{2\pi}\left[\arg \frac{\operatorname{sym} B^+(x)}{\operatorname{sym} B^-(x)}\right]_{t_0}^{T_0} + \frac{1}{2\pi}\left[\arg \operatorname{sym} B(t_0,\lambda)\right]_{-\infty}^{+\infty} \\ &\quad -\frac{1}{2\pi}\left[\arg \operatorname{sym} B(T_0,\xi)\right]_{-\infty}^{+\infty}, \end{aligned}$$

where the point $t_0 \in \mathbb{R}_+$ is close enough to 0, and such that

$$\inf_{0<x\leq t_0,\ \lambda\in\mathbb{R}} |\operatorname{sym} B(x,\lambda)| > 0,$$

and $T_0 \in \mathbb{R}_+$ is a large enough number, such that

$$\inf_{T_0\leq x<\infty,\ \xi\in\mathbb{R}} |\operatorname{sym} B(x,\xi)| > 0.$$

Proof. Follows the standard method of separations of singularities [9]. □

Theorem 5.13. *The C^*-algebra $\mathcal{R}$ is the norm–closure of the set of all operators of the form*

$$B = \sum_{j=1}^{N} a_j(x)A_j + K,$$

where $a_j \in SV^\infty$, each A_j is of the form (5.1), and K is a compact operator.

Proof. In fact, by Proposition 5.4 the commutator $[a(x)I, A]$ is compact for each $a \in SV^\infty$ and A of the form (5.1). Thus to finish the proof, it is sufficient to check that the composition of two operators of the form (5.1) is an operator of the same form, modulo a compact operator.

Let A_1, A_2 be two operators of the form (5.1), and let

$$\operatorname{sym} A_1 = \begin{cases} \sigma_1(0,\lambda), & x=0 \\ (\sigma_1^+,\sigma_1^-), & x\in(0,+\infty) \\ \sigma_1(+\infty,\xi), & x=+\infty \end{cases},$$

$$\operatorname{sym} A_2 = \begin{cases} \sigma_2(0,\lambda), & x=0 \\ (\sigma_2^+,\sigma_2^+), & x\in(0,+\infty) \\ \sigma_2(+\infty,\xi), & x=+\infty \end{cases}$$

be their symbols. Then by Theorem 5.10 we have $A_1A_2 = A + K$, where the operator A is of the form (5.1) with

$$\begin{aligned} k(x) &= (F^{-1}(\sigma_1(+\infty,\xi)\sigma_2(+\infty,\xi)))(x), \\ b(x,y) &= i\sqrt{\frac{\pi}{2}}\frac{(y-x)}{y}\left(M^{-1}\left(\frac{2\sigma_1(0,\lambda)\sigma_2(0,\lambda)-(\sigma_1^+\sigma_2^+ + \sigma_1^-\sigma_2^-)}{\sigma_1^+\sigma_2^+ - \sigma_1^-\sigma_2^-}\right)\right)\left(\frac{x}{y}\right), \end{aligned}$$

and K is a compact operator. Here $M^{-1} : L_2(\mathbb{R}) \longrightarrow L_2(\mathbb{R}_+)$ is the inverse Mellin transform

$$(M^{-1}u)(t) = \frac{1}{\sqrt{2\pi}} \int_{\mathbb{R}_+} t^{i\lambda - 1/2} u(\lambda)\, d\lambda.$$

□

Consider the subalgebra $\mathcal{R}_0$ of the algebra $\mathcal{R}$ which is generated by multiplication operators $a(x)I$, with $a \in C(\overline{\mathbb{R}}_+)$, and the integral operators of the form (5.1) with the property (5.6). Introduce

$$\Gamma = [0, +\infty] \times [-\infty, +\infty] = \{(x, \xi) : x \in \overline{\mathbb{R}}_+,\ \xi \in \overline{\mathbb{R}}\},$$

and orient Γ starting with the standard (from 0 to $+\infty$) orientation of the side $+\infty \times [0, +\infty]$. Then from Theorem 5.10 we have

Theorem 5.14. *The Fredholm symbol algebra* $\operatorname{Sym} \mathcal{R}_0 = \mathcal{R}_0/\mathcal{K} = \widehat{\mathcal{R}_0}$ *of the algebra* $\mathcal{R}_0$ *is isomorphic and isometric to the algebra* $C(\Gamma)$*. Under their identification the symbol homomorphism*

$$\operatorname{sym} : \mathcal{R}_0 \longrightarrow \operatorname{Sym} \mathcal{R}_0 = C(\Gamma)$$

is generated by the following mapping of the generators of the algebra $\mathcal{R}_0$*:*

$$\operatorname{sym} : a(x)I \longmapsto \begin{cases} a(0), & (x,\xi) \in \{0\} \times [-\infty, +\infty] \\ a(x), & (x,\xi) \in (0, +\infty) \times \{\pm\infty\} \\ a(+\infty), & (x,\xi) \in \{+\infty\} \times [-\infty, +\infty] \end{cases},$$

$$\operatorname{sym} : A \longmapsto \begin{cases} \frac{1}{2}[(\widehat{k}_+ + \widehat{k}_-) + (\widehat{k}_+ - \widehat{k}_-) q_A(\lambda)], & (x,\xi) \in \{0\} \times [-\infty, +\infty] \\ \hat{k}_\pm, & (x,\xi) \in (0, +\infty) \times \{\pm\infty\}, \\ \hat{k}(\xi), & (x,\xi) \in \{+\infty\} \times [-\infty, +\infty] \end{cases}$$

where $a(x) \in C(\overline{\mathbb{R}}_+)$*, and* A *is of the form (5.1).*

An operator B *from the algebra* $\mathcal{R}_0$ *is Fredholm if and only if its symbol* $\operatorname{sym} B$ *is invertible in the algebra* $\operatorname{Sym} \mathcal{R}_0 = C(\Gamma)$*.*

The index of a Fredholm operator $B \in \mathcal{R}$ *is given by*

$$\operatorname{Ind} B = \frac{1}{2\pi} [\arg \operatorname{sym} B]_\Gamma.$$

6. Algebra generated by Bergman-Toeplitz operators

As follows from Proposition 2.5, if $b = b(x)$ ($\in L_\infty(\mathbb{R})$) does not depend on y, then the Bergman-Toeplitz operator T_b is unitary equivalent to the operator $A_b = RT_bR^*$, where

$$(A_b)u(x) = \frac{1}{\sqrt{2\pi}} \int_{\mathbb{R}_+} \frac{2\sqrt{xt}}{x+t}\, k_b(x-t) u(t) dt \tag{6.1}$$

and $k_b(x) = (Fb)(x)$ is the Fourier transform of $b(x)$ regarded in the sense of distribution theory. Note, that $\hat{k}_b(\xi) = (Fk_b)(\xi) = b(-\xi)$.

The operator A_b is of the form (5.1). Calculate

$$q_{A_b}(\lambda) = \frac{1}{i\pi}\int_{\mathbb{R}_+} t^{1/2-i\lambda}\frac{2\sqrt{t}}{(1+t)(1-t)}\frac{dt}{t} = \frac{2}{i\pi}\int_{\mathbb{R}_+}\frac{t^{-i\lambda}dt}{(1+t)(1-t)}, \quad \lambda \in \mathbb{R}.$$

First, suppose that $0 < \operatorname{Im}\lambda < 1$, then

$$q_{A_b}(\lambda) = \frac{1}{i\pi}\int_{\mathbb{R}_+}\frac{t^{-i\lambda}dt}{t+1} - \frac{1}{i\pi}\int_{\mathbb{R}_+}\frac{t^{-i\lambda}dt}{t-1}, \quad 0 < \operatorname{Im}\lambda < 1.$$

The second integral is well-known and is equal to $\coth(-\pi\lambda)$. Calculating the first integral by means of the residue theory, we have

$$\frac{1}{i\pi}\int_{\mathbb{R}_+}\frac{t^{-i\lambda}dt}{t+1} = \frac{2e^{\pi\lambda}}{1-e^{2\pi\lambda}}.$$

Hence,

$$q_{A_b}(\lambda) = \frac{2e^{\pi\lambda}}{1-e^{2\pi\lambda}} - \coth(-\pi\lambda) = \frac{e^{\pi\lambda}+e^{-\pi\lambda}-2}{e^{\pi\lambda}-e^{-\pi\lambda}} = \tanh\frac{\pi\lambda}{2}, \quad 0 < \operatorname{Im}\lambda < 1. \tag{6.2}$$

Passing to the limit on the left and right hand sides of formula (6.2), we deduce that formula (6.2) holds in the strip $0 \le \operatorname{Im}\lambda < 1$. Hence,

$$q_{A_b}(\lambda) = \tanh\frac{\pi\lambda}{2}, \quad \text{for } \lambda \in \mathbb{R}. \tag{6.3}$$

Proposition 6.1. *Let $b(x) \in C(\overline{\mathbb{R}})$, then $A_b \in \mathcal{R}$.*

Proof. Indeed, each function $b(x) \in C(\overline{\mathbb{R}})$ can be approximated in the sup-norm by a sequence $\{b_m(x)\}$ of functions satisfying the conditions (5.2), (5.3). This implies that

$$\|A_b - A_{b_m}\| = \|T_b - T_{b_m}\| \le \|b - b_m\|_{L_\infty(\mathbb{R})} \to 0, \quad \text{if } \ m \to \infty.$$

□

Now, as follows from Proposition 2.4, if a function $a = a(y)$ ($\in L_\infty(\mathbb{R}_+)$) does not depend on x, then the Bergman-Toeplitz operator T_a is unitary equivalent to the multiplication operator

$$RT_aR^* = \gamma_a(x)I,$$

where the function $\gamma_a(x)$ is given by (2.1).

Proposition 6.2. *Let $a(y) \in C_b^1(\mathbb{R}_+)$, and*

$$\lim_{y\to\infty} ya'(y) = 0, \tag{6.4}$$

then $\gamma_a(x) \in SV$, and thus $\gamma_a(x)I \in \mathcal{R}$.

Proof. We have

$$\gamma_a'(x) = -\frac{1}{x}\int_{\mathbb{R}_+} \frac{t}{2x}\, a'(\frac{t}{2x})\, e^{-t}dt.$$

The function $ya'(y)$ is bounded on $\mathbb{R}_+$, and

$$|\gamma_a'(x)| \leq \frac{1}{x} \sup_{y\in\mathbb{R}_+} |ya'(y)|,$$

thus

$$\lim_{x\to\infty} \gamma_a'(x) = 0.$$

Hence $\gamma_a(x)$ is slowly varying if $x \to \infty$. Consider now the behavior of $\gamma_a(x)$ near the point 0. We have

$$x\gamma_a'(x) = -\int_{\mathbb{R}_+} \frac{t}{2x} a'(\frac{t}{2x}) e^{-t}dt, \tag{6.5}$$

and the integral (6.5) converges uniformly with respect to the parameter $1/2x$, when $x \to 0$. Passing to the limit when $x \to 0$, we obtain that

$$\lim_{x\to 0} x\gamma_a'(x) = 0.$$

□

Let $\widetilde{SV}(\mathbb{R}_+)$ be the C^*-algebra, generated by functions $a = a(y)$ satisfying the conditions of Proposition 6.2. Denote by $\mathcal{T} = \mathcal{T}(C(\overline{\mathbb{R}}), \widetilde{SV}(\mathbb{R}_+))$ the C^*-algebra generated by all Bergman-Toeplitz operators

$$T_b, \text{ where } b = b(x) \in C(\overline{\mathbb{R}}), \quad \text{and} \quad T_a, \text{ where } a = a(y) \in \widetilde{SV}(\mathbb{R}_+),$$

acting on the space $\mathcal{A}^2(\Pi)$.

Now as a direct corollary from Propositions 6.1, 6.2 and Theorems 5.10, 5.12 we have

Theorem 6.3. *The Fredholm symbol algebra* $\operatorname{Sym}\mathcal{T} = \mathcal{T}/\mathcal{K}$ *of the Toeplitz operator algebra* $\mathcal{T} = \mathcal{T}(C(\overline{\mathbb{R}}), \widetilde{SV}(\mathbb{R}_+))$ *is isomorphic and isometric to the algebra* $\mathfrak{S}$. *Under their identification the symbol homomorphism*

$$\text{sym} : \mathcal{T} \longrightarrow \operatorname{Sym}\mathcal{T} = \mathfrak{S}$$

is generated by the following mapping of the generators of the algebra $\mathcal{R}$:

$$\text{sym} : T_a \longmapsto \begin{cases} \gamma_a + C_0^0(\mathbb{R}_+), & x = 0 \\ (\gamma_a(x), \gamma_a(x)), & x \in (0, +\infty) \\ \gamma_a + C_0^{+\infty}(\mathbb{R}_+), & x = +\infty \end{cases},$$

$$\text{sym} : T_b \longmapsto \begin{cases} \frac{1}{2}[(b(-\infty) + b(+\infty)) + (b(-\infty) - b(+\infty)) \tanh\frac{\pi\lambda}{2}], & x = 0 \\ (b(-\infty), b(+\infty)), & x \in (0, +\infty), \\ b(-\xi), & x = +\infty \end{cases}$$

where $a(y) \in \widetilde{SV}(\mathbb{R}_+)$, $b(x) \in C(\overline{\mathbb{R}})$.

An operator T *from the algebra* $\mathcal{T}$ *is Fredholm if and only if its symbol* $\operatorname{sym} T$ *is invertible in the algebra* $\operatorname{Sym}\mathcal{T} = \mathfrak{S}$.

The index of a Fredholm operator $T \in \mathcal{T}$ *is given by*

$$\begin{aligned} \operatorname{Ind} T &= -\frac{1}{2\pi}\left[\arg\frac{\operatorname{sym} T^+(x)}{\operatorname{sym} T^-(x)}\right]_{r_0}^{R_0} + \frac{1}{2\pi}\left[\arg\operatorname{sym} T(r_0,\lambda)\right]_{-\infty}^{+\infty} \\ &\quad -\frac{1}{2\pi}\left[\arg\operatorname{sym} T(R_0,\xi)\right]_{-\infty}^{+\infty}, \end{aligned}$$

where the point $r_0 \in \mathbb{R}_+$ *is close enough to* 0*, and such that*

$$\inf_{0<x\le r_0,\ \lambda\in\mathbb{R}} |\operatorname{sym} T(x,\lambda)| > 0,$$

and $R_0 \in \mathbb{R}_+$ *is a large enough number, that*

$$\inf_{R_0\le x<\infty,\ \xi\in\mathbb{R}} |\operatorname{sym} T(x,\xi)| > 0.$$

Let T be an operator from the algebra $\mathcal{T}$, and $\operatorname{sym} T$ be its (Fredholm) symbol. Again, consider a function $\sigma_0(x,\lambda) \in SV \otimes C(\overline{\mathbb{R}})$, whose image in $\mathcal{R}(0)$ coincides with $\operatorname{sym} T(0)$, and a function $\sigma_{+\infty}(x,\xi) \in SV \otimes C(\overline{\mathbb{R}})$, whose image in $\mathcal{R}(+\infty)$ coincides with $\operatorname{sym} T(+\infty)$.

Denote by $\{\widetilde{\sigma}_0(\lambda)\}$ the set of all partial limits of the function $\sigma_0(x,\lambda)$, when $x \to 0$, and by $\{\widetilde{\sigma}_{+\infty}(\xi)\}$ the set of all partial limits of the function $\sigma_{+\infty}(x,\xi)$, when $x \to +\infty$.

Corollary 6.4. *The essential spectrum of an operator* T *of the algebra* $\mathcal{T}$ *is given by*

$$\begin{gathered} \operatorname{ess\text{-}sp} T = \operatorname{Im}\operatorname{sym} T = \\ \bigcup\{z\in\mathbb{C} : z = \widetilde{\sigma}_0(\lambda),\ \lambda\in\overline{\mathbb{R}}\}\bigcup\{z\in\mathbb{C} : z=\widetilde{\sigma}_{+\infty}(\xi),\ \xi\in\overline{\mathbb{R}}\} \bigcup_{x\in(0,+\infty)} \operatorname{sym} T(x), \end{gathered}$$

where the first union is taken over the all partial limits of the function $\sigma_0(x,\lambda)$*, when* $x \to 0$*, and the second union is taken over the all partial limits of the function* $\sigma_{+\infty}(x,\xi)$*, when* $x \to +\infty$.

Note that for operators from $\mathcal{T}$ their essential spectra are, in general, massive sets, i.e., have positive Lebesgue plain measure. This effect is due to the oscillation of functions γ_a for $a \in \widetilde{SV}(\mathbb{R}_+)$.

In particular, the essential spectrum of the operator $T_a \cdot T_b$ (with a function $a = a(y)$, such that the corresponding function γ_a has oscillation either in 0, or in $+\infty$) is massive, while the essential spectrum of each operator T_a and T_b is always not.

References

[1] S. Axler, *Bergman spaces and their operators,* Surveys on Some Recent Results in Operator Theory, Vol I, Pitman Research Notes in Math. **171,** (1988) 1–50.

[2] S. Axler, J. Conway and G. McDonald, *Toeplitz operators on Bergman spaces,* Can. J. Math., **34,** (1982) 466–483.

[3] V. Bergman, *On a Hilbert space of analytic functions,* Comm. Pure Appl. Math., **3,** (1961) 215–228.

[4] F. A. Berezin, *Wick and anti-Wick symbol of operators,* Math. USSR Sbornik, **84,** (1971) 578–610.

[5] F. A. Berezin, *Method of second Quantization,* Nauka, Moscow (1988).

[6] H. O. Cordes, *Pseudo-differential operators on a half-line,* J. Math. Mech., **18,** (9) (1969) 893–908.

[7] J. Danus and K. H. Hofmann, *Representation of Rings by Sections,* Memoirs Amer. Math. Soc, **83,** (1968).

[8] R. G. Douglas, Banach Algebras Techniques in Operator Theory, Academic Press, (1972).

[9] I. Gohberg and N. Krupnik, One-dimensional linear sigular equations, Vol. 2, General theory and applications, Birkhauser Verlag, Basel (1992).

[10] K. H. Hofmann, *Representation of algebras by continuous functions,* Bull. Amer. Math. Soc, **78,** (3) (1972) 291–373.

[11] G. McDonald and C. Sundberg, *Toeplitz operators on the disc,* Indiana Math. J., **28,** (1979) 595–611.

[12] V. S. Rabinovich, *Singular integral operators on complicated contours and pseudodiferential operators,* Math. Notes, **58,** (1995) 772–734.

[13] M. E. Taylor, Pseudodiferential operators, Princeton University Press, New Jersey, (1981).

[14] J. Varela, *Duality on C^*-algebras,* Meomories Amer. Math. Soc, **148,** (1974) 97–108.

[15] N. L. Vasilevski, *Banach Algebras generated by two-dimensional integral operators with Bergman kernel and pice-wise continuous coefficientes, I,* Soviet Math. (Izv. VUZ), **30,** (3) (1986) 14–24.

[16] N. L. Vasilevski, *Convolution operators on standard CR-manifold, II. Algebras of convolution operators on the Heisenberg group,* Integr. Equat, Oper. Th., **19,** (1994) 327–348.

[17] N. L. Vasilevski, *On Bergman-Toepliz operators with commutative symbol algebras,* Integr. Equat, Oper, Th., (to appear).

[18] N. L. Vasilevski, *On the structure of Bergman and poly-Bergman spaces,* Integr. Equat, Oper, Th., (to appear).

[19] V. S. Vladimirov, *Equations of Mathematical Physics,* Nauka, Moskow, Russia (1981) 327–348.

[20] Kehe Zhu, *VMO, ESV, and Toeplitz operators on the Bergman space,* Trans. AMer, Math. Soc., **302,** (1987) 617–646.

[21] Kehe Zhu, Operator Theory in Function Spaces, Marcel Dekker, Ins., (1990) 617–646.

V. Rabinovich
Departamento de Telecomunicaciones,
ESIME del I.P.N. Unidad Zacatenco,
Av. I.P.N. s/n, Ed. 1, 07738
México, D.F., México

E-mail address: rabinov@maya.esimez.ipn.mx

N. Vasilevski
Departamento de Matemáticas,
CINVESTAV del I.P.N.
Apartado Postal 14-740, 07000
México, D.F., México
E-mail address: nvasilev@math.cinvestav.mx

Operator Theory:
Advances and Applications, Vol. 114

The small Hankel operator in several complex variables

Bernard Russo

Abstract. A survey of known results and open problems concerning boundedness, compactness, and trace ideal membership of the small Hankel operator. The setting is either the Bergman or Hardy space over a bounded symmetric domain or a strongly pseudoconvex domain in several complex variables, with special attention to the unit polydisk and multivariable harmonic analysis.

Acknowledgements: The author thanks Song-Ying Li for numerous discussions on Hankel operators.

This expository paper attempts to give the status of research on small Hankel operators in several complex variables insofar as it is concerned with certain basic problems for operators associated with a symbol. Although it is primarily about small Hankel operators on Hardy spaces, the Bergman spaces and big Hankel operators are also mentioned. Hankel operators, Toeplitz operators, and composition operators are at the center of the study of certain aspects of contemporary operator theory in function spaces. For any of these operators, one can consider the following problems: characterize the symbols for which the operator with that symbol is bounded, compact, or in a Schatten p-class. In this note we consider these problems for the small Hankel operator.

Hankel operators are of interest in pure and applied operator theory. They appear in the following contexts, to name a few (for the first three, see [39]):

- H^∞ control theory (engineering)
- interpolation problems (Nevanlinna-Pick,Caratheodory-Fejer)
- approximation theory
- noncommutative geometry (quantum Hall effect,[8])
- $\overline{\partial}_b$ equation ([1])

This paper contains three sections. Section 1 gives the background on the types of function spaces, domains, and operators of interest, and poses the problems to be discussed in later sections. The literature for the Bergman space versions of our problems is discussed here. In Section 2, the state of affairs regarding the Hardy spaces of the unit ball is discussed. Also in that section, the known results for the Hardy spaces of the unit disk are given. Most of these, as well as references to results on more general domains, are given in the monograph [62] so they are only mentioned briefly here without much discussion. Section 3 is an exposition of the multiparameter harmonic analysis as it applies to the study of the small Hankel operator on the polydisk, which is also presented there.

1. Preliminaries

1.1. Bergman and Hardy spaces

Let Ω be a domain in $\mathbf{C}^n$. The *Bergman space* is the set of all holomorphic functions on Ω which are p-integrable with respect to Lebesgue volume measure dV on $\mathbf{C}^n = \mathbf{R}^{2n}$:

$$A^p(\Omega) \subset L^p(\Omega, dV) \qquad 0 < p < \infty.$$

$A^p(\Omega)$ is a closed subspace of $L^p(\Omega, dV)$. When $n = 1$, we use the notation dA for dV. The Hardy space $\mathcal{H}^p(\Omega)$, $0 < p < \infty$, as well as the embedding $\mathcal{H}^p(\Omega) \subset L^p(\partial\Omega)$ are a little more complicated. We begin with three familiar cases. For any function f and $r > 0$, let $f_r(z) = f(rz)$.

In the following, $d\theta/2\pi$ denotes normalized Lebesgue measure on the unit circle $\mathbf{T} = \partial\Delta$, σ denotes a unique rotation invariant measure on the unit sphere ∂B, where B is the unit ball in $\mathbf{C}^n$, and in the case of the unit polydisk Δ^n, $r = (r_1, \cdots, r_n)$ and $\theta = (\theta_1, \cdots, \theta_n)$, with obvious meanings for $0 < r < 1$ and $e^{i\theta}$ in this case.

- $\Omega =$ the unit disk: for a holomorphic function f on the unit disk $\Delta = \{z \in \mathbf{C} : |z| < 1\}$ and $0 < p < \infty$, $f \in \mathcal{H}^p(\Delta)$ if

$$\|f\|^p_{\mathcal{H}^p} = \sup_{0<r<1} \int_0^{2\pi} |f_r(e^{i\theta})|^p \, d\theta/2\pi < \infty.$$

- $\Omega =$ the unit ball: for a holomorphic function f on the unit ball $B = \{z = (z_1, \cdots, z_n) \in \mathbf{C}^n : \sum |z_j|^2 < 1\}$ and $0 < p < \infty$, $f \in \mathcal{H}^p(B)$ if

$$\|f\|^p_{\mathcal{H}^p} = \sup_{0<r<1} \int_{\partial B} |f_r(\zeta)|^p \, d\sigma(\zeta) < \infty.$$

- $\Omega =$ the unit polydisk: for a holomorphic function f on the unit polydisk $\Delta^n \subset \mathbf{C}^n$ and $0 < p < \infty$, $f \in \mathcal{H}^p(\Delta^n)$ if

$$\|f\|^p_{\mathcal{H}^p} = \sup_{0<r<1} \int_{\mathbf{T}^n} |f_r(e^{i\theta})|^p \, d\theta_1 \cdots d\theta_n/(2\pi)^n < \infty.$$

In each of the above cases, any $\mathcal{H}^p$ function f has nontangential boundary values f^* almost everywhere, which belong to $L^p(\partial\Omega)$, and the map

$$\mathcal{H}^p \ni f \mapsto f^* \in L^p(\partial\Omega)$$

is norm preserving ([28], [56], [55]). In fact, for any bounded domain in $\mathbf{C}^n$ with C^2-boundary, f^* exists, see [33, Ch. 8]. Moreover, in the two cases considered below, that is, bounded symmetric domains and strongly pseudoconvex domains, the embedding $\mathcal{H}^p(\Omega) \subset L^p(\partial\Omega)$ is an isometry.

1.2. Domains of Interest

We shall limit our attention in this paper to two types of domains, namely, *bounded symmetric domains* and especially the unit polydisk, and *strongly pseudoconvex domains.* Before discussing the definitions, we show how the Hardy spaces are defined in each case. For a summary of an algebraic approach to bounded symmetric domains, see the survey paper [57].

A bounded symmetric domain can be defined as a domain in $\mathbf{C}^n$ which is the open unit ball of a certain Banach space structure on $\mathbf{C}^n$. This is because in finite dimensions, the bounded symmetric domains have been classified, first using Lie theory [12], and afterwards using Jordan theory [32], [48]. Namely, the underlying Banach spaces of all finite dimensional bounded symmetric domains are contained in the following list. We shall not specify the norms in the last three cases. For a fuller discussion, see [48] or [57].

- $M_{m,n}(\mathbf{C})$: rectangular m by n complex matrices with the operator norm
- $S_n(\mathbf{C})$: symmetric n by n complex matrices with the operator norm
- $A_n(\mathbf{C})$: anti-symmetric n by n complex matrices with the operator norm
- Spin_n: the complex "spin factor" of dimension n
- I_{16}: the "exceptional" complex Jordan triple system of dimension 16
- I_{27}: the "exceptional" complex Jordan algebra of dimension 27

In particular, we obtain the unit disk, unit ball, and unit polydisk, from $M_{1,1}$, $M_{1,n}$, and $M_{1,1} \times M_{1,1} \times \cdots \times M_{1,1}$, respectively.

For any bounded symmetric domain Ω, there is a unique probability measure σ on the Silov boundary $\partial^* = \partial^*\Omega$, which is invariant under the action of the compact group of linear automorphisms of Ω. Since Ω is the open unit ball for a norm on $\mathbf{C}^n$, the following definition makes sense for a holomorphic function f on Ω ([27]).

For $0 < p < \infty$, $f \in \mathcal{H}^p(\Omega)$ if

$$\|f\|^p_{\mathcal{H}^p} = \sup_{0<r<1} \int_{\partial^*} |f_r(\zeta)|^p \, d\sigma(\zeta) < \infty.$$

A strongly pseudoconvex domain Ω is given by a defining function $\rho : \mathbf{C}^n \to (0, \infty)$ with certain properties which will not be mentioned here: $\Omega = \{z \in \mathbf{C}^n : \rho(z) < 0\}$. With Ω_ϵ defined by $\{\rho < -\epsilon\}$, the conditions on ρ guarantee the existence of a surface area probability measure σ_ϵ on $\partial\Omega_\epsilon$ so the following definition makes sense ([33, Ch. 8]). For $0 < p < \infty$, $f \in \mathcal{H}^p(\Omega)$ if

$$\|f\|^p_{\mathcal{H}^p} = \sup_{\epsilon>0} \int_{\partial\Omega_\epsilon} |f(z)|^p \, d\sigma_\epsilon(z) < \infty.$$

The unit ball is an example of a bounded symmetric domain and of a strongly pseudoconvex domain, the defining function given by $\rho(z) = |z|^2 - 1$.

1.3. Operators of Interest

The Bergman space $A^2(\Omega)$ is a closed subspace of the Hilbert space $L^2(\Omega)$ and its orthogonal projection (the Bergman projection) is given as an integral operator with kernel $K(z,w)$ (the Bergman kernel). We shall denote this projection by P,

$$Pf(z) = \int_\Omega f(w)K(z,w)\,dV(w) \quad , \quad f \in L^2(\Omega), \quad z \in \Omega.$$

Similarly, the Hardy space $\mathcal{H}^2(\Omega)$ is a closed subspace of the Hilbert space $L^2(\partial\Omega)$ and its orthogonal projection (the Szegö projection) is given as an integral operator with kernel $S(z,w)$ (the Szegö kernel). We shall denote this projection by S,

$$Sf(z) = \int_{\partial\Omega} f(w)S(z,w)\,d\sigma(w), \quad f \in L^2(\partial\Omega), \quad z \in \Omega.$$

Let $f : \Omega \to \mathbf{C}$ (say $f \in L^2$) and define formally the following:

Hankel operator $H_f : A^2 \to A^{2\perp}$: ; $H_f g = (I-P)(fg), \quad g \in A^2, \; fg \in L^2$

Small Hankel operator $h_f : A^2 \to A^2$: ; $h_f g = P(f\overline{g}), \quad g \in A^2, \; f\overline{g} \in L^2$

We make several remarks in connection with these definitions. The definitions above are for the operators on the Bergman space with $p = 2$. There is a corresponding Hardy space operator in each case; replace Ω by $\partial\Omega$, A^2 by $\mathcal{H}^2$ and P by S. Although these operators can also be defined on A^p and $\mathcal{H}^p$ for $0 < p \leq \infty$, we shall restrict our attention to the Hilbert space case of $p = 2$. Both of these operators are densely defined, and the small Hankel operators are conjugate linear. The small Hankel operator is essentially the same as the Hankel operator only in the case of $\mathcal{H}^2(\Delta)$, because $\mathcal{H}^2(\Delta)^\perp$ is one dimension away from $\overline{\mathcal{H}^2(\Delta)}$. It is sometimes convenient to consider these operators as acting from L^2 into L^2.

The Bergman and Szegö projections are important tools in the study of operator theory in function spaces, and indeed are instrumental in the very definition of Hankel operators. Let's give some explicit formulas for the Bergman and Szegö kernels in the cases of interest to us.

unit ball: ([33, p.60,66]),

$$K(z,w) = \frac{n!}{\pi^n}\frac{1}{(1-z\cdot\overline{w})^{n+1}},$$

$$S(z,w) = \frac{(n-1)!}{2\pi^n}\frac{1}{(1-z\cdot\overline{w})^n}$$

strongly pseudoconvex domain: In this case, there is no explicit formula, but an asymptotic expansion due to C. Fefferman [22] allows one to transfer techniques known for the unit ball to this setting.

unit polydisk: ([33, p.61,67]),

$$K(z,w) = \frac{1}{\pi^n} \prod_{j=1}^{n} \frac{1}{(1 - z_j \overline{w_j})^2},$$

$$S(z,w) = \frac{1}{(2\pi)^n} \prod_{j=1}^{n} \frac{1}{(1 - z_j \overline{w_j})}$$

bounded symmetric domain: In this case, the Bergman kernel can be expressed in terms of the Jordan algebraic structure associated with bounded symmetric domains as follows:

$$K(z,w) = c \det B(z,w)^{-1}$$

where $B(x,y)$ is the "Bergman operator". This description of the Bergman kernel can be found in [48] and [21], see also [57]. Formulas for these kernels can be found in [29], see [15].

1.4. Problems of Interest

For small Hankel operators on the Hardy space $\mathcal{H}^2$ or the Bergman space A^2 we shall be interested in the following natural questions.

1. For which symbols f is h_f bounded?
2. For which symbols f is h_f compact?
3. For which symbols f does h_f belong to some Schatten-von Neumann class $\mathcal{S}_q$, $0 < q < \infty$?

For a given domain, the above list implies that there are six questions of interest, three for the Hardy space and three for the Bergman space. In the case of the Bergman space, all of these problems have been essentially solved except for the third one in the case of a strongly pseudoconvex domain, see Table 2 below. In particular, all three questions have been answered for the Bergman space of a bounded symmetric domain and therefore for the Bergman space of the unit ball and of the unit polydisk.

On the contrary, all three problems are completely open in the case of the Hardy space of a bounded symmetric domain, and one of them (see Table 1) is open in the particular cases of the unit polydisk and a strongly pseudoconvex domain.

The above discussion is summarized in the following two tables, whose entries show the appropriate authors and year of publication for the solution of the problem associated with the entry. We have used the abbreviation SPCD for strongly pseudoconvex domain and BSD for bounded symmetric domain. An entry in parentheses means it was initially proved in a more general context. The author apologizes if there are some other references that should have been included here which have been overlooked.

Table 1: Problems on the Hardy space

	bounded	compact	Schatten
unit disk	Nehari 57	Hartman 58	Peller 80
unit ball	Coifman &Rochberg & Weiss 76	Coifman &Rochberg & Weiss 76	Feldman & Rochberg 90 Zhang 91
unit polydisk	Lin-Russo 95	Lin-Russo 95	OPEN
SPCD	Krantz-Li 95	Krantz-Li 95	OPEN
BSD	OPEN	OPEN	OPEN

Table 2: Problems on the Bergman space

	bounded	compact	Schatten
unit disk	Janson & Rochberg & Peetre 87	Janson & Rochberg & Peetre 87	Janson,Rochberg & Peetre 87; Arazy Fisher & Peetre 88
unit ball	Coifman &Rochberg & Weiss 76	Coifman &Rochberg & Weiss 76	Feldman & Rochberg 90 Burbea 87–unpub.
unit polydisk	(Zhu 95)	(Zhu 95)	(Zhu 95)
SPCD	Coupet 89	Coupet 89	OPEN
BSD	Zhu 95	Zhu 95	Zhu 95

In contrast to the Hardy space case, the theory of the small Hankel operator on the Bergman space is fairly complete. The following is an elaboration of Table 2.

unit disk: Boundedness and compactness have been characterized in terms of Bloch and little Bloch spaces in [31], and trace ideal criteria were worked out in terms of Besov spaces in [3],[31] and in [10].

unit ball: Boundedness and compactness criteria have been established in [17]. Trace ideal criteria are established in the unpublished paper [11], and are obtained as a consequence of the Hardy space theory in [24].

strongly pseudoconvex domain: Boundedness and compactness have been characterized in terms of Bloch and little Bloch spaces in [20]. The trace ideal criteria have not been done up to now, but there are some sufficient conditions in this case, as well as in the case of finite type domains in $\mathbf{C}^n$ (convex if $n > 2$). Since the small Hankel operator is "dominated" by the big Hankel operator (see for example [36]), the work in the 1990s on the latter, for example [7],[40],[42], [38],[50], [51],[43],[44],[58], automatically give sufficient conditions for each of the three problems of interest. Finding conditions which are both necessary and sufficient for the small Hankel operator, and for the big Hankel operator for $p < 2$ has proved difficult to achieve, however see [44]. In this context there is also a a useful relation between Bergman space results and Hardy space results in one higher dimension, see [37] and [58] for example.

bounded symmetric domain: A complete theory of boundedness, compactness, and trace ideal criteria have been established in [65].

The theory of the big Hankel operator differs significantly from that of the small Hankel operator. For example, there are cut-off phenomenon, going back to the setting of $\mathbf{R}^n$ in [30]. In more than one variable, the references below represent work which appeared in print after 1990.

The big Hankel operator on the *Hardy space* has been studied in at least two contexts, the unit ball [24] and the unit polydisk [19]. There does not seem to be any other references which study the big Hankel operator on Hardy spaces over domains more general than the unit ball and unit polydisk.

On the Bergman space there is more activity. The types of problems considered in this paper for the small Hankel operator have been studied for the big Hankel operator in the following works, which however will not be discussed here. In some cases, the operator in question is more general than the Hankel operator defined here. The author apologizes if some relevant references have been overlooked. In addition to the above references for strongly pseudoconvex domains, we also have [5],[3] for the unit disk, [4],[63],[59],[50] for the unit ball, [41],[64] for the unit polydisk, and [6],[61],[2] for bounded symmetric domains.

2. The small Hankel operator on the Hardy space of the unit ball

2.1. The unit disk

Let Δ be the open unit disk in $\mathbf{C}$ with normalized Lebesgue measure dA, and $\mathbf{T} = \partial\Delta$ the unit circle with normalized arc length measure $d\sigma$. Let $\mathcal{H}^p = \mathcal{H}^p(\Delta)$ be the Hardy space for $p \geq 1$, and let $S : L^2(\mathbf{T}, d\sigma) \to \mathcal{H}^2$ be the orthogonal projection. For holomorphic f, the *small Hankel operator* h_f on $\mathcal{H}^2$ is defined by

$$h_f g = S(f\overline{g}), \quad g \in \mathcal{H}^2, \quad f\overline{g} \in L^2(\mathbf{T}, d\sigma).$$

We know by the theorems of Nehari and Hartman respectively, that h_f is bounded or compact if and only if $f \in BMOA$ or $f \in VMOA$ (see [62, Chapter 9] for details or [47] for a summary of this). Let $\mathcal{S}_p$ be the Schatten class of operators on $\mathcal{H}^2$. The following well known theorem due to Peller [49] characterizes those holomorphic functions f for which $h_f \in \mathcal{S}_p$.

Theorem 2.1. *Let f be a holomorphic function on Δ, and $p \geq 1$. Then $h_f \in \mathcal{S}_p$ if and only if*

$$\int_D |f''(z)|^p (1 - |z|^2)^{2p-2}\, dA(z) < \infty.$$

For a detailed proof, see [62, Chapter 9].

2.2. Boundedness and Compactness

Let $B =$ be the unit ball in $\mathbf{C}^n$, and let σ denote Lebesgue area measure on ∂B. Recall that the Hardy space $\mathcal{H}^p(B)$ consists of all holomorphic functions $F : B \to \mathbf{C}$ satisfying

$$\|F\|_p^p = \sup_{0<r<1} \int_{\partial B} |F(rz)|^p \, d\sigma(z) < \infty.$$

The space $BMO(B)$ is defined as the space of functions $b : \partial B :\to \mathbf{C}$ such that

$$\|b\|_{BMO} = \sup_S \frac{1}{|S|} \int_S |b(y) - m_S(b)| \, d\sigma < \infty$$

where S runs over all spheres in ∂B with respect to the metric $|1 - z \cdot \overline{w}|^{1/2}$, and $m_S(b) = \int_S b \, d\sigma / |S|$.

The following three theorems form a pattern which can be used in various contexts. For the unit ball, they are contained in [17]

Theorem 2.2 (Factorization). *Every $F \in \mathcal{H}^1(B)$ can be written $F = \sum_i G_i H_i$, where $G_i, H_i \in \mathcal{H}^2(B)$ and $\sum \|G_i\|_2 \|H_i\|_2 \leq c\|F\|_1$.*

Theorem 2.3 (Boundedness). *For $f \in \mathcal{H}^2(B)$, if h_f denotes the small Hankel operator, then*

$$h_f \in B(\mathcal{H}^2(B)) \Leftrightarrow f \in BMOA(B).$$

The space $VMO = VMO(B)$ consists of those functions $b \in BMO(B)$ for which

$$\lim_{|S| \to 0} \frac{1}{|S|} \int_S |b - m_S(b)| \, d\sigma = 0,$$

and the space $VMOA(B)$ denotes the BMO-closure of the analytic polynomials. We have the duality relations:

$$VMOA^* = \mathcal{H}^1 \quad , \quad \mathcal{H}^{1^*} = BMOA.$$

Theorem 2.4 (Compactness). *For $f \in \mathcal{H}^2(B)$, if h_f denotes the small Hankel operator, then*

$$h_f \text{ is compact} \Leftrightarrow f \in VMOA(B).$$

The results of this subsection have been proved in the setting of a bounded strongly pseudoconvex domain in $\mathbf{C}^n$ with smooth boundary, as well as a bounded pseudoconvex domain of finite type in $\mathbf{C}^2$ in [34]. This work, as well as [35] has established the foundation for harmonic analysis on domains in several complex variables, and the results obtained, about Hardy spaces, BMO, Hankel operators, had been sought for fifteen years or more.

2.3. Trace Ideal Criteria

The problem of Schatten class membership has a history going back to 1980, but when restricted to the small Hankel operator on Hardy space, it has only been solved for the unit ball in one or several complex variables. More precisely, Peller, in [49] proved that the Hankel operator on the Hardy space of the unit disk belongs to the Schatten p-class $\mathcal{S}_p$ if and only if the symbol belongs to the Besov space $B^p, 1 \leq p < \infty$ (see 2.4). A similar theorem was obtained for the upper half plane in $\mathbf{C}$ by Coifman and Rochberg [16] for $p = 1$ and by Rochberg [52] for $p > 1$.

In more than one variable, there are two papers which prove the corresponding result for the open unit ball, namely Feldman and Rochberg [24] and Zhang [60]. The former involves the techniques of harmonic analysis on the Heisenberg group as well as the notion of nearly weakly orthonormal sequences [53], [54]. The latter involves duality of Bergman spaces and complex interpolation theory. These two papers show the richness of the problem and provide ideas for generalizations to domains other than the unit ball. For completeness, we state the result here.

Theorem 2.5. *Let f be a holomorphic function on the unit ball B in $\mathbf{C}^n$ and let $p \geq 1$. Then the small Hankel operator h_f belongs to the Schatten class $\mathcal{S}_p$ over the Hardy space $\mathcal{H}^2(B)$ if and only if*

$$\sum_{|\alpha|=n+1} \int_B \left| \frac{\partial^{|\alpha|} f}{\partial z^\alpha} \right|^p (1 - |z|^2)^{(p-1)(n+1)} \, dV(z) < \infty.$$

2.4. Hankel operators in the Dixmier class

Let H be a Hilbert space over $\mathbf{C}$. For $0 < p < \infty$, recall that $T \in \mathcal{S}_p(H)$ (Schatten-von Neumann p-class) if $\{\mu_n(T)\}_{n=1}^\infty \in \ell^p$, where the $\mu_n(T)$ are the eigenvalues of $|T| = (T^*T)^{1/2}$.

An important class of operators which lies between $\mathcal{S}_1(H)$ and $\mathcal{S}_{1+\epsilon}(H)$ is the Macaev ideal $\mathcal{S}_1^+(H)$ (also denoted by $\mathrm{L}^{(1,\infty)}(H)$), which we shall call the Dixmier class, cf. [18]. We say that $T \in \mathcal{S}_1^+(H)$ if $\{\sigma_n/\log n\}_{n=2}^\infty \in \ell^\infty$ where $\sigma_n = \sum_{j=1}^n \mu_j(T)$. This class was used in 1966 by Dixmier, see [18, p. 303] or [8, p. 5408], to settle in the negative the question of the uniqueness of the trace on $\mathcal{L}(H)$. We mention that $\mathcal{S}_1^+$ is a Banach space under the norm: $\|T\|_{\mathcal{S}_1^+} = \sup_{n\geq 2}\{\sigma_n(T)/\log n\}$.

More recently, J. Bellissard and co-workers have connected Hankel operators on the Hardy space of the unit disk with their study of the quantum Hall effect [8], thereby proposing the following question: What is the holomorphic function space which consists of precisely the symbols of Hankel operators belonging to the Dixmier class $\mathcal{S}_1^+$? An answer to this question is given in [45] as follows.

Recall that for $1 \leq p < \infty$, $B^p(\Omega)$ denotes the holomorphic Besov space over a domain Ω in $\mathbf{C}^n$, with the seminorm $\|\cdot\|_{B^p}$ defined as follows:

$$\|f\|_{B^p}^p = \int_\Omega |f^{(n+1)}(z)|^p K(z,z)^{1-p} \, dV(z), \quad |f^{(n+1)}(z)| = \sum_{|\beta|=n+1} \left| \frac{\partial^{n+1} f}{\partial z^\beta} \right|.$$

For each $\alpha > 0$, we let $dV_\alpha(z) = c_\alpha(1-|z|^2)^{\alpha-1}dV(z)$ where dV is Lebesgue volume measure, and $\int_\Omega dV_\alpha = 1$. Let $A^2_\alpha(\Omega)$ denote the weighted Bergman space on Ω and $P_\alpha : L^2(\Omega, dV_\alpha) \to A^2_\alpha(\Omega)$ the Bergman projection with Bergman kernel $K^\alpha_w(z) = K^\alpha(z,w) = c_\alpha(1 - z\cdot\overline{w})^{-n-\alpha}$. Note that, as a limiting case, $\alpha = 0$ gives rise to the Hardy space.

For a domain $\Omega \subset \mathbf{C}^n$, we say that a holomorphic function f over Ω belongs to $B^1_+(\Omega)$ if,

$$\|f\|_{B^1_+(\Omega)} = \int_\Omega \frac{|f^{(n+1)}(z)|}{1+|\log F(f)|} dV(z) < \infty, \quad F(f) = \frac{1+|f^{(n+1)}(z)|}{K_z(z)}.$$

Then we have the following theorem.

Theorem 2.6. *Let $\alpha \geq 0$ and let $f \in \mathcal{H}^2(B)$, where B is the unit ball. Then*

(i) $h^\alpha_f \in S^+_1(A^2_\alpha(B))$ if and only if $\sup_{1<p\leq 2}\left\{(p-1)\|f\|^p_{B^p}\right\} < \infty$.

(ii) If $f \in B^1_+(B)$, then $h^\alpha_f \in S^+_1(A^2_\alpha(B))$.

(iii) If $h^\alpha_f \in S^+_1(A^2_\alpha(B))$, then for any $p \in (1,2)$,

$$\int_B |f^{(n+1)}(z)|(1+|\log F(f)|)(1+\log(1+|\log F(f)|))^{-p} dV(z) < \infty.$$

By using the results on the boundedness and compactness of h_f in [34], [58] and the asymptotic expansion of the Bergman and Szegö kernels given in [22], one can prove Theorem 2.6 in the case of a smoothly bounded strictly pseudoconvex domain in $\mathbf{C}^n$. This remark can also apply to other domains in $\mathbf{C}^n$, such as bounded symmetric domains, by using the results proved in [65].

3. The small Hankel operator on the Hardy space of the bidisk

The problem of boundedness and compactness in this setting has been discussed in [47], where sufficient conditions are given, based on the study of multiparameter Fourier analysis done in [13]. A survey of these theorems on the unit disk (Theorems of Nehari and Hartman) as well as on the unit ball in several complex variables (Theorems of Coifman-Rochberg-Weiss) is given in [47, 1.1]. As in [17], a proof of boundedness could be based on a factorization theorem and a proof of compactness could be based on factorization and a duality between H^1 and $VMOA$ on the bidisc. The work in [46, Theorem 2.3] proves a factorization theorem for atoms but the proof for a general H^1 function, as stated there, is incomplete. The author is grateful to Aline Bonami for bringing this to his attention. Thus, the subtle question of whether the sufficient conditions for boundedness and compactness are also necessary is, as with H^1-factorization, still an open problem.

The unit ball in $\mathbf{C}^n$ is an example of a bounded symmetric domain and of a strongly pseudoconvex domain. For the small Hankel operator on the Hardy space in these contexts (other than for the unit disk or unit ball), the only criteria on boundedness and compactness are those of Krantz-Li for strongly pseudoconvex domains [34]. It should be noted that some sufficient conditions for the boundedness, compactness, and belonging to a Schatten class are proved in [58] in the case of a bounded pseudoconvex domain of finite type in $\mathbf{C}^2$ with smooth boundary. Moreover, for a general strongly pseudoconvex domain, necessary and sufficient conditions are proved in [9], which also considers the problem in the setting of complex ellipsoids.

To summarize then, there are three ingredients needed for a theorem of Nehari type on the Hardy space of a domain. Namely,

- duality of $\mathcal{H}^1$ with BMO
- atomic decomposition of $\mathcal{H}^1$
- factorization in $\mathcal{H}^1$

As already noted, these results are known for the unit disk (Fefferman, Coifman), unit ball (Coifman-Rochberg-Weiss), and strongly pseudoconvex domains (Krantz-Li). For the polydisk, the first two are known (Chang-Fefferman). In this section we give an exposition in the case of the polydisk, and in the process correct some inaccuracies in [47] and [46].

3.1. Multiparameter Harmonic Analysis

Hardy spaces of the bidisc

We denote by $\Gamma_j(\theta_j)$ a standard cone in the unit disc Δ with vertex at $e^{i\theta_j} \in \mathbf{T}$, that is, for $j = 1, 2$,

$$\Gamma_j(\theta_j) = \{z_j \in \Delta : |1 - z_j e^{-i\theta_j}| < 1 - |z_j|\},$$

and we set, for $\theta = (\theta_1, \theta_2) \in \mathbf{T}^2$,

$$\Gamma(\theta) = \Gamma_1(\theta_1) \times \Gamma_2(\theta_2)$$

For a measurable function u on Δ^2, let $N(u)(\theta)$ be the unrestricted nontangential maximal function,

$$N(u)(\theta) = \sup_{z \in \Gamma(\theta)} |u(z)|,$$

and let $A(u)$ denote the area integral of u, that is,

$$A^2(u) = A_{12}^2(u) + A_1^2(u) + A_2^2(u) + |u(0)|^2,$$

where

$$A_{12}^2(u)(\theta) = \int_{\Gamma(\theta)} |\nabla_1 \nabla_2 u(z)|^2 \, dz,$$

$$A_1^2(u)(\theta) = \int_{\Gamma_1(\theta_1)} |\nabla_1 u(z_1, 0)|^2 \, dz_1,$$

$$A_2^2(u)(\theta) = \int_{\Gamma_2(\theta_2)} |\nabla_2 u(0, z_2)|^2 \, dz_2,$$

and dz_j is Lebesgue measure on Δ, $dz = dz_1 dz_2$.

We shall be dealing with functions u which are harmonic in each variable: $\Delta_1 u = \Delta_2 u = 0$. For such a function, it is known ([26, Th.1]) that for $0 < p < \infty$, $N(u) \in L^p(\mathbf{T})$ if and only if $A(u) \in L^p(\mathbf{T})$, and the space $H^p(\Delta^2)$ is defined to be the set of functions u harmonic in each variable, such that this condition is satisfied. The space H^p is normed as follows:

$$\|u\|_{H^p} = \|N(u)\|_{L^p} \text{ or } \|A(u)\|_{L^p},$$

which are equivalent. We let f denote the boundary distribution of u and identify u with f when convenient. It is noteworthy that the usual holomorhic Hardy spaces defined in section 1.2 and denoted here by $H^p_A(\Delta^2)$, are included in these spaces; $H^p_A(\Delta^2) \subset H^p(\Delta^2)$.

There is a companion result which deals with the bi-upper half-plane $D = \mathbf{R}^2_+ \times \mathbf{R}^2_+$. For $x = (x_1, x_2) \in \mathbf{R}^2$, let

$$\Gamma(x) = \Gamma(x_1) \times \Gamma(x_2) = \{(y_1, t_1, y_2, t_2) : |x_1 - y_1| < t_1, |x_2 - y_2| < t_2\}.$$

Let $u(x,t)$ be harmonic in each variable (x_j, t_j) $(j = 1, 2)$ and denote by u^* the nontangential maximal function

$$u^*(x_1, x_2) = \sup_{(y,t)\in\Gamma(x)} |u((y_1, t_1, y_2, t_2)|,$$

and Su the square function

$$S^2(u)(x) = \int_{\Gamma(x)} |\nabla_1 \nabla_2 u(y,t)|^2 \, dy_1 dy_2 dt_1 dt_2.$$

For a function $f : \mathbf{R}^2 \to \mathbf{C}$, let $u(x,t) = P[f](x,t)$ be its bi-Poisson integral. Then, for $0 < p < \infty$, by definition, f belongs to $H^p(D)$ if $u^* \in L^p(\mathbf{R}^2)$. It is known that this is the case if and only if $S(u) \in L^p(\mathbf{R}^2)$ ([23, pp.103–109]). As above, $H^p_A(D) \subset H^p(D)$ and $\|f\|_{H^p}$ is given by either of the equivalent norms $\|u^*\|_{L^p}$ or $\|Su\|_{L^p}$.

The space $H^1(D)$ will be of special interest to us. It is defined as

$$\begin{aligned} H^1(D) &= \{u : u \text{ harmonic on } D, u^* \in L^1(\mathbf{R}^2)\} \\ &= \{f : f \text{ defined on } \mathbf{R}^2, u = \text{Poisson integral of } f, f^* = u^* \in L^1(\mathbf{R}^2)\}. \end{aligned}$$

It is proved in [13, Th.1] that

$$H^1(D) = \{f = \sum \lambda_k a_k, a_k \text{ atoms }, \sum |\lambda_k| \le A\|f\|_{H^1}\}. \tag{1}$$

Atoms will be defined below in subsection 3.2. Equation (1) is the atomic decomposition of H^1 and will be discussed further below.

Duality of H^1 with BMO on the bidisc

By the combined efforts of A. Chang and R. Fefferman, the following three conditions serve as criteria for a function φ to belong to BMO in the multiparameter setting. We state only the bidisc version.

Theorem 3.1. *Let $\varphi : \mathbf{T}^2 \to \mathbf{C}$. Then $\varphi \in BMO(\mathbf{T}^2)$ if one of the following equivalent conditions holds:*

(i): $\varphi \in (H^1)^*$, *that is*

$$\left|\int f\varphi\right| \leq \|\varphi\|_{BMO}\|f\|_{H^1}.$$

(ii): $\varphi \in L^\infty + H_1 L^\infty + H_2 L^\infty + H_1 H_2 L^\infty$, *where H_j is the Hilbert transform in the variable z_j.*

(iii): *If $u = P[\varphi]$ and $\Omega \subset \mathbf{T}^2$ is an open set, then*

$$\int_{S(\Omega)} |\nabla_1 \nabla_2 u|^2 \log\frac{1}{|z_1|}\log\frac{1}{|z_2|}\, dA(z_1)dA(z_2) \leq C|\Omega|,$$

where $S(\Omega) = \{(z_1, z_2) \in \Delta^2 : I_{z_1} \times I_{z_2} \subset \Omega\}$.

It is now a simple matter to obtain a holomorphic duality theorem, correcting an omission in [47]. Let $BMOA(\Delta^2) = H^2_A(\Delta^2) \cap BMO(\mathbf{T}^2)$.

Theorem 3.2. $BMOA = (H^1_A)^*$.

Proof. Let $\ell \in (H^1_A)^*$. Then $\ell(f) = \int f\bar{g}$ for some $g \in L^\infty$. For $f \in H^2_A$, $\ell(f) = \int f\bar{g} = \int f\overline{Sg}$, where S is the Szegö projection from L^2 onto H^2_A. Now $Sg \in S(L^\infty) \cap H^2_A = BMOA$, since $S(L^\infty) \subset BMO$ by virtue of the relation of the Szegö projection and Hilbert transform in one variable: $Sf = (iHf + f - \hat{f}(0))/2$.

Conversely, if $g \in BMOA$, then for $f \in H^2_A \subset H^1_A$,

$$\left|\int f\bar{g}\right| \leq C\|f\|_{H^1}\|g\|_{BMO}.$$

Let $\pi : BMOA \to (H^1_A)^*$ be the map $\pi(g) = \ell_g$. By the above arguments, π is linear and onto. To show that it is one-to-one, suppose $\pi(g_1) = \pi(g_2)$. Then for any $f \in L^2(\mathbf{T}^2)$,

$$\int g\overline{(g_1 - g_2)} = (f, g_1 - g_2)_{L^2} = (Sf, g_1 - g_2)_{L^2} = 0$$

since $Sf \in H^2_A$. Thus $g_1 = g_2$, and since $BMOA$ is complete, by the open mapping theorem and the inequality $\|\pi(g)\| \leq C\|g\|_{BMO}$, the norms $\|\pi(g)\|$ and $\|g\|_{BMO}$ are equivalent. □

3.2. Factorization of an atom on the bidisc

In this subsection we first elaborate on the atomic decomposition (1).

We work in the context first of the bi-upper half plane D. An *atom* is a function $a = a(x_1, x_2)$ on the Shilov boundary $\mathbf{R}^2$, supported in an open set Ω of finite measure, which satisfies the following conditions:

1. $\|a\|_{L^2} \leq |\Omega|^{-1/2}$
2. a has mean zero over every component interval of every x_j-cross section of Ω
3. a is further decomposed into "elementary particles" $a = \sum_R a_R$ where

(a): Each a_R is supported on a rectangle $R \subset \Omega$ with $R \not\subset 3R'$ for any $R \neq R'$ in the sum

(b): $\int_{I_j} a_R(x)\, dx_j = 0$, $R = I_1 \times I_2$

(c): a_R satisfies

- $\|a_R\|_\infty \leq c_R|R|^{-1/2}$
- $\|\partial a_R/\partial x_j\|_\infty \leq c_R|I_j|^{-1}|R|^{-1/2}$
- $\|\partial^2 a_R/\partial x_1 \partial x_2\|_\infty \leq c_R|R|^{-3/2}$

where $\sum_R c_R^2 \leq A|\Omega|^{-1}$.

With this definition of atom, one has the atomic decomposition (1) of Chang and Fefferman. It is natural to expect that (1) and Theorem 3.3 below would lead to a factorization theorem for an arbitrary element of H^1, but as of this writing, this has not been proved.

Theorem 3.3 (Theorem 2.3 of [46]). *Let a be an atom. Then for each R in the decomposition $a = \sum_R a_R$, there exist $B_j, C_j \in H^2(D)$ such that*

$$S(a_R) = \sum_1^4 B_j C_j$$

and

$$\sum_1^4 \|B_j\|_2 \|C_j\|_2 \leq c c_R |R|^{1/2}.$$

Because all functions involved are holomorphic, the multivariable Cayley transform can be used to transfer Theorem 3.3 to the setting of the bidisc. Moreover, the factorization can be done for any p-atom for $0 < p \leq 1$, as defined in [14], and in this case, the holomorphic component functions B_j, C_j belong to H^{2p}. The corresponding factorization theorem for arbitrary elements of H^p for this range of p, for the unit ball and for strongly pseudoconvex domains, were proved in [25] and [34] respectively.

References

[1] E. Amar, *Big Hankel operator and $\overline{\partial}_b$-equation*, J. Operator Theory 33 (1995), 223–233.

[2] J. Arazy, *Boundedness and compactness of generalized Hankel operators on bounded symmetric domains*, J. Funct. Anal. 137 (1996), 97–151.

[3] J. Arazy, S.D. Fisher and J. Peetre *Hankel operators on weighted Bergman spaces*, Amer. J. Math. 110 (1988), 989–1053.

[4] J. Arazy, S.D. Fisher, S. Janson and J. Peetre, *Membership of Hankel operators on the ball in unitary ideals*, J. London Math. Soc 43 (1991), 485–508.

[5] S. Axler, *The Bergman space, the Bloch space, and commutators of multiplication operators*, Duke Math. J. 53 (1986), 315–332.

[6] D. Bekolle, C. Berger, L. Coburn, and K. Zhu, *BMO in the Bergman metric of bounded symmetric domains*, J. Funct. Anal. 93 (1990), 310–350.

[7] F. Beatrous and S.Y. Li, *Trace ideal criteria for operators of Hankel type*, Ill. J. Math. 39 (1995), 723–754.

[8] J. Bellissard, A. van Elst, H. Schulz-Baldes, *The noncommutative geometry of the quantum Hall effect*, J. Math. Phys. 35 (1994), 5373–5451.

[9] A. Bonami, M. Peloso, and F. Symesak, *Powers of the Szegö kernel and Hankel operators on Hardy spaces*, preprint 1998.

[10] F. F. Bonsall, *Hankel operators on the Bergman space for the disc*, J. London Math. Soc. 33 (1986), 355–364.

[11] J. Burbea, Trace ideal criteria for Hankel operators in the ball, preprint 1987.

[12] E. Cartan, *Sur les domaines bornés homogènes de l'espace de n variables complexes*, Abh. Math. Sem. Univ. Hamburg 11 (1935), 116–162.

[13] S.-Y. A. Chang and R. Fefferman, *A continuous version of duality of H^1 with BMO on the bidisc*, Ann. of Math. 112 (1980), 179–201.

[14] S.-Y. A. Chang and R. Fefferman, *The Calderon-Zygmund decomposition on product domains*, Amer. J. Math. 104 (1982), 455–468.

[15] J. M. Cohen and F. Colonna, *Bounded holomorphic functions on bounded symmetric domains*, Trans. Amer. Math. Soc. 343 (1994), 135–156.

[16] R. Coifman and R. Rochberg, *Representation theorems for holomorphic and harmonic functions in L^p*, Astérisque 77 (1980), 11–66.

[17] R. Coifman, R. Rochberg and G. Weiss, *Factorization theorems for Hardy spaces in several variables*, Ann. of Math. 103 (1976), 611–635.

[18] A. Connes, *Noncommutative geometry*, Academic Press, 1994.

[19] M. Cotlar and C. Sadosky, *Nehari and Nevanlinna-Pick problems and holomorphic extensions in terms of restricted BMO*, J. Funct. Anal. 124 (1994), 205–210.

[20] B. Coupet, *Décomposition atomique des espaces de Bergman*, Indiana Univ. Math. J. 38 (1989), 917–941.

[21] J. Faraut and A. Koranyi, *Analysis on symmetric cones*, Oxford University Press, 1994.

[22] C. Fefferman, *The Bergman kernel and biholomorphic mappings of pseudoconvex domains*, Invent. Math 26 (1974), 1–65.

[23] R. Fefferman, *Multiparameter Fourier Analysis*, In: Beijing Lectures, Annals of Mathematics Studies 112, Princeton Univ. Press (1986), 47–130.

[24] M. Feldman and R. Rochberg, *Singular value estimates for commutators and Hankel operators on the unit ball and the Heisenberg group*, 121–160, Analysis and PDE (C. Sadosky, ed.), A collection of papers dedicated to M. Cotlar, Dekker, New York 1990.

[25] J. Garnett and R. Latter, *The atomic decomposition for Hardy spaces in several complex variables*, Duke Math. J. 45 (1978), 815–845.

[26] R. F. Gundy and E. M. Stein, *H^p theory for the poly-disc*, Proc. Natl. Acad. Sci. USA 76 (1979), 1026–1029.

[27] K. T. Hahn and J. Mitchell, *Representation of linear functionals in H^p spaces over bounded symmetric domains in* $\mathbf{C}^n$, J. Math. Anal. Appl. 56 (1976), 379–396.

[28] Hoffman, K., *Banach spaces of analytic functions*, Prentice-Hall 1962

[29] L. K. Hua, *Harmonic analysis of functions of several complex variables in the classical domains*, Translation of Mathematical Monographs, 6. Amer. Math. Soc. 1979.

[30] S. Janson and T. Wolfe, *Schatten classes and commutators ofsingular integral operators*, Ark. Mat. 20 (1982), 301–310.

[31] S. Janson, J. Peetre, R. Rochberg, *Hankel forms and the Fock space*, Revista Math. Iberoamericano 3 (1987), 61–138.

[32] M. Koecher, *An elementary approach to bounded symmetric domains*, Lecture Notes, Rice University, 1969.

[33] S. Krantz, *Function theory of several complex variables*, Wadsworth and Brooks/Cole, 1992.

[34] S. Krantz and S.-Y. Li, *On decomposition theorems for Hardy spaces on domains in* $\mathbf{C}^n$ *and applications*, J. Fourier Anal. Appl. 2 (1995), 65–107.

[35] S. G. Krantz and S.-Y. Li, *Duality theorems for Hardy and Bergman spaces on convex domains of finite type in* $\mathbf{C}^n$, Ann. Inst. Fourier (Grenoble) 45 (1995), 1305–1327.

[36] S. G. Krantz, S.-Y. Li, P. Lin and R. Rochberg, *The effect of boundary regularity on the singular numbers of Friedrichs operators on Bergman spaces*, Michigan Math. J. 43 (1996), 337–348.

[37] S. G. Krantz, S.-Y. Li and R. Rochberg, *Analysis of some function spaces associated to Hankel operators*, Ill. J. Math. 41 (1997), 398–411.

[38] S. G. Krantz, S.-Y. Li and R. Rochberg, *The effect of boundary geometry on Hankel operators belonging to the trace ideals of Bergman spaces*, Int. Eq. Operator Theory 28 (1997), 196–213.

[39] S. V. Khrushchev and V. V. Peller, *Hankel operators, best approximations and stationary Gaussian processes*, Uspekhi Mat. Nauk 37 (1982), 53–124. English translation: Russian Math. Surveys 37 (1982), 61–144.

[40] H. Li, *Schatten class Hankel operators on the Bergman spaces of strongly pseudoconvex domains*, Proc. Amer. Math. Soc. 119 (1993), 1211–1221.

[41] H. Li, *Hankel operators on the Bergman space of the unit polydisc*, Proc. Amer. Math. Soc. 120 (1994), 1113–1121.

[42] H. Li, Hankel operators on the Bergman space of strongly pseudoconvex domains, Int. Eq. Oper. Theory 19 (1994), 458–476.

[43] H. Li and D. Luecking, *BMO on strongly pseudoconvex domains: Hankel operators, duality, and* $\overline{\partial}$*-estimates*, Trans. Amer. Math. Soc. 346 (1994), 661–691.

[44] H. Li and D. Luecking, *Schatten class of Hankel and Toeplitz operators on the Bergman space of strongly pseudoconvex domains*, Contemporary Math. 185 (1995), 237–257.

[45] S.Y. Li and B. Russo *Hankel operators in the Dixmier class*, C. R. Acad. Sci. Paris, 325 (1997), 21–26.

[46] I.-J. Lin, *Factorization theorems for Hardy spaces of the bidisc,* $0 < p \leq 1$, Proc. Amer. Math. Soc. 124 (1996), 549–560.

[47] I.-J. Lin and B. Russo, *Applications of factorization in the Hardy spaces of the polydisk*, In: Interactions between functional analysis, harmonic analysis, and probability (Columbia Missouri), edited by N. Kalton, E. Saab, S. Montgomery-Smith, Marcel Dekker 1995 , pp. 131-146.

[48] O. Loos, *Bounded symmetric domains and Jordan pairs*, University of California, Irvine, 1977.

[49] V. V. Peller, *Hankel operators of class* S_p *and applications*, Mat. Sb. 113 (1980), 538–581; translation in Math. USSR Sbornik 4 (1982), 443–479.

[50] M. Peloso, *Besov spaces, mean oscillation, and generalized Hankel operators*, Pacific J. Math. 161 (1993), 155–184.

[51] M. Peloso, *Hankel operators on weighted Bergman spaces on strongly pseudoconvex domainsk* Ill. J. Math. 38 (1994), 223–249.

[52] R. Rochberg, *Trace ideal criteria for Hankel operators and commutators*, Indiana U. Math. J. 31 (1982), 913–925.

[53] R. Rochberg and S. Semmes, *A decomposition theorem for BMO and applications*, J. Funct. Anal. 67 (1986), 228–263.

[54] R. Rochberg and S. Semmes, *Nearly weakly orthonormal sequences, singular value estimates, and Calderon-Zygmund operators*, J. Funct. Anal. 86 (1989), 237–306.

[55] W. Rudin, *Function Theory in Polydisks*, W. A. Benjamin 1969

[56] W. Rudin, *Function Theory in the unit ball of* $\mathbf{C}^n$, Springer-Verlag 1980

[57] B. Russo, *Composition operators in several complex variables*, In: Composition Operators, Proceedings of the Rocky Mountain Mathematics Consortium Conference 1996, Eds: C. Cowen, F. Jafari, B. MacCluer, D. Porter, American Mathematical Society, Contemporary Mathematics, to appear.

[58] F. Symesak, *Hankel operators on pseudoconvex domains of finite type in* $\mathbf{C}^2$, Can. J. Math. 50 (1998), 658–672.

[59] R. Wallsten, *Hankel operators between weighted Bergman spaces in the ball*, Ark. Mat. 28 (1990), 183–192.

[60] G. Zhang, *Hankel operators on Hardy spaces and Schatten classes*, Chinese Ann. of Math. 3 (1991), 282–293.

[61] D. Zheng, *Schatten class Hankel operators on the Bergman space*, Int. Eq. Operator Theory 13 (1990), 442–459.

[62] K. Zhu, *Operator Theory in Function Spaces*, Marcel Dekker, Inc., 1990.

[63] K. Zhu, *Schatten class Hankel operators on the Bergman space of the unit ball*, Amer. J. Math. 113 (1991), 147–167.

[64] K. Zhu, *Hankel operators on the Bergman space of bounded symmetric domains*, Trans. Amer. Math. Soc. 324 (1991), 707–730.

[65] K. Zhu, *Holomorphic Besov spaces on bounded symmetric domains. II*, Indiana U. Math. J. 44 (1995), 1017–1031.

University of California
Irvine, CA 92697-3875
E-mail address: brusso@math.uci.edu

Operator Theory:
Advances and Applications, Vol. 114

The reproducing kernel Hilbert space and its multiplication operators

Franciszek Hugon Szafraniec

To commemorate the 90th anniversary of the introduction of the reproducing kernel property by Stanisław Zaremba

Abstract. This exposition collects some facts concerning the reproducing kernel Hilbert space and its multiplication operators and is oriented towards applications, especially to modelling *unbounded* Hilbert space operators as in [5]. It, by the way, reflects the author's personal view of the RKHS approach as well as his interest in clarifying the circumstances.

Prelude

Let X be a set and $\boldsymbol{f} = \{f_\alpha\}_{\alpha\in A}$ be a family of complex functions on X. If $\mathcal{F}(X)$ denote the linear space of all linear combinations of the members of $\boldsymbol{f}$, define the inner product on $\mathcal{F}(X)$ by extending

$$\langle f_\alpha, f_\beta\rangle_{\boldsymbol{f}} = \delta_{\alpha\beta}$$

provided $\boldsymbol{f}$ is linearly independent. Completing the inner product space $\mathcal{F}(X)$ gives us a Hilbert space in which the sequence $\boldsymbol{f}$, when properly imbedded, forms an orthonormal basis. This fast (apparently somewhat lazy though sometimes the only possible) way of defining a Hilbert space in which a prescribed space of functions is "dense" is too rough; so simple an analytic object as the operator of multiplication by Z when $\mathcal{F}(X) = \mathbb{C}[Z]$, the space of all analytic polynomials, after passing to the Hilbert space becomes rather vague. However, making a more fortunate choice of $\boldsymbol{f}$ one can have $\mathcal{F}(X)$ imbedded as a dense subspace of a Hilbert space of functions. This is usually done by means of the reproducing kernel construction, which we present in detail and with some care.

The reproducing kernel Hilbert space

Let X be a set. Suppose we are given a couple $(\mathcal{H}, K)$ where $\mathcal{H}$ is a Hilbert space of complex functions on X (with inner product denoted by $\langle\cdot,\cdot\rangle$) and K is a complex

1991 *Mathematics Subject Classification.* Primary 47B20, 47B60, 46B20, 46B22; Secondary 47B38, 30C40.
Key words and phrases. reproducing kernel Hilbert space, the multiplication operator, multiplier, cyclic operator, joint cyclic operators, joint eigenvector and joint point spectrum.

function on $X \times X$. The function K is called a *reproducing kernel of* $\mathcal{H}$ and the space $\mathcal{H}$ a *reproducing kernel Hilbert space with respect to* K if

$$f(x) = \langle f, K_x \rangle, \; x \in X, \; f \in \mathcal{H}, \tag{1}$$

where $K_x \stackrel{\text{df}}{=} K(\cdot, x)$, is sometimes called a kernel function. We refer to $(\mathcal{H}, K)$ as a *RKHS couple on* X and this is what we deliberately intend to start with; on many occasions one finds the key word "RKHS" used either for consequences or elements of a construction, very often mixed up together. At any rate, the formula (1) is usually known to as a *reproducing kernel property*, in our presentation, *of the couple* $(\mathcal{H}, K)$.

The immediate consequences are:

(**α**) The kernel K must necessarily be *positive definite*; that is,

$$\sum_{i,j=0}^{N} K(x_i, x_j)\lambda_i\bar{\lambda}_j \geq 0, \quad x_1, \ldots, x_N \in X, \; \lambda_1, \ldots \lambda_N \in \mathbb{C}. \tag{2}$$

(**β**) The linear functionals

$$\phi_x : \mathcal{H} \ni f \mapsto f(x) \in \mathbb{C}$$

are *continuous* for every $x \in X$

(**γ**) The kernel is uniquely determined by the space in the sense that if $(\mathcal{H}, K_1)$ and $(\mathcal{H}, K_2)$ are two RKHS couples, then $K_1 = K_2$. The kernel can be obtained, in particular, from the formula

$$K(x, y) = \sum_{\alpha \in A} f_\alpha(x)\overline{f_\alpha(y)}, \; x, y \in X \tag{3}$$

where $\{\alpha\}_{\alpha \in A}$ is any[1] orthonormal basis of $\mathcal{H}$.

(**δ**) The set $\{K_x; \; x \in X\}$ is total in $\mathcal{H}$ and, consequently, the space is uniquely determined by the kernel in the sense that if $(\mathcal{H}_1, K)$ and $(\mathcal{H}_2, K)$ are RKHS couples, then $\mathcal{H}_1 = \mathcal{H}_2$.

Moreover, the following *RKHS test* (see [6] for instance) gives a characterization of those complex function on X which belong to $\mathcal{H}$:

(**η**) f is in $\mathcal{H}$ if and only if there exists $C > 0$ such that

$$|\sum_{i=0}^{N} f(x_i)\lambda_i|^2 \leq C^2 \sum_{i,j=1}^{N} K(x_i, x_j)\lambda_i\bar{\lambda}_j, \quad x_1, \ldots, x_N \in X, \; \lambda_1, \ldots \lambda_N \in \mathbb{C}. \tag{4}$$

If this happens, then

$$\|f\| = \inf\{C : \text{(4) holds}\}. \tag{5}$$

---

[1]For any x, y there is always at most countable number of α's different from zero which makes such sums possible.

There are in fact two standard ways, corresponding to (α) and (β) respectively, in which a RKHS couple $(\mathcal{H}, K)$ can be introduced (any other construction sooner or later leads to one of these two)[2]:

(a) *Coupling the functional space with a kernel.* Suppose we are given a kernel $K : X \times X \mapsto \mathbb{C}$. Set $\mathcal{D}_K \overset{\mathrm{df}}{=} \operatorname{lin}\{K_x;\ x \in X\}$, the linear span of the functions K_x. If K is *positive definite*, then

$$\langle K_x, K_y \rangle \overset{\mathrm{df}}{=} K(y, x) \tag{6}$$

defines an inner product in $\mathcal{D}_K$ and the completion $\mathcal{H}_K$ of $\mathcal{D}_K$ can still be realized as a space of functions. The resulting space $\mathcal{H}_K$ is a RKHS with the kernel K (for details see [1]).

(b) *Coupling the kernel with a functional space.* Suppose we are given a Hilbert space $\mathcal{H}$ of functions on X. If the linear functional $\phi_x : \mathcal{H} \ni f \mapsto f(x) \in \mathbb{C}$ is *continuous* for any $x \in X$, then $K_{\mathcal{H}}(x, y) \overset{\mathrm{df}}{=} \phi_y^* \phi_x$, where ϕ_y^* stands for the adjoint of the Hilbert space operator ϕ_x, becomes a kernel of $\mathcal{H}$.

A RKHS from its prospective basis

Suppose we are given a family $\boldsymbol{f} = \{f_\alpha\}_{\alpha \in A}$ of functions on X such that

$$\sum_{\alpha \in A} |f_\alpha(x)|^2 < +\infty, \ x \in X. \tag{7}$$

Then $K^{\boldsymbol{f}}$ defined by

$$K^{\boldsymbol{f}}(x, y) \overset{\mathrm{df}}{=} \sum_{\alpha \in A} f_\alpha(x) \overline{f_\alpha(y)}, \ x, y \in X \tag{8}$$

is positive definite and we can follow the construction as in **(a)** resulting in the RKHS $\mathcal{H}^{\boldsymbol{f}}$. Because this approach is not very popular we take the opportunity to prove some simple facts to make it useful.

Fact A. *Let $\xi = \{\xi_\alpha\}_{\alpha \in A} \in \ell^2(A)$. Then for every $x \in X$, the series $\sum_{\alpha \in A} \xi_\alpha f_\alpha(x)$ converges absolutely and the function $f_\xi : x \mapsto \sum_{\alpha \in A} \xi_\alpha f_\alpha(x)$ is in $\mathcal{H}^{\boldsymbol{f}}$; the series $\sum_{\alpha \in A} \xi_\alpha f_\alpha$ converges in $\mathcal{H}^{\boldsymbol{f}}$ to f_ξ. In particular, any function f_α, $\alpha \in A$ belongs to $\mathcal{H}^{\boldsymbol{f}}$ and $\sum_{\alpha \in A} \overline{f_\alpha(x)} f_\alpha$ converges in $\mathcal{H}^{\boldsymbol{f}}$ to $K_x^{\boldsymbol{f}}$.*

[2]While often one takes one of them as a definition of RKHS, so as to avoid roundabout proofs, it seems to be appropriate to make the definition independent of a particular construction.

Proof. It is clear that, due to (7), for any $\{\xi_\alpha\}_{\alpha\in A}\in \ell^2(A)$ and any $x\in X$ the series $\sum_{\alpha\in A}\xi_\alpha f_\alpha(x)$ converges absolutely. Thus

$$|\sum_{i=0}^{N}(\sum_{\alpha\in A}\xi_\alpha f_\alpha(x_i))\lambda_i|^2$$

$$=|\sum_{\alpha\in A}\xi_\alpha\sum_{i=0}^{N}f_\alpha(x_i)\lambda_i|^2\le\sum_{\alpha\in A}|\xi|^2\|\sum_{i=0}^{N}f_\alpha(x_i)\lambda_i\|_{\ell^2(A)}=$$

$$\|\xi\|_{\ell^2(A)}\sum_{i,j=0}^{N}K^{\boldsymbol{f}}(x_i,x_j)\lambda_i\bar{\lambda}_j. \quad (9)$$

This implies by (4) that the function f_ξ is in $\mathcal{H}^{\boldsymbol{f}}$. Applying the RKHS test (5), we infer that $\sum_{\alpha\in A}\xi_\alpha f_\alpha$ converges in $\mathcal{H}^{\boldsymbol{f}}$, necessarily to f_ξ. The rest of the conclusion choices comes from particular choices of $\{\xi_\alpha\}_{\alpha\in A}$. □

Since, as we have already noticed the sequence $\{\sum_{i=0}^{N}f_\alpha(x_i)\lambda_i\}_{\alpha\in a}$ is in $\ell^2(A)$ for $\lambda_0,\dots,\lambda_N\in\mathbb{C}$, using (3) we get

$$\|\{\sum_{i=0}^{N}\overline{f_\alpha(x_i)}\lambda_i\}_{\alpha\in A}\|_{\ell^2(A)}=\|\sum_{i=0}^{N}\lambda_i K^{\boldsymbol{f}}_{x_i}\|_{\mathcal{H}^{\boldsymbol{f}}}. \quad (10)$$

This enables us, after setting $\mathcal{E}\overset{\text{df}}{=}\operatorname{clolin}\{\{\overline{f_\alpha(x)}\}_{\alpha\in A};\ x\in X\}$, to define a unitary operator

$$V:\mathcal{E}\ni\{\sum_{i=0}^{N}\overline{f_\alpha(x_i)}\lambda_i\}_{\alpha\in A}\mapsto\sum_{i=0}^{N}\lambda_i K^{\boldsymbol{f}}_{x_i}\in\mathcal{H}^{\boldsymbol{f}}. \quad (11)$$

On the other hand, due to Fact A, the operator

$$W:\ell^2(A)\ni\xi\mapsto\sum_{\alpha\in A}\xi_\alpha f_\alpha\in\mathcal{H}^{\boldsymbol{f}} \quad (12)$$

is well defined and, because of (5), the inequality (9) implies $\|\sum_{\alpha\in A}\xi_\alpha f_\alpha\|_{\mathcal{H}^{\boldsymbol{f}}}\le\|\{\xi_\alpha\}_{\alpha\in A}\|_{\ell^2(A)}$, which means that W is a contraction. It is a matter of direct verification that

$$Wf=Vf,\quad f\in\mathcal{E}, \quad (13)$$

Indeed, due to Fact A,

$$W(\{\sum_{i=0}^{N}\overline{f_\alpha(x_i)}\lambda_i\}_{\alpha\in A})=\sum_{\alpha\in A}(\sum_{i=0}^{N}\overline{f_\alpha(x_i)}\lambda_i)f_\alpha=\sum_{i=0}^{N}\lambda_i\sum_{\alpha\in A}\overline{f_\alpha(x_i)}f_\alpha=\sum_{i=0}^{N}\lambda_i K^{\boldsymbol{f}}_{x_i}.$$

This implies W (which is bounded) maps onto $\mathcal{H}^{\boldsymbol{f}}$. Let $\xi\in\ell^2(A)$ be orthogonal to $\mathcal{E}$. Then, by the reproducing kernel property (1),

$$0=\langle\xi,\{f_\alpha(x)\}_{\alpha\in A}\rangle=\sum_{\alpha\in A}\xi_\alpha f_\alpha(x)=\langle\sum_{\alpha\in A}\xi_\alpha f_\alpha,K^{\boldsymbol{f}}_x\rangle=\langle W\xi,K^{\boldsymbol{f}}_x\rangle,$$

which means that $W\xi = 0$. Consequently, W is a partial isometry from $\ell^2(A)$ onto $\mathcal{H}^{\boldsymbol{f}}$ with initial space $\mathcal{E}$.

Fact B. *The family $\boldsymbol{f} = \{f_\alpha\}_{\alpha\in A}$ is always complete in $\mathcal{H}^{\boldsymbol{f}}$. Moreover, the following conditions are equivalent*[3]

(i) $\xi \in \ell^2(A)$ *and* $\sum_{\alpha\in A} \xi_\alpha f_\alpha(x) = 0$ *for all* $x \in X$ *implies* $\xi = 0$;

(ii) $\{f_\alpha\}_{\alpha\in A}$ *is orthonormal in* $\mathcal{H}^{\boldsymbol{f}}$.

Proof. Since W is onto, every f in $\mathcal{H}^{\boldsymbol{f}}$ is of the form $\sum_{\alpha\in A} \xi_\alpha f_\alpha$ and this implies that $\boldsymbol{f}$ is complete.
(i)⇒(ii). Condition (i) implies that that the set $\{\{\overline{f_\alpha(x)}\}_{\alpha\in A};\ x \in X\}$ is total in $\ell^2(A)$ and, consequently, $\mathcal{E} = \ell^2(A)$. This implies, in turn, that W is unitary (use (13)) and, because

$$W\epsilon_\alpha = f_\alpha, \tag{14}$$

where $\epsilon_\alpha \overset{\text{df}}{=} \{\delta_{\alpha\beta}\}_{\beta\in A}$, $\{f_\alpha\}_{\alpha\in A}$ is orthonormal.
(ii)⇒(i). Condition (ii), because of (14), implies W is unitary and this forces its initial space $\mathcal{E}$ as a partial isometry to be the whole of $\ell^2(A)$. This, in turn, means (i) holds. □

Conclusion. *Let $\mathcal{F}(X)$ be a linear space of complex functions on X. Suppose there is $\{f_\alpha\}_{\alpha\in A} \subset \mathcal{F}(X)$ the linear span of which is $\mathcal{F}(X)$ and such that* (7) *holds. Then the RKHS couple $(\mathcal{H}, K)$ defined above is such that $\mathcal{H}$ includes $\mathcal{F}(X)$ as a dense subspace*[4]. *The RKHS couple $(\mathcal{H}, K)$ is uniquely determined regardless the choice of $\{f_\alpha\}_{\alpha\in A}$ satisfying* (7).

Proof. According to Fact A the family $\{f_\alpha\}_{\alpha\in A}$ is contained in $\mathcal{H}$ and because it linearly spans the space $\mathcal{F}(X)$, the latter is contained in and, due to Fact B, dense in $\mathcal{H}$.
Suppose there are given two RKHS couples $(\mathcal{H}_i, K_i)$, $i = 1, 2$. Then, if V_i and W_i are the operators defined by (11) and (12) and corresponding to $i = 1$ and $i = 2$, resp., then $U \overset{\text{df}}{=} V_2V_1^{-1}$ maps $\mathcal{H}_1$ unitarily onto $\mathcal{H}_2$. Since $Uf_\alpha = W_2(W_1|_{\mathcal{E}})^{-1}f_\alpha = f_\alpha$, the unitary W becomes the identity on $\mathcal{F}(X)$ and, because the latter is dense in both $\mathcal{H}_1$ as well as $\mathcal{H}_2$, it establishes the identity of $\mathcal{H}_1$ and $\mathcal{H}_2$. □

This conclusion fulfils what we were looking for in the Prelude and creates the environment for modelling cyclic operators, which, in turn, goes back to problems of classical Analysis such as orthogonality of polynomials, especially of analytic ones (cf., [8]) or interpolation problems (cf. for instance, [7] or [4]).

---

[3] Equivalence of (i) and of the conjunction of (ii) and of completenes of $\{f_\alpha\}_{\alpha\in A}$ is stated, after a somewhat narrative proof, as a Theorem in [2].

[4] Because we pre-suppose no norm in $\mathcal{F}(X)$, this has nothing in common, at least at this stage, with so called functional completion (cf. [2] for a discussion).

A RKHS of transforms

Let X be a set. Suppose we are given a mapping $\tau : X \ni x \mapsto h_x \in \mathcal{H}$. Set

$$K^\tau(x,y) \stackrel{\text{df}}{=} \langle h_y, h_x\rangle, \quad x,y \in X \tag{15}$$

and define $\mathcal{H}^\tau$ corresponding to the positive definite kernel K as in **(a)**. Set

$$f^\tau(x) \stackrel{\text{df}}{=} \langle f, h_x\rangle. \quad x \in X. \tag{16}$$

Then, according to the RKHS test ($\boldsymbol{\eta}$), the transform f^τ of f is in $\mathcal{H}^\tau$. Define two linear mappings

$$V : \operatorname{clolin}\{h_x;\ x \in X\} \ni h_x \mapsto K_x^\tau \in \mathcal{H}^\tau$$

and

$$W : \mathcal{H} \ni f \mapsto f^\tau \in \mathcal{H}^\tau. \tag{17}$$

Then evidently V is a unitary mapping of $\operatorname{clolin}\{h_x;\ x \in X\}$ onto $\mathcal{H}^\tau$ and, because $f^\tau \perp \operatorname{clolin}\{h_x;\ x \in X\}$ forces $f = 0$, $(0 = \langle f, h_x\rangle = f^\tau(x) \Rightarrow\ f = 0)$ and because $Wf = Vf$ for $f \in \operatorname{clolin}\{h_x;\ x \in X\}$, W is a partial isometry with the initial space $\operatorname{clolin}\{h_x;\ x \in X\}$.

Fact C. *Suppose $\{e_\alpha\}_{\alpha\in A}$ is an orthonormal basis*[5] *in $\mathcal{H}$ and*

$$K^\tau(x,y) = \sum_{\alpha\in A} e_\alpha^\tau(x)\overline{e_\alpha^\tau(y)}, \quad x,y \in X.$$

Then the following conditions are equivalent.

(i) *$\{e_\alpha^\tau\}_{\alpha\in A}$ is an orthonormal basis of $\mathcal{H}^\tau$,*
(ii) $\operatorname{clolin}\{h_x;\ x \in X\} = \mathcal{H}$.

Proof. (i)⇒(ii). Take $f = \sum_{\alpha\in A} \xi_\alpha e_\alpha$ such that $\langle f, h_x\rangle = 0$ for any $x \in X$. Then $0 = \langle f, h_x\rangle = \sum_{\alpha\in A} \xi_\alpha \langle e_\alpha, h_x\rangle = \sum_{\alpha\in A} \xi_\alpha e_\alpha^\tau(x)$. Now, by Fact B, (i) implies that $\xi_\alpha = 0$ for all α. This implies $f = 0$ which gives (ii).

(ii)⇒(i). Since $\operatorname{clolin}\{h_x;\ x \in X\} = \mathcal{H}$, $W = V =$ unitary (both W and V were defined just above) and (i) follows. □

The multiplication operator in a RKHS

Let φ be a complex function on X. Given an RKHS couple $(\mathcal{H}, K)$, define the operator $M_{\varphi,\max}$ as

$$\mathcal{D}(M_{\varphi,\max}) \stackrel{\text{df}}{=} \{f \in \mathcal{H};\ \varphi f \in \mathcal{H}\}, \quad M_{\varphi,\max} f \stackrel{\text{df}}{=} \varphi f,\ f \in \mathcal{D}(M_{\varphi,\max})$$

Fact D. *$M_{\varphi,\max}$ is a closed operator.*

Proof. This goes straightforwardly if one remembers that $\mathcal{H}$-norm convergence implies pointwise (simple) convergence. □

Set $\mathcal{D}_K = \operatorname{lin}\{K_x;\ x \in X\}$. Recall that $\mathcal{D}_K$ is dense in $\mathcal{H}$.

[5]In fact it is enough to assume $\{e_\alpha\}_{\alpha\in A}$ is a Riesz basis in $\mathcal{H}$.

Fact E. (i) *If $M_{\varphi,\max}$ is densely defined, then*

$$R_\varphi : \sum_{i=1}^{N} \lambda_i K_{x_i} \mapsto \sum_{i=1}^{N} \lambda_i \overline{\varphi(x_i)} K_{x_i} \tag{18}$$

is a well defined operator with domain $\mathcal{D}_K$ and $R_\varphi \subset M^*_{\varphi,\max} \overset{\mathrm{df}}{=} (M_{\varphi,\max})^*$.
(ii) *Conversely, if* (18) *leads to a well defined operator R_φ (with $\mathcal{D}(R_\varphi) = \mathcal{D}_K$), then $R^*_\varphi = M_{\varphi,\max}$.*
(iii) *If $M_{\varphi,\max}$ is densely defined then R_φ is closable and $M^*_{\varphi,\max} = R^-_\varphi$.*
(iv) *For any $x \in X$, K_x is always an eigenvector of R_φ and the corresponding eigenvalue is $\overline{\varphi(x)}$.*

Proof. Notice that, due to the reproducing kernel property,

$$\langle \varphi f, \sum_{i=1}^{N} \lambda_i K_{x_i} \rangle = \langle f, \sum_{i=1}^{N} \lambda_i \overline{\varphi(x_i)} K_{x_i} \rangle \tag{19}$$

for $f \in \mathcal{D}(M_{\varphi,\max})$.
Suppose $M_{\varphi,\max}$ is densely defined and take $f \in \mathcal{D}(M_{\varphi,\max})$. If $\sum_{i=1}^{N} \lambda_i K_{x_i} = 0$, then, by (19), $\sum_{i=1}^{N} \lambda_i \overline{\varphi(x_i)} K_{x_i} = 0$ and then the formula (18) defines an operator and implies that

$$R_\varphi \subset M^*_{\varphi,\max}. \tag{20}$$

Suppose (18) defines an operator. Then $f \in \mathcal{D}(R^*_\varphi)$ is equivalent to

$$|\langle f, \sum_{i=1}^{N} \lambda_i R_\varphi K_{x_i} \rangle|^2 \le C \| \sum_{i=1}^{N} \lambda_i K_{x_i} \|^2.$$

The left hand side of this inequality is, due to (19), equal to $|\sum_{i=1}^{N} \varphi(x_i) f(x_i) \lambda_i|^2$ and this, in turn, is equivalent, due to the RKHS test (4), to the fact that $f \in \mathcal{D}(M_{\varphi,\max})$. Consequently,

$$R^*_\varphi = M_{\varphi,\max} \tag{21}$$

Since (20) implies R_φ is closable, (20) toghether with (21) concludes in

$$R^-_\varphi \subset M^*_{\varphi,\max} = R^{**}_\varphi = R^-_\varphi.$$

The last statement (iv) follows directly from (18). □

Suppose we are given a family $\boldsymbol{f} = \{f_\alpha\}_{\alpha \in A}$ of complex functions on X such that

$$\sum_{\alpha \in A} |f_\alpha(x)|^2 < +\infty, \; x \in X. \tag{22}$$

Construct the couple $(\mathcal{H}, K)$ as in the preceding Section.

Fact F. *If*

$$\varphi f_\alpha \in \mathcal{H}, \quad \alpha \in A, \tag{23}$$

then $\mathcal{D}(M_{\varphi,\max})$ *is dense in* $\mathcal{H}$ *and* $\mathcal{D} \stackrel{\text{df}}{=} \operatorname{lin}\{f_\alpha;\ \alpha \in A\}$ *is core of* $M_{\varphi,\max}$.

Proof. Because $\mathcal{D}$ is dense in $\mathcal{H}$, $M_{\varphi,\max}$ is densely defined.

For any $f \in \mathcal{D}(M_{\varphi,\max})$, since $\boldsymbol{f}$ is complete, f can be aproximated by finite sums of elements in $\{f_\alpha\}_{\alpha\in A}$ both norm and pointwisely, according to Fact A. This implies that $(M_{\varphi,\max}|_{\mathcal{D}})^- = M_{\varphi,\max}$ which is the other part of the conclusion. □

Remark. If $\mathcal{D}(M_{\varphi,\max}) = \mathcal{H}$, φ is usually called a *multiplier* of $\mathcal{H}$. If this happens for $(\mathcal{H}, K)$ being a RKHS couple, $M_{\varphi,\max}$ must necessarily be a bounded operator [3]. So one may distinguish bounded and unbounded multipliers according to whether $\mathcal{D}(M_{\varphi,\max})$ is equal to $\mathcal{H}$ or is dense in it.

Operators of multplication by independent variables in $\mathbb{C}^N$

Suppose we are given a sequence $\boldsymbol{p} = \{p_n\}_{n=0}^\infty \subset \mathbb{C}[Z_1, \ldots, Z_N]$ such that

$$\omega_{\boldsymbol{p}} \stackrel{\text{df}}{=} \{z \in \mathbb{C}^N;\ \sum_{n=0}^\infty |p_n(z)|^2 < \infty\} \neq \varnothing. \tag{24}$$

Define the kernel $K^{\boldsymbol{p}}$ as in (8) with $X = \omega_{\boldsymbol{p}}$, $A = \mathbb{N}$ and $f_n = p_n$. Let $\mathcal{H}^{\boldsymbol{p}}$ be the corresponding RKHS.

Referring to the previous Section set $M_{i,\max} \stackrel{\text{df}}{=} M_{Z_i,\max}$ and $R_i \stackrel{\text{df}}{=} M(Z_i)$. Since, due to Fact A, every p is in $\mathcal{H}^{\boldsymbol{p}}$, we can define one more operator M_i by

$$\mathcal{D}(M_i) \stackrel{\text{df}}{=} \mathbb{C}[Z_1, \ldots, Z_N]|_{\omega_{\boldsymbol{p}}}, \quad M_i p \stackrel{\text{df}}{=} Z_i p.$$

Specifying Fact E and using Fact F we get

Fact G. $M_i^- = M_{i,\max} = R_i^*$ *and* $M_i^* = M_{i,\max}^* = R_i^-$, $i = 1, \ldots, N$.

Introduce a shorthand notation:

$$\boldsymbol{M} \stackrel{\text{df}}{=} (M_1, \ldots, M_N), \boldsymbol{M}_{\max} \stackrel{\text{df}}{=} (M_{1,\max}, \ldots, M_{N,\max})$$

and

$$\boldsymbol{R} \stackrel{\text{df}}{=} (R_1, \ldots, R_N).$$

Now we can improve conclusion (iv) of Fact E by

Fact H. *Suppose* h *is a joint eigenvector of* $\boldsymbol{R}$. *Then it must be necessarily of the form* $K_z^{\boldsymbol{p}}$ *for some* z *in* $\omega_{\boldsymbol{p}}$.

Proof. Let h be eigenvector in question. Then $\langle 1, h\rangle$ is different from zero (if not, $\langle 1, \overline{p(z)}h\rangle = \langle p(z), h\rangle = 0$ implies, due to density of polynomials in $\mathcal{H}^{\boldsymbol{p}}$, that $h = 0$), so say $\langle 1, h\rangle = 1$. Then

$$\langle p, h\rangle = p(z)$$

and, consequently,

$$\|h^2\| = \sum_{n=0}^\infty |\langle h, p_n\rangle|^2 = \sum_{n=0}^\infty |p(z)|^2$$

which implies $z \in \omega_{\boldsymbol{p}}$. So $\langle p, h\rangle = p(z) = \langle p, K_z^{\boldsymbol{p}}\rangle$ and density of polynomials in $\mathcal{H}^{\boldsymbol{p}}$ used again leads to the conclusion. □

Models of some classes of unbounded operators

Suppose $\mathcal{H}$ is a Hilbert space. Let $\boldsymbol{A} \stackrel{\text{df}}{=} (A_1, \ldots, A_N)$ be an n-tuple of densely defined operators such that $\mathcal{D}(A_i) = \{p(\boldsymbol{A})f;\ p \in \mathbb{C}[Z_1, \ldots, Z_N]\}$, $i = 1, \ldots, N$, with some $f \in \mathcal{H}$ and the operators A_i commute on $\mathcal{D}(\boldsymbol{A}) \stackrel{\text{df}}{=} \mathcal{D}(A_i)$ [6]. Suppose $\boldsymbol{p} = \{p_n\}_{n=0}^{\infty}$ is such that the vectors $p_n(\boldsymbol{A})f$, $n = 0, 1, \ldots$, are linearly independent and (24) holds. Introduce the RKHS couple $(\mathcal{H}^{\boldsymbol{p}}, K^{\boldsymbol{p}})$ as in the preceeding Section. Define a linear map

$$U^{\boldsymbol{p}} : \mathcal{D}(\boldsymbol{A}) \ni p(\boldsymbol{A})f \mapsto p \in \mathcal{H}^{\boldsymbol{p}}.$$

where M_i are as in the preceeding Section.

Fact I. *$U^{\boldsymbol{p}}A_i = M_iU^{\boldsymbol{p}}$, $i = 1, \ldots, N$. If $\{p_n\}_{n=0}^{\infty}$ is an orthonormal basis in $\mathcal{H}$ and if $\xi = \{\xi_n\}_{n=0}^{\infty} \in \ell^2$, $\sum_{n=0}^{\infty} \xi_n p_n(z) = 0$, $z \in \omega_{\boldsymbol{p}}$ implies always $\xi = 0$, then $U^{\boldsymbol{p}}$ is unitary.*

Proof. The first conclusion is direct. Fact B tells us that $U^{\boldsymbol{p}}$ defined above extends from $\mathcal{D}(\boldsymbol{A})$ to $\mathcal{H}^{\boldsymbol{p}}$ as a unitary operator. □

Let $\boldsymbol{B} \stackrel{\text{df}}{=} (B_1, \ldots, B_N)$ be another N-tuple of densely defined operators. Suppose the joint point spectrum $\sigma_{\text{jp}}(\boldsymbol{B}) \neq \varnothing$. Denote by h_z the common eigenvector of $\boldsymbol{B}$ corresponding to $z \in \sigma_{\text{jp}}(\boldsymbol{B})$. Suppose that the multiplicity of any h_z is 1. Then we can define $\tau : \sigma_{\text{jp}}(\boldsymbol{B}) \ni z \mapsto h_z \in \mathcal{H}$ and subsequently define the kernel K^{τ} by (15) as well as the corresponding Hilbert space $\mathcal{H}^{\tau}$. Denote the partial isometry given by (17) by $U_{\boldsymbol{B}}$.

Fact J. *$U^{\tau}R_i = R_iU_{\tau}$, $i = 1, \ldots, N$, where R_i are defined as in the preceeding Section.*

Proof. Since $M_{i,\max}$ are densely defined, Fact E part (i) implies that R_i are well defined. The rest is direct. □

Set $(\boldsymbol{B})^{-} \stackrel{\text{df}}{=} (B_1^{-}, \ldots, B_N^{-})$ and $\boldsymbol{A}^{*} \stackrel{\text{df}}{=} (A_1^{*}, \ldots, A_N^{*})$. Putting all the pieces together we can end up with the following two statments.

Theorem 1. *Suppose $\boldsymbol{A}$ is jointly cyclic with a cyclic vector f and $\sigma_{jp}(\boldsymbol{A}^{*}) \neq \varnothing$. If $\boldsymbol{p} = \{p_n\}_{n=0}^{\infty}$ is such that $\{p_n(\boldsymbol{A})f\}_{n=0}^{\infty}$ is an orthonormal basis of $\mathcal{H}$ [7] then there is a unitary operator $U : \mathcal{H} \mapsto \mathcal{H}^{\boldsymbol{p}}$ such that*

$$UA_i^{-} = M_{i,\max}U,\ UA_i^{*} = R_i^{-}U,\ i = 1, \ldots, N$$

provided either

$$\xi = \{\xi_n\}_{n=0}^{\infty} \in \ell^2,\ \sum_{n=0}^{\infty} \xi_n p_n(z) = 0,\ z \in \omega_{\boldsymbol{p}} \text{ implies } \xi = 0,$$

[6]Such an N-tuple of operators is said to be *jointly cyclic* and f is its *joint cyclic vector*.

[7]This can be always obtained by Gram-Schmidt orthonormalization

or

$$\operatorname{clolin}\{h_z;\ z \in \sigma_{jp}(\boldsymbol{B})\} = \mathcal{H}$$

And *viceversa.*

Theorem 2. *Suppose we are given two N-tuples $\boldsymbol{A}$ and $\boldsymbol{B}$ of densely defined operators such that*

1° *$\boldsymbol{A}$ is jointly cyclic with the joint cyclic vector f,*

2° *there is a sequence of polynomials $\boldsymbol{p} = \{p_n\}_{n=0}^{\infty}$ such that*

$$\{p_n(\boldsymbol{A})f\}_{n=0}^{\infty} \text{ is an orthonormal basis of } \mathcal{H}$$

3° *$\boldsymbol{B}$ has one dimensional joint eigenspaces exclusively.*

If either either

$$\xi = \{\xi_n\}_{n=0}^{\infty} \in \ell^2,\ \sum_{n=0}^{\infty} \xi_n p_n(z) = 0,\ z \in \omega_{\boldsymbol{p}} \text{ implies } \xi = 0,$$

or

$$\operatorname{clolin}\{h_z;\ z \in \sigma_{jp}(\boldsymbol{B})\} = \mathcal{H}$$

and

$$\omega_{\boldsymbol{p}} = \sigma_{jp}(\boldsymbol{B}) \neq \varnothing \text{ as well as } K^{\boldsymbol{p}} = K^{\tau}$$

after suitable normalization of joint eigenvectors h_z of $\boldsymbol{B}$, then

$$(\boldsymbol{B})^{-} = \boldsymbol{A}^{*}.$$

References

1. N. Aronszajn, Theory of reproducing kernels, *Trans. Amer. Math. Soc.* **68** (1950), 337–404.

2. W. F. Donoghue, Orthonormal sets in reproducing kernel spaces and functional completion, *Note Mat.* 10, suppl. **1** (1990) 223-227.

3. P.R. Halmos, *A Hilbert space problem book*, van Nostrand, Toronto, London, 1967.

4. P. Quiggin, For which reproducing kernel Hilbert spaces is Pick's theorem true?, *Integr. Equat. Op. Th.* **16** (1993), 244-266.

5. Jan Stochel, F.H. Szafraniec, On normal extensions of unbounded operators. III. Spectral properties, *Publ. RIMS, Kyoto Univ.* **25** (1989), 105-139.

6. F.H. Szafraniec, Interpolation and domination by positive definite kernels, *Complex Analysis - Fifth Romanian-Finish Seminar, Part 2* Proc., Bucarest (Romania), 1981, (C. Andrean Cazacu, N. Boboc, M. Jurchescu and I. Suciu, eds.); *Lecture Notes in Math.* vol. 1014, pp. 291-295. Springer, Berlin-Heidelberg, 1983.

7. ———, On bounded holomorphic interpolation in several variables, *Monatsh. Mat.* **101** (1986), 59-66.

8. ———, A (little) step towards orthogonality of analytic polynomials, *J. Comput. Appl. Math.* **49** (1993), 225–261.

9. S. Zaremba, L'équation biharmonique et une class remarquable de functions fondamentales harmoniques, *Bulletin International de l'Académie des Sciences de Cracovie* (1907), 147-196.

10. ______, Sur le calcul numérique des fonctions demandées dans le problème de Dirichlet et le problème hydrodynamique, *ibidem* (1909), 125–195.

Instytut Matematyki,
Uniwersytet Jagielloński,
ul.Reymonta 4, PL-30059 Kraków
E-mail address: fhszafra@im.uj.edu.pl

Operator Theory:
Advances and Applications, Vol. 114

Lie algebras in Fock space

A. Turbiner

Abstract. A catalogue of explicit realizations of representations of Lie (super) algebras and quantum algebras in Fock space is presented.

This article is an attempt to present a catalogue of known representations of (super) Lie algebras and quantum algebras acting on different Fock spaces. Of course, we do not have so ambitious goal as to present a complete list of all representations of all possible algebras, but we plan to present some of them, those we consider important for applications, mainly restricting ourselves by those possessing finite-dimensional representations. Many representations are known in the folklore spread throughout the literature under often different names[1]. Therefore, we provide references according to our taste and knowledge, often quite arbitrarily. This work does not pretend to be totally original. Throughout the text we usually consider complex algebras.

Lie algebras

1. sl_2.

Take two operators a and b obeying the commutation relation

$$[a,b] \equiv ab - ba \;=\; 1, \tag{A.1.1}$$

with the identity operator on the r.h.s. – they span the three-dimensional Heisenberg algebra. By definition the universal enveloping algebra of the Heisenberg algebra is the algebra of all normal-ordered polynomials in a, b: any monomial is taken to be of the form $b^k a^m$ [2]. If, besides the polynomials, all entire functions in a, b are considered, then the *extended* universal enveloping algebra of the Heisenberg algebra appears or, in other words, the extended Heisenberg-Weyl algebra. In the (extended) Heisenberg-Weyl algebra one can find the non-trivial embeddings of the Heisenberg algebra [3], whose can be treated as a certain type of quantum canonical transformations We say that the (extended) Fock space appears if we take the (extended) universal enveloping algebra of the Heisenberg algebra and add to it the vacuum state $|0>$ such that

$$a|0> \;=\; 0\ . \tag{A.1.2}$$

[1]For instance, in nuclear physics some of them are known as boson representations.

[2]Sometimes this is called the Heisenberg-Weyl algebra

[3]This means that there exists a family of pairs of the non-trivial elements of the Heisenberg-Weyl algebra obeying the commutation relations (A.1.1)

One of the most important realizations of (A.1.1) is the coordinate-momentum representation:

$$a \;=\; \frac{d}{dx} \equiv \partial_x \;,\; b \;=\; x \;, \tag{A.1.3}$$

where x stands for the multiplication operator on x in a space of functions $f(x)$. In this case the vacuum is a constant and without a loss of generality we put $|0> = 1$. Recently, a finite-difference analogue of (A.1.3) has been found [1],

$$a \;=\; \mathcal{D}_+,\; b \;=\; x(1-\delta\mathcal{D}_-) \;, \tag{A.1.4}$$

where

$$\mathcal{D}_+ f(x) = \frac{f(x+\delta)-f(x)}{\delta} \;,$$

is the finite-difference operator, $\delta \in \mathbf{C}$ and $\mathcal{D}_+ \to \mathcal{D}_-$, if $\delta \to -\delta$.

(a). It is easy to check that if the operators a, b obey (A.1.1), then the following three operators

$$J_n^+ = b^2 a - nb \;,$$
$$J_n^0 = ba - \frac{n}{2} \;, \tag{A.1.5}$$
$$J_n^- = a \;,$$

span the sl_2-algebra with the commutation relations:

$$[J^0, J^\pm] = \pm J^\pm \;,\; [J^+, J^-] = -2J^0 \;,$$

where $n \in \mathbf{C}$. For the representation (A.1.5) the quadratic Casimir operator is equal to

$$C_2 \equiv \frac{1}{2}\{J^+, J^-\} - J^0 J^0 = -\frac{n}{2}\left(\frac{n}{2}+1\right) \;, \tag{A.1.6}$$

where $\{\;,\;\}$ denotes the anticommutator. If n is a non-negative integer, then (A.1.5) possesses a finite-dimensional, irreducible representation in the Fock space leaving invariant the space

$$\mathcal{P}_n(b) \;=\; \langle 1, b, b^2, \ldots, b^n\rangle|0\rangle, \tag{A.1.7}$$

of dimension $\dim \mathcal{P}_n = (n+1)$.

Substituting of (A.1.3) into (A.1.5) leads to a well-known realization of the sl_2-algebra as an algebra of differential operators of the first order [4]

$$J_n^+ = x^2\partial_x - nx \;,$$
$$J_n^0 = x\partial_x - \frac{n}{2} \;, \tag{A.1.8}$$
$$J^- = \partial_x \;,$$

[4]This representation was known to Sophus Lie.

where the finite-dimensional representation space (A.1.7) becomes the space of polynomials of degree not higher than n

$$\mathcal{P}_n(x) \ = \ \langle 1, x, x^2, \ldots, x^n \rangle \ . \tag{A.1.9}$$

(b). The existence of a non-trivial embedding of the Heisenberg algebra into its extended universal enveloping algebra, namely, $[\hat{a}(a,b), \hat{b}(a,b)] \ = \ [a,b] \ = \ 1$ allows to construct different representations of the algebra sl_2 by $a \to \hat{a}, b \to \hat{b}$ in (A.1.5). In particular, such an embedding of the Heisenberg algebra into its extended universal enveloping algebra is realized by the following two operators,

$$\hat{a} = \frac{(e^{\delta a} - 1)}{\delta} \ ,$$

$$\hat{b} = b e^{-\delta a} \ , \tag{A.1.10}$$

where δ is any complex number. If δ goes to zero then $\hat{a} \to a, \hat{b} \to b$. In other words, (A.1.10) is a 1-parameter quantum canonical transformation of the deformation type of the Heisenberg algebra (A.1.1). It is one of the (several) possible quantum analogies of a point-to-point canonical transformation. The substitution of the representation (A.1.10) into (A.1.5) results in the following representation of the sl_2-algebra

$$J_n^+ = (\frac{b}{\delta} - 1) b e^{-\delta a} (1 - n - e^{-\delta a}) \ ,$$

$$J_n^0 = \frac{b}{\delta}(1 - e^{-\delta a}) - \frac{n}{2} \ , \ J^- = \frac{1}{\delta}(e^{\delta a} - 1) \ . \tag{A.1.11}$$

If n is a non-negative integer, then (A.1.11) possesses a finite-dimensional irreducible representation of dimension $\dim \mathcal{P}_n = (n+1)$ coinciding with (A.1.7). It is worth noting that the vacuum for (A.1.10) remains the same, for instance (A.1.2). Also the value of the quadratic Casimir operator for (A.1.11) coincides with that given by (A.1.6).

The operator $\hat{a}$ in the particular representation (A.1.4) becomes the well-known translationally-covariant finite-difference operator

$$\hat{a} f(x) = \frac{(e^{\delta \partial_x} - 1)}{\delta} f(x) \ = \ \mathcal{D}_+ f(x) \tag{A.1.12}$$

while $\hat{b}$ takes the form

$$\hat{b} f(x) = x e^{-\delta \partial_x} f(x) \ = \ x f(x - \delta) = x(1 - \delta \mathcal{D}_-) f(x) \ . \tag{A.1.13}$$

After substitution of (A.1.12)–(A.1.13) into (A.1.11) we arrive at a representation of the sl_2-algebra by finite-difference operators,

$$J_n^+ = x(\frac{x}{\delta} - 1) e^{-\delta \partial_x} (1 - n - e^{-\delta \partial_x}) \ ,$$

$$J_n^0 = \frac{x}{\delta}(1 - e^{-\delta \partial_x}) - \frac{n}{2} \ , \ J^- = \frac{1}{\delta}(e^{\delta \partial_x} - 1) \ , \tag{A.1.14}$$

or, equivalently,

$$J_n^+ = x(1-\frac{x}{\delta})(\delta^2\mathcal{D}_-\mathcal{D}_- - (n+1)\delta\mathcal{D}_- + n) ,$$

$$J_n^0 = x\mathcal{D}_- - \frac{n}{2} , \quad J^- = \mathcal{D}_+. \tag{A.1.15}$$

The finite-dimensional representation space for (A.1.14)–(A.1.15) for integer values of n is again given by the space (A.1.9) of polynomials of degree not higher than n.

(c). Another example of quantum canonical transformation is given by the oscillator representation

$$\hat{a} = \frac{b+a}{\sqrt{2}} ,$$

$$\hat{b} = \frac{b-a}{\sqrt{2}} . \tag{A.1.16}$$

Inserting (A.1.16) into (A.1.5) it is easy to check that the following three generators form a representation of the sl_2-algebra,

$$J_n^+ = \frac{1}{2^{3/2}}[b^3 + a^3 - b(b+a)a - (2n+1)(b-a) - 2b] ,$$

$$J_n^0 = \frac{1}{2}(b^2 - a^2 - n - 1) , \tag{A.1.17}$$

$$J^- = \frac{b+a}{\sqrt{2}} ,$$

where $n \in C$. In this case the vacuum state

$$(b+a)|0> = 0, \tag{A.1.18}$$

differs from (A.1.2). If n is a non-negative integer, then (A.1.17) possesses a finite-dimensional irreducible representation in a subspace of the Fock space

$$\mathcal{P}_n(b) = \langle 1, (b-a), (b-a)^2, \ldots, (b-a)^n \rangle |0\rangle , \tag{A.1.19}$$

of dimension $\dim \mathcal{P}_n = (n+1)$.

Taking a, b in the realization (A.1.3) and substituting them into (A.1.17), we obtain

$$J_n^+ = \frac{1}{2^{3/2}}[x^3 + \partial_x^3 - x(x+\partial_x)\partial_x - (2n+1)(x-\partial_x) - 2\partial_x] ,$$

$$J_n^0 = \frac{1}{2}(x^2 - \partial_x^2 - n - 1) , \tag{A.1.20}$$

$$J^- = \frac{x+\partial_x}{\sqrt{2}} ,$$

which represents the sl_2-algebra by means of differential operators of finite order (but not of first order as in (A.1.8)). The operator J_n^0 coincides with the Hamiltonian of the harmonic oscillator (with the reference point for eigenvalues changed). The vacuum state is

$$|0> = e^{-\frac{x^2}{2}} , \tag{A.1.21}$$

and the representation space is

$$\mathcal{P}_n(x) = \langle 1, x, x^2, \ldots, x^n \rangle e^{-\frac{x^2}{2}} , \tag{A.1.22}$$

(cf.(A.1.19)).

(d). The following three operators

$$\begin{aligned} J^+ &= \frac{a^2}{2} , \\ J^0 &= -\frac{\{a, b\}}{4} , \\ J^- &= \frac{b^2}{2} , \end{aligned} \tag{A.1.23}$$

are generators of the sl_2-algebra and the quadratic Casimir operator for this representation is

$$C_2 = \frac{3}{16} .$$

This is the so-called metaplectic representation of sl_2 (see, for example, [2]). This representation is infinite-dimensional. Taking the realization (A.1.2) or (A.1.4) of the Heisenberg algebra we get the well-known representation

$$J^+ = \frac{1}{2}\partial_x^2 , \; J^0 = -\frac{1}{2}(x\partial_x - \frac{1}{2}) , \; J^- = \frac{1}{2}x^2 \tag{A.1.24}$$

in terms of differential operators, or

$$J^+ = \frac{1}{2}\mathcal{D}_+^2 , \; J^0 = -\frac{1}{2}(x\mathcal{D}_- - \frac{1}{2}) ,$$

$$J^- = \frac{1}{2}x(x-\delta)(1 - 2\delta\mathcal{D}_- - \delta^2\mathcal{D}_-^2) , \tag{A.1.25}$$

in terms of finite-difference operators, correspondingly.

(e). Take two operators a and b from the Clifford algebra s_2

$$\{a, b\} \equiv ab + ba = 0 , \; a^2 = b^2 = 1 . \tag{A.1.26}$$

Then the operators

$$J^1 = a , \; J^2 = b , \; J^3 = ab , \tag{A.1.27}$$

form the sl_2-algebra.

(f). Take the $(2p + 1)$-dimensional Heisenberg algebra H_{2p+1}

$$[a_i, b_j] = \delta_{ij}, \; i, j = 1, 2, \ldots, p , \tag{A.1.28}$$

where δ_{ij} is the Kronecker symbol. For $p=2$ we get the algebra H_5. The operators

$$J^1 = b_1 a_2 \ , \ J^2 = b_2 a_1 \ , \ J^3 = b_1 a_1 - b_2 a_2 \ , \qquad \text{(A.1.29)}$$

form the sl_2-algebra. This representation is reducible. If (A.1.28) is given in the coordinate-momentum representation

$$a_i \ = \ \frac{d}{dx_i} \equiv \partial_i \ , \ b_i \ = \ x_i \ , \qquad \text{(A.1.30)}$$

where x_i stands for the multiplication by x_i in $C[x_1, x_2]$, the representation (A.1.29) becomes the well-known vector-field representation. The vacuum is a constant. Finite-dimensional representations appear if a linear space of homogeneous polynomials of fixed degree is taken.

2. sl_3

(a). Take the Fock space associated with the five-dimensional Heisenberg algebra H_5 with vacuum

$$a_i|0> \ = \ 0 \ , \ i=1,2 \qquad \text{(A.2.1)}$$

One can show that the following operators are a set of generators of the sl_3-algebra

$$J_1^+ \ = \ b_1(b_1 a_1 \ + \ b_2 a_2 - \ n) \ , \ J_2^+ \ = \ b_2(b_1 a_1 \ + \ b_2 a_2 - \ n) \ ,$$

$$J_1^- \ = \ a_1 \ , \ J_2^- \ = \ a_2 \ ,$$

$$J_{21}^0 \ = \ b_2 a_1 \ , \ J_{12}^0 \ = \ b_1 a_2 \ ,$$

$$J_1^0 \ = \ b_1 a_1 - b_2 a_2 \ , \ J_2^0 \ = \ b_1 a_1 + b_2 a_2 \ - \ \frac{2}{3} n \ , \qquad \text{(A.2.2)}$$

where n is a complex number. If n is a non-negative integer, (A.2.2) possesses a finite-dimensional representation and its representation space is given by the inhomogeneous polynomials of degree not higher than n in the Fock space:

$$\mathcal{P}_n = \langle b_1^{n_1} b_2^{n_2} \mid 0 \le (n_1 + n_2) \le n \rangle \ . \qquad \text{(A.2.3)}$$

In the coordinate-momentum representation (A.1.30) the representation (A.2.2) becomes

$$J_1^+ \ = \ x_1(x_1\partial_1 \ + \ x_2\partial_2 - \ n) \ , \ J_2^+ \ = \ x_2(x_1\partial_1 \ + \ x_2\partial_2 - \ n) \ ,$$

$$J_1^- \ = \ \partial_1 \ , \ J_2^- \ = \ \partial_2 \ ,$$

$$J_{21}^0 \ = \ x_2\partial_1 \ , \ J_{12}^0 \ = \ x_1\partial_2 \ ,$$

$$J_1^0 \ = \ x_1\partial_1 - x_2\partial_2 \ , \ J_2^0 \ = \ x_1\partial_1 + x_2\partial_2 \ - \ \frac{2}{3} n \ , \qquad \text{(A.2.4)}$$

where the vacuum $|0>=1$ and for non-negative integer n the space of the finite-dimensional representation is given by

$$\mathcal{P}_n = \langle x_1^{n_1} x_2^{n_2} \mid 0 \le (n_1 + n_2) \le n \rangle \ . \qquad \text{(A.2.5)}$$

(b). An important example of a quantum canonical transformation of the five-dimensional Heisenberg algebra (A.1.28) is a generalization of (A.1.10) and has the form

$$\hat{a}_i = \frac{(e^{\delta_i a_i} - 1)}{\delta_i} ,$$

$$\hat{b}_i = b_i e^{-\delta_i a_i} , \quad i = 1,2 , \qquad \text{(A.2.6)}$$

where $\delta_{1,2}$ are complex numbers. Under this transformation the vacuum remains the same (A.2.1). Finally, we are led to the following representation of the sl_3-algebra

$$J_1^+ = \hat{b}_1(\hat{b}_1\hat{a}_1 + \hat{b}_2\hat{a}_2 - n) , \quad J_2^+ = \hat{b}_2(\hat{b}_1\hat{a}_1 + \hat{b}_2\hat{a}_2 - n) ,$$

$$J_1^- = \hat{a}_1 , \quad J_2^- = \hat{a}_2 ,$$

$$J_{21}^0 = \hat{b}_2\hat{a}_1 , \quad J_{12}^0 = \hat{b}_1\hat{a}_2 ,$$

$$J_1^0 = \hat{b}_1\hat{a}_1 - \hat{b}_2\hat{a}_2 , \quad J_2^0 = \hat{b}_1\hat{a}_1 + \hat{b}_2\hat{a}_2 - \frac{2}{3}n , \qquad \text{(A.2.7)}$$

As in previous case (a) for a non-negative integer n the representation (A.2.7) becomes finite-dimensional with the corresponding representation space given by (A.2.3). We should mention that in the coordinate-momentum representation the operators $\hat{a}, \hat{b}$ can be rewritten in terms of finite-difference operators (A.1.12-A.1.13), $\mathcal{D}_\pm^{(x,y)}$ and, finally, the generators become

$$J_1^+ = x(1 - \delta_1\mathcal{D}_-^{(x)})(x\mathcal{D}_-^{(x)} + y\mathcal{D}_-^{(y)} - n) ,$$

$$J_2^+ = y(1 - \delta_2\mathcal{D}_-^{(y)})(x\mathcal{D}_-^{(x)} + y\mathcal{D}_-^{(y)} - n) ,$$

$$J_1^- = \mathcal{D}_+^{(x)} , \quad J_2^- = \mathcal{D}_+^{(y)} ,$$

$$J_{21}^0 = y(1 - \delta_2\mathcal{D}_-^{(y)})\mathcal{D}_+^{(x)} , \quad J_{12}^0 = x(1 - \delta_1\mathcal{D}_-^{(x)})\mathcal{D}_+^{(y)} ,$$

$$J_1^0 = x\mathcal{D}_-^{(x)} - y\mathcal{D}_-^{(y)} , \quad J_2^0 = x\mathcal{D}_-^{(x)} + y\mathcal{D}_-^{(y)} - \frac{2n}{3} . \qquad \text{(A.2.8)}$$

(c). Another representation of the sl_3-algebra is related to the seven-dimensional Heisenberg algebra (A.1.28) for $p = 3$. The generators are

$$J_1^+ = -(b_2 - b_1b_3)a_1 - b_2b_3a_2 - b_3^2a_3 + nb_3 ,$$

$$J_2^+ = -b_1(b_2 - b_1b_3)a_1 - b_2^2a_2 - b_2b_3a_3 - mb_1b_3 + (n+m)b_2,$$

$$J_1^- = a_2 , \quad J_2^- = a_3 ,$$

$$J_{32}^0 = a_1 + b_3a_2 , \quad J_{23}^0 = -b_1^2a_1 + b_2a_3 + mb_1,$$

$$J_1^0 = -b_1a_1 + b_2a_2 + 2b_3a_3 - n ,$$

$$J_2^0 \;=\; 2b_1a_1 \;+\; b_2a_2 \;-\; b_3a_3 \;-\; m\;, \qquad \text{(A.2.9)}$$

where m, n are real numbers. In the coordinate-momentum representation of Heisenberg algebra the algebra (A.2.9) becomes the sl_3-algebra of first order differential operators in the regular representation (on the flag manifold)

$$J_1^+ \;=\; -(y-xz)\partial_x \;-\; yz\partial_y \;-\; z^2\partial_z \;+\; nz\;,$$

$$J_2^+ = -x(y-xz)\partial_x - y^2\partial_y - yz\partial_z - mxz + (n+m)y\;,$$

$$J_1^- \;=\; \partial_y\;,\; J_2^- \;=\; \partial_z\;,$$

$$J_{32}^0 \;=\; \partial_x \;+\; z\partial_y\;,\; J_{23}^0 \;=\; -x^2\partial_x \;+\; y\partial_z \;+\; mx,$$

$$J_1^0 \;=\; -x\partial_x \;+\; y\partial_y \;+\; 2z\partial_z \;-\; n\;,$$

$$J_2^0 \;=\; 2x\partial_x \;+\; y\partial_y \;-\; z\partial_z \;-\; m\;. \qquad \text{(A.2.10)}$$

Using the realization (A.2.6) of the generators of the Heisenberg algebra H_7 and the coordinate-momentum representation, a realization of the sl_3-algebra emerges in terms of finite-difference operators acting on C^3 functions, which is similar to (A.2.8).

3. $gl_2 \ltimes \mathbf{C}^{r+1}$

Among the subalgebras of the (extended) universal enveloping algebra of the Heisenberg algebra H_5 there is the one-parameter family of non-semi-simple algebras $gl_2 \ltimes \mathbf{C}^{r+1}$:

$$J^1 \;=\; a_1\;,$$

$$J^2 \;=\; b_1a_1 \;-\; \frac{n}{3}\;,\; J^3 \;=\; b_2a_2 \;-\; \frac{n}{3r}\;,$$

$$J^4 \;=\; b_1^2a_1 \;+\; rb_1b_2a_2 \;-\; nb_1\;,$$

$$J^{5+k} \;=\; b_1^ka_2\;,\; k=0,1,\ldots,r\;, \qquad \text{(A.3.1)}$$

where $r = 1,2,\ldots$ and n is a complex number. Here the generators J^{5+k}, $k = 0,1,\ldots,r$ span the $(r+1)$-dimensional abelian subalgebra $\mathbf{C}^{r+1}$. If n is a non-negative integer, a finite-dimensional *reducible* representation in the corresponding Fock space occurs,

$$\mathcal{P}_n = \langle b_1^{n_1}b_2^{n_2} \mid 0 \le (n_1 + rn_2) \le n\rangle\;. \qquad \text{(A.3.2)}$$

It is easy to see that (A.3.2) has an invariant subspace

$$\tilde{\mathcal{P}}_n = \langle b_1^{n_1} \mid 0 \le n_1 \le n\rangle\;, \qquad \text{(A.3.3)}$$

with respect to action of the generators (A.3.1). One can show that (A.3.2) becomes a finite-dimensional *irreducible* representation space if an extra operator

$$T \;=\; b_2a_1^r\;, \qquad \text{(A.3.4)}$$

is added to the operators (A.3.1).

Taking a concrete realization of the Heisenberg algebra in terms of differential or finite-difference operators in two variables similar to (A.1.2) or (A.1.4), respectively, in the generators (A.3.1) we arrive at the $gl_2 \ltimes \mathbf{C}^{r+1}$-algebra realized as the algebra of first-order differential operators or finite-difference operators, respectively [5]. Explicitly, it has a form

$$J^1 = \partial_x ,$$

$$J^2 = x\partial_x - \frac{n}{3} , \quad J^3 = ry\partial_y - \frac{n}{3} ,$$

$$J^4 = x^2\partial_x + rxy\partial_y - nx ,$$

$$J^{5+i} = x^i\partial_y , \quad i = 0, 1, \dots, r , \tag{A.3.5}$$

while the operator (A.3.4) becomes

$$T = y\partial_x^r . \tag{A.3.6}$$

Since the vacuum of Fock space in the coordinate-momentum representation of H_5 is the constant, the finite-dimensional representation spaces (A.3.2,A.3.3) become spaces of polynomials

$$\mathcal{P}_n = \langle x^{n_1} y^{n_2} \mid 0 \le (n_1 + rn_2) \le n \rangle . \tag{A.3.7}$$

and

$$\tilde{\mathcal{P}}_n = \langle x^{n_1} \mid 0 \le n_1 \le n \rangle , \tag{A.3.8}$$

correspondingly. The algebra $gl_2 \ltimes \mathbf{C}^{r+1}$ geometrically related with $r-$th Hirzebruch surface Σ_r and the modules are sections of holomorphic line bundles over this surface [?].

For the case of finite-difference operators the algebra (A.3.1) becomes

$$J_1^- = \mathcal{D}_+^{(x)},$$

$$J^2 = x\mathcal{D}_-^{(x)} - \frac{n}{3} , \quad J^3 = ry\mathcal{D}_-^{(y)} - \frac{n}{3} ,$$

$$J^4 = x(1 - \delta_1\mathcal{D}_-^{(x)})(x\mathcal{D}_-^{(x)} + ry\mathcal{D}_-^{(y)} - n) ,$$

$$J^{5+r} = x(1 - \delta_1\mathcal{D}_-^{(x)})^k \, \mathcal{D}_+^{(y)} , \tag{A.3.9}$$

however, the finite-dimensional representation space (A.3.2) converts to the same space of polynomials (A.3.7) as in the algebra of differential operators (A.3.5).

4. gl_k

[5] The algebra of vector fields acting on functions of two complex variables was found first by Sophus Lie and, recently, it has been extended to the algebra of first order differential operators [3].

There are different Fock spaces where the gl_k-algebra can act. We consider the action of gl_k on Fock space associated with the $(2k-1)$-dimensional Heisenberg algebra H_{2k-1}, which is the minimal dimension of the Heisenberg algebra, where gl_k acts. The explicit formulas for the generators are given by

$$J_i^- = a_i \ , \quad i = 2, 3, \dots, k \ ,$$

$$J_{i,j}^0 = b_i J_j^- = b_i a_j \ , \quad i, j = 2, 3, \dots, k \ ,$$

$$J^0 = n - \sum_{p=2}^{k} b_p a_p \ ,$$

$$J_i^+ = b_i J^0 \ , \quad i = 2, 3, \dots, k \ , \tag{A.4.1}$$

where the parameter n is a complex number. The generators $J_{i,j}^0$ span the algebra gl_{k-1}. If n is a non-negative integer, the representation (A.4.1) becomes the finite-dimensional representation acting on the space of polynomials

$$V_n(t) = \text{span}\{b_2^{n_2} b_3^{n_3} b_4^{n_4} \dots b_k^{n_k} \mid 0 \le \sum n_i \le n\} \ . \tag{A.4.2}$$

Substituting the a, b-generators of the Heisenberg algebra in the coordinate-momentum representation into (A.4.1) and using the vacuum, $|0>= 1$, we get a representation of the gl_k-algebra in terms of first-order differential operators (see, for example, [4])

$$J_i^- = \frac{\partial}{\partial x_i} \ , \quad i = 2, 3, \dots, k \ ,$$

$$J_{i,j}^0 = x_i J_j^- = x_i \frac{\partial}{\partial x_j} \ , \quad i, j = 2, 3, \dots, k \ ,$$

$$J^0 = n - \sum_{p=2}^{k} x_p \frac{\partial}{\partial x_p} \ ,$$

$$J_i^+ = x_i J^0 \ , \quad i = 2, 3, \dots, k \ , \tag{A.4.3}$$

which acts on functions of $x \in \mathbf{C}^{\mathbf{k-1}}$. One of the generators, namely, $J^0 + \sum_{p=2}^{k} J_{p,p}^0$ is proportional to a constant and, if it is taken out, we end up with the sl_k-subalgebra of the original algebra. The generators $J_{i,j}^0$ form the sl_{k-1}-algebra of the vector fields. If n is a non-negative integer, the representation (A.4.3) becomes the finite-dimensional representation acting on the space of polynomials

$$V_n(x) = \text{span}\{x_2^{n_2} x_3^{n_3} x_4^{n_4} \dots x_k^{n_k} \mid 0 \le \sum n_i \le n\} \ . \tag{A.4.4}$$

This representation corresponds to a Young tableau of one row with n blocks and is irreducible.

If the a, b-generators of the Heisenberg algebra are taken in the form of finite-difference operators (A.1.4) and are inserted into (A.4.1), the gl_k-algebra appears as the algebra of the finite-difference operators:

$$J_i^- = \mathcal{D}_+^{(i)} \ , \quad i = 2, 3, \ldots, k \ ,$$

$$J_{i,j}^0 = x_i(1 - \delta_i \mathcal{D}_-^{(i)}) \ , \ J_j^- = x_i(1 - \delta_i \mathcal{D}_-^{(i)})\mathcal{D}_+^{(j)}, \ i, j = 2, 3, \ldots k,$$

$$J^0 = n - \sum_{p=2}^{k} x_p \mathcal{D}_-^{(p)} \ ,$$

$$J_i^+ = x_i(1 - \delta_i \mathcal{D}_-^{(i)})J^0 \ , \quad i = 2, 3, \ldots, k \ , \tag{A.4.5}$$

where $\mathcal{D}_\pm^{(i)}$ denote the finite-difference operators acting in the direction x_i (see (A.1.4)).

Lie Super-Algebras

In order to work with superalgebras we must introduce the super Heisenberg algebra. This is the $(2k + 2r + 1)$-dimensional algebra which contains the H_{2k+1}-Heisenberg algebra (A.1.28) as a subalgebra and also the Clifford algebra s_r :

$$\{a_i^{(f)}, a_j^{(f)}\} = \{b_i^{(f)}, b_j^{(f)}\} = 0 \ ,$$

$$\{a_i^{(f)}, b_j^{(f)}\} = \delta_{ij} \ , \ i, j = 1, 2, \ldots, r \ , \tag{S.1}$$

as another subalgebra. There are two widely used realizations of the Clifford algebra (S.1):

(i) The fermionic analogue of the coordinate-momentum representation (A.1.30):

$$a_i^{(f)} = \theta_i^+ \ , \ b_i^{(f)} = \theta_i \ , \ i = 1, 2, \ldots, r \ , \tag{S.2}$$

or, differently,

$$a_i^{(f)} = \frac{\partial}{\partial \theta_i} \ , \ b_i^{(f)} = \theta_i \ , \ i = 1, 2, \ldots, r \ , \tag{S.3}$$

and

(ii) The matrix representation

$$a_i^{(f)} = \underbrace{\sigma^0 \otimes \ldots \otimes \sigma^0}_{i-1} \otimes \sigma^+ \otimes \underbrace{\mathbf{1} \otimes \ldots \otimes \mathbf{1}}_{r-i} \ ,$$

$$b_i^{(f)} = \underbrace{\sigma^0 \otimes \ldots \otimes \sigma^0}_{i-1} \otimes \sigma^- \otimes \underbrace{\mathbf{1} \otimes \ldots \otimes \mathbf{1}}_{r-i} \ , \ i = 1, 2, \ldots, r \ , \tag{S.4}$$

where the $\sigma^{\pm,0}$ are Pauli matrices in standard notation,

$$\sigma^+ = \begin{pmatrix} 0 & 1 \\ 0 & 0 \end{pmatrix} \ , \ \sigma^- = \begin{pmatrix} 0 & 0 \\ 1 & 0 \end{pmatrix} \ , \ \sigma^0 = \begin{pmatrix} 1 & 0 \\ 0 & -1 \end{pmatrix} \ .$$

In what follows we will consider the Fock space and also the realizations of the superalgebras assuming that the Clifford algebra generators are taken in the fermionic representation (S.3) or the matrix representation (S.4).

1. $osp(2,2)$

Let us define a spinorial Fock space as a linear space of all 2-component spinors with normal ordered polynomials in a, b as components and with a definition of the vacuum

$$|0> = \begin{pmatrix} |0>_1 \\ |0>_2 \end{pmatrix}$$

such that any component is annihilated by the operator a:

$$a|0>_i = 0 \ , \ i = 1,2 \tag{S.1.1}$$

(a). Take the Heisenberg algebra (A.1.1). Then consider the following two sets of 2×2 matrix operators:

$$\begin{aligned} T^+ &= b^2 a - nb + b\sigma^-\sigma^+ , \\ T^0 &= ba - \frac{n}{2} + \frac{1}{2}\sigma^-\sigma^+ , \\ T^- &= a , \\ J &= -\frac{n}{2} - \frac{1}{2}\sigma^-\sigma^+ , \end{aligned} \tag{S.1.2}$$

called bosonic (even) generators and

$$Q = \begin{bmatrix} \sigma^+ \\ b\sigma^+ \end{bmatrix} , \ \bar{Q} = \begin{bmatrix} (ba-n)\sigma^- \\ -a\sigma^- \end{bmatrix} , \tag{S.1.3}$$

called fermionic (odd) generators. The explicit matrix form of the even generators is given by:

$$T^+ = \begin{pmatrix} J_n^+ & 0 \\ 0 & J_{n-1}^+ \end{pmatrix} , T^0 = \begin{pmatrix} J_n^0 & 0 \\ 0 & J_{n-1}^0 \end{pmatrix} , T^- = \begin{pmatrix} J^- & 0 \\ 0 & J^- \end{pmatrix} ,$$

$$J = \begin{pmatrix} -\frac{n}{2} & 0 \\ 0 & -\frac{n+1}{2} \end{pmatrix} ,$$

and of the odd ones by

$$Q_1 = \begin{pmatrix} 0 & 1 \\ 0 & 0 \end{pmatrix} , \ Q_2 = \begin{pmatrix} 0 & b \\ 0 & 0 \end{pmatrix} ,$$

$$\overline{Q}_1 = \begin{pmatrix} 0 & 0 \\ ba-n & 0 \end{pmatrix} , \ \overline{Q}_2 = \begin{pmatrix} 0 & 0 \\ -a & 0 \end{pmatrix} , \tag{S.1.4}$$

where the $J_n^{\pm,0}$ are the generators of sl_2 given by (A.1.10).

The above generators span the superalgebra $osp(2,2)$ with the commutation relations:

$$[T^0, T^{\pm}] = \pm T^{\pm} \quad , \quad [T^+, T^-] = -2T^0 \quad , \quad [J, T^{\alpha}] = 0 \quad , \alpha = \pm, 0$$

$$\{Q_1, \overline{Q}_2\} = -T^- \quad , \quad \{Q_2, \overline{Q}_1\} = T^+ ,$$

$$\frac{1}{2}(\{\overline{Q}_1, Q_1\} + \{\overline{Q}_2, Q_2\}) = J \; , \; \frac{1}{2}(\{\overline{Q}_1, Q_1\} - \{\overline{Q}_2, Q_2\}) = T^0 ,$$

$$\{Q_1, Q_1\} = \{Q_2, Q_2\} = \{Q_1, Q_2\} = 0 ,$$

$$\{\overline{Q}_1, \overline{Q}_1\} = \{\overline{Q}_2, \overline{Q}_2\} = \{\overline{Q}_1, \overline{Q}_2\} = 0 ,$$

$$[Q_1, T^+] = Q_2 \; , \; [Q_2, T^+] = 0 \; , \; [Q_1, T^-] = 0 \; , \; [Q_2, T^-] = -Q_1 ,$$

$$[\overline{Q}_1, T^+] = 0 \; , \; [\overline{Q}_2, T^+] = -\overline{Q}_1 \; , \; [\overline{Q}_1, T^-] = \overline{Q}_2 \; , \; [\overline{Q}_2, T^-] = 0 ,$$

$$[Q_{1,2}, T^0] = \pm\frac{1}{2}Q_{1,2} \quad , \quad [\overline{Q}_{1,2}, T^0] = \mp\frac{1}{2}\overline{Q}_{1,2} ,$$

$$[Q_{1,2}, J] = -\frac{1}{2}Q_{1,2} \quad , \quad [\overline{Q}_{1,2}, J] = \frac{1}{2}\overline{Q}_{1,2} . \qquad \text{(S.1.5)}$$

If, in the expressions (S.1.2)–(S.1.3), the parameter n is a non-negative integer, then (S.1.2)–(S.1.3) possess a finite-dimensional representation in the spinorial Fock space

$$\mathcal{P}_{n,n-1} = \left\langle \begin{array}{c} 1, b, b^2, \ldots, b^n \\ 1, b, b^2, \ldots, b^{n-1} \end{array} \right\rangle = \left\langle \begin{array}{c} \mathcal{P}_n \\ \mathcal{P}_{n-1} \end{array} \right\rangle . \qquad \text{(S.1.6)}$$

If we take a representation (A.1.3) of the Heisenberg algebra, the generators (S.1.2) become 2×2 matrix differential operators, where the bosonic generators are [5]

$$T^+ = x^2\partial_x - nx + x\sigma^-\sigma^+,$$

$$T^0 = x\partial_x - \frac{n}{2} + \frac{1}{2}\sigma^-\sigma^+ , \qquad \text{(S.1.7)}$$

$$T^- = \partial_x ,$$

$$J = -\frac{n}{2} - \frac{1}{2}\sigma^-\sigma^+ ,$$

and the fermionic generators

$$Q = \begin{bmatrix} \sigma^+ \\ x\sigma^+ \end{bmatrix} , \; \bar{Q} = \begin{bmatrix} (x\partial_x - n)\sigma^- \\ -\partial_x\sigma^- \end{bmatrix} . \qquad \text{(S.1.8)}$$

The finite-dimensional representation space for non-negative integer values of the parameter n in (S.1.7-S.1.8) becomes a linear space of 2-component spinors with polynomial components:

$$\mathcal{P}_{n,n-1}(x) = \left\langle \begin{array}{c} 1, x, x^2, \ldots, x^n \\ 1, x, x^2, \ldots, x^{n-1} \end{array} \right\rangle = \left\langle \begin{array}{c} \mathcal{P}_n(x) \\ \mathcal{P}_{n-1}(x) \end{array} \right\rangle \tag{S.1.9}$$

(b). Taking the quantum canonical transformation (A.1.9) and substituting it into (S.1.2) we arrive at the $osp(2,2)$-algebra analogue of the representation (A.1.11) for the sl_2-algebra,

$$\begin{aligned} T^+ &= (\frac{b}{\delta} - 1) b e^{-\delta a}(1 - n - e^{-\delta a} + \sigma^-\sigma^+) , \\ T^0 &= \frac{b}{\delta}(1 - e^{-\delta a}) - \frac{n}{2} + \frac{\sigma^-\sigma^+}{2} , \\ T^- &= \frac{1}{\delta}(e^{\delta a} - 1) , \\ J &= -\frac{1}{2} - \frac{\sigma^-\sigma^+}{2} , \end{aligned} \tag{S.1.10}$$

and,

$$Q = \begin{bmatrix} \sigma^+ \\ b e^{-\delta a}\sigma^+ \end{bmatrix} , \ \bar{Q} = \begin{bmatrix} \frac{b - b e^{-\delta a} - n}{\delta}\sigma^- \\ \frac{1 - e^{\delta a}}{\delta}\sigma^- \end{bmatrix} . \tag{S.1.11}$$

Taking for the generators a, b the coordinate-momentum realization (A.1.3), we obtain a representation of the algebra $osp(2,2)$ in terms of finite-difference operators

$$\begin{aligned} T^+ &= (\frac{x}{\delta} - 1) x e^{-\delta \partial_x}(1 - n - e^{-\delta \partial_x} + \sigma^-\sigma^+) , \\ T^0 &= \frac{x}{\delta}(1 - e^{-\delta \partial_x}) - \frac{n}{2} + \frac{\sigma^-\sigma^+}{2} , \\ T^- &= \frac{1}{\delta}(e^{\delta \partial_x} - 1) , \\ J &= -\frac{1}{2} - \frac{\sigma^-\sigma^+}{2} , \end{aligned} \tag{S.1.12}$$

and,

$$Q = \begin{bmatrix} \sigma^+ \\ x e^{-\delta \partial_x}\sigma^+ \end{bmatrix} , \ \bar{Q} = \begin{bmatrix} \frac{x - x e^{-\delta \partial_x} - n}{\delta}\sigma^- \\ \frac{1 - e^{\delta \partial_x}}{\delta}\sigma^- \end{bmatrix} . \tag{S.1.13}$$

Or, in terms of the operators $\mathcal{D}_\pm$, their explicit matrix forms are the following

$$T^+ = \begin{pmatrix} J_n^+ & 0 \\ 0 & J_{n-1}^+ \end{pmatrix}, T^0 = \begin{pmatrix} x\mathcal{D}_- - \frac{n}{2} & 0 \\ 0 & x\mathcal{D}_- - \frac{n-1}{2} \end{pmatrix}, T^- = \begin{pmatrix} \mathcal{D}_+ & 0 \\ 0 & \mathcal{D}_+ \end{pmatrix},$$

$$J = \begin{pmatrix} -\frac{n}{2} & 0 \\ 0 & -\frac{n+1}{2} \end{pmatrix} ,$$

for the bosonic generators and

$$Q_1 = \begin{pmatrix} 0 & 1 \\ 0 & 0 \end{pmatrix} , \; Q_2 = \begin{pmatrix} 0 & x(1-\delta\mathcal{D}_-) \\ 0 & 0 \end{pmatrix} ,$$

$$\overline{Q}_1 = \begin{pmatrix} 0 & 0 \\ x\mathcal{D}_- - n & 0 \end{pmatrix} , \; \overline{Q}_2 = \begin{pmatrix} 0 & 0 \\ -\mathcal{D}_+ & 0 \end{pmatrix} , \qquad \text{(S.1.14)}$$

for the fermionic generators, where the generator J_n^+ is given by (A.1.15).

(c). The super-metaplectic representation of the $osp(2,2)$-algebra can be easily constructed and has the following form. The even generators are given by

$$T^+ = \begin{pmatrix} \frac{a^2}{2} & 0 \\ 0 & \frac{a^2}{2} \end{pmatrix} , T^0 = \begin{pmatrix} -\frac{\{a,b\}}{4} & 0 \\ 0 & -\frac{\{a,b\}}{4} \end{pmatrix} , T^- = \begin{pmatrix} \frac{b^2}{2} & 0 \\ 0 & \frac{b^2}{2} \end{pmatrix} ,$$

$$J = \begin{pmatrix} \frac{1}{4} & 0 \\ 0 & -\frac{1}{4} \end{pmatrix} ,$$

while the odd ones are

$$Q_1 = \begin{pmatrix} 0 & -\frac{b}{\sqrt{2}} \\ 0 & 0 \end{pmatrix} , \; Q_2 = \begin{pmatrix} 0 & \frac{a}{\sqrt{2}} \\ 0 & 0 \end{pmatrix} ,$$

$$\overline{Q}_1 = \begin{pmatrix} 0 & 0 \\ \frac{a}{\sqrt{2}} & 0 \end{pmatrix} , \; \overline{Q}_2 = \begin{pmatrix} 0 & 0 \\ \frac{b}{\sqrt{2}} & 0 \end{pmatrix} . \qquad \text{(S.1.15)}$$

Taking the realization of the Heisenberg algebra H_3 in terms of the differential or finite-difference operators (A.1.2), (A.1.4), respectively, and inserting it into (S.1.15) we end up with a realization of the super-metaplectic representation of the $osp(2,2)$-algebra in terms of differential or finite-difference operators.

2. $gl(k+1, r+1)$

One of the simplest representations of the $gl(k+1, r+1)$-superalgebra can be written as follows

$$T_i^- = a_i \, , \quad i = 1, 2, \ldots , k \, ,$$

$$T_{i,j}^0 = b_i T_j^- = b_i a_j \, , \quad i, j = 1, 2, \ldots , k \, ,$$

$$T^0 = n - \sum_{p=1}^{k} b_p a_p - \sum_{p=1}^{r} \theta_p \frac{\partial}{\partial \theta_p} \, ,$$

$$T_i^+ = b_i T^0 \, , \quad i = 1, 2, \ldots , k \, , \qquad \text{(S.2.1)}$$

$$\overline{Q_i} = \frac{\partial}{\partial \theta_i} \, , \quad i = 1, 2, \ldots , r \, ,$$

$$\overline{Q}_i^+ = \theta_i T^0 , \quad i = 1,2,\dots,r ,$$

$$Q_{ij}^- = \theta_i T_j^- = \theta_i a_j , \quad i,j = 1,2,\dots,r ,$$

$$Q_{ij}^+ = b_i \overline{Q}_j^- = b_i \frac{\partial}{\partial \theta_i} , \quad i = 1,2,\dots,k , \; j = 1,2,\dots,r ,$$

$$J_{i,j}^0 = \theta_i \overline{Q}_i^- = \theta_i \frac{\partial}{\partial \theta_j} , \quad i,j = 1,2,\dots,r ,$$

These generators can be represented by the following $(k+p) \times (k+p)$ matrix,

$$\left(\begin{array}{c|c} k \times k & p \times k \\ BB & BF \\ \hline k \times p & p \times p \\ FB & FF \end{array} \right) , \tag{S.2.2}$$

where the notation $B(F)B(F)$ means the product of a bosonic operator B (fermionic F) with a bosonic operator B (fermionic F). Correspondingly, the operators T in (S.2.1) are of BB-type (mixed with FF-type), J are of FF-type, while the rest operators are of BF-type. The algebra is defined by the (anti)commutation relations

$$\{[E_{IJ}, E_{KL}]\} = \delta_{IL} E_{JK} \pm \delta_{JK} E_{IL} ,$$

where the generalized indices I, J, K, L are in $\{B, F\}$. Anticommutators are taken for generators of FB, BF types only, while for all other cases the defining relations are given by commutators. The dimension of the algebra is $(k+p)^2$.

The generators $J_{i,j}^0$ span the sl_k-algebra of the vector fields. The parameter n in (S.2.1) can be any complex number. However, if n is a non-negative integer, the representation (S.2.1) becomes the finite-dimensional representation acting on a subspace of the Fock space,

$$V_n(b) = \mathrm{span}\{b_1^{n_1} b_2^{n_2} b_3^{n_3} \dots b_k^{n_k} \theta_1^{m_1} \theta_2^{m_2} \dots \theta_r^{m_r} | 0 \le \sum n_i + \sum m_j \le n\}. \tag{S.2.3}$$

Taking the coordinate-momentum realization of the Heisenberg algebra (A.1.30) in the generators (S.2.1), we obtain the $gl(k+1, r+1)$-superalgebra realized in terms of first order differential operators (see, for example, [4]):

$$T_i^- = \frac{\partial}{\partial x_i} , \quad i = 1,2,\dots,k ,$$

$$T_{i,j}^0 = x_i T_j^- = x_i \frac{\partial}{\partial x_j} , \quad i,j = 1,2,\dots,k ,$$

$$T^0 = n - \sum_{p=1}^{k} x_p \frac{\partial}{\partial x_p} - \sum_{p=1}^{r} \theta_p \frac{\partial}{\partial \theta_p} ,$$

$$T_i^+ = x_i T^0\ , \quad i = 1, 2, \ldots, k\ , \tag{S.2.4}$$

$$\overline{Q_i^-} = \frac{\partial}{\partial\theta_i}\ , \quad i = 1, 2, \ldots, r\ ,$$

$$\overline{Q_i^+} = \theta_i T^0\ , \quad i = 1, 2, \ldots, r\ ,$$

$$Q_{ij}^- = \theta_i T_j^- = \theta_i \frac{\partial}{\partial x_j}\ , \quad i = 1, 2, \ldots, r; j = 1, 2, \ldots, k\ ,$$

$$Q_{ij}^+ = x_i \overline{Q_j^-} = x_i \frac{\partial}{\partial\theta_i}\ , \quad i = 1, 2, \ldots, k; j = 1, 2, \ldots, r\ ,$$

$$J_{i,j}^0 = \theta_i \overline{Q_i^-} = \theta_i \frac{\partial}{\partial\theta_j}\ , \quad i, j = 1, 2, \ldots, r\ ,$$

which acts on functions in $\mathbf{C}^k \otimes \mathbf{G}^r$.

A combination of the generators $J^0 + \sum_{p=1}^k T_{p,p}^0 + \sum_{p=1}^r J_{p,p}^0$, is proportional to a constant and, if it is taken out, we end up with the superalgebra $sl(k+1, r+1)$. The generators $T_{i,j}^0$, $J_{p,q}^0$, $i, j = 1, 2, \ldots, k$, $p, q = 1, 2, \ldots, r$ span the algebra of the vector fields $gl(k, r)$. The parameter n in (S.2.4) can be any complex number. If n is a non-negative integer, the representation (S.2.1) becomes the finite-dimensional representation acting on the space of polynomials

$$V_n(t) = \text{span}\{x_1^{n_1} x_2^{n_2} x_3^{n_3} \ldots x_k^{n_k} \theta_1^{m_1} \theta_2^{m_2} \ldots \theta_r^{m_r} | 0 \le \sum n_i + \sum m_j \le n\}. \tag{S.2.5}$$

This representation corresponds to a Young tableau of one row with n blocks in the bosonic direction and is irreducible.

If the a, b-generators of the Heisenberg algebra are taken in the form of finite-difference operators (A.1.4) and are inserted into (A.4.1), the $gl(k+1, r+1)$-algebra appears as the algebra of the finite-difference operators:

$$T_i^- = \mathcal{D}_+^{(i)}\ , \quad i = 1, 2, \ldots, k\ ,$$

$$T_{i,j}^0 = x_i(1 - \delta_i \mathcal{D}_-^{(i)}) T_j^- = x_i(1 - \delta_i \mathcal{D}_-^{(i)}) \mathcal{D}_+^{(j)}\ , \quad i, j = 1, 2, \ldots, k\ ,$$

$$T^0 = n - \sum_{p=1}^k x_p \mathcal{D}_-^{(p)} - \sum_{p=1}^r \theta_p \frac{\partial}{\partial\theta_p}\ ,$$

$$T_i^+ = x_i(1 - \delta_i \mathcal{D}_-^{(i)}) T^0\ , \quad i = 1, 2, \ldots, k\ , \tag{S.2.6}$$

$$\overline{Q_j^-} = \frac{\partial}{\partial\theta_j}\ , \quad j = 1, 2, \ldots, r\ ,$$

$$\overline{Q_j^+} = \theta_j T^0\ , \quad j = 1, 2, \ldots, r\ ,$$

$$Q_{ij}^- = \theta_i T_j^- = \theta_i \mathcal{D}_+^{(j)}\ , \quad i = 1, 2, \ldots, r; j = 1, 2, \ldots, k\ ,$$

$$Q^{+}_{ij} = x_i(1-\delta_i\mathcal{D}^{(i)}_{-})\overline{Q}^{-}_{j} = x_i(1-\delta_i\mathcal{D}^{(i)}_{-})\frac{\partial}{\partial\theta_i},\ i=1,2,\dots k; j=1,2,\dots r,$$

$$J^{0}_{i,j} = \theta_i\overline{Q}^{-}_{i} = \theta_i\frac{\partial}{\partial\theta_j}\ ,\quad i,j=1,2,\dots r\ ,$$

It is worth mentioning that for the integer n, the algebra (S.2.6) has the same finite-dimensional representation (S.2.5) as the algebra of the first order differential operators (S.2.4).

Quantum Algebras

sl_{2q}

Take two operators $\tilde{a}$ and $\tilde{b}$ obeying the commutation relation

$$\tilde{a}\tilde{b} - q\tilde{b}\tilde{a} \ = \ 1\ , \tag{Q.1}$$

with the identity operator on the right hand side. They define the so-called q-deformed Heisenberg algebra. Here $q \in C$. One can define a q-deformed analogue of the universal enveloping algebra by taking all ordered monomials $\tilde{b}^k\tilde{a}^m$. Introducing a vacuum

$$\tilde{a}|0> \ = \ 0\ , \tag{Q.2}$$

in addition to the q-deformed analogue of the universal enveloping algebra we arrive at a construction which is a q-analogue of Fock space.

It can be easily checked that the q-deformed Heisenberg algebra is a subalgebra of the extended universal enveloping Heisenberg algebra. This can be shown explicitly as follows. For any $q \in C$, two elements of the extended universal enveloping Heisenberg algebra

$$\tilde{a} \ = \ \left(\frac{1}{b}\right)\left(\frac{q^{ba}-1}{q-1}\right)\ ,\ \tilde{b} \ = \ b\ , \tag{Q.3}$$

obey the commutation relations (Q.1). It can be shown that the universal enveloping Heisenberg algebra does not contain the q-deformed Heisenberg algebra as a subalgebra. The formula (Q.3) allows us to construct different realizations of the the q-deformed Heisenberg algebra. One of them is a q-analogue of the coordinate-momentum representation (A.1.3):

$$\tilde{a} \ = \ \tilde{D}_x\ ,\ \tilde{b} \ = \ x\ , \tag{Q.4}$$

where

$$\tilde{D}_x f(x) \ = \ \frac{f(qx)-f(x)}{x(q-1)}\ , \tag{Q.5}$$

is the so-called Jackson symbol or the Jackson derivative.

Another realization of the (Q.1) appears if a quantum canonical transformation of the Heisenberg algebra (A.1.10) is taken:

$$\tilde{a} \ = \ \left(\frac{1}{b+\delta}\right)e^{\delta a}\left(\frac{q^{\frac{b}{\delta}(1-e^{-\delta a})}-1}{q-1}\right)\ ,\ \tilde{b} \ = \ be^{-\delta a}\ , \tag{Q.6}$$

where δ is any complex number. In terms of translationally-covariant finite-difference operators $\mathcal{D}_\pm$ the realization has the form

$$\tilde{a} = \left(\frac{1}{x+\delta}\right)(\delta\mathcal{D}_+ + 1)\left(\frac{q^{x\mathcal{D}_-} - 1}{q-1}\right), \quad \tilde{b} = x(1-\delta\mathcal{D}_-) . \tag{Q.7}$$

In these cases the vacuum is a constant, say, $|0> = 1$, as in the non-deformed case.

The following three operators

$$\begin{aligned} \tilde{J}^+_\alpha &= \tilde{b}^2\tilde{a} - \{\alpha\}\tilde{b} , \\ \tilde{J}^0_\alpha &= \tilde{b}\tilde{a} - \hat{\alpha} , \\ \tilde{J}^- &= \tilde{a}, \end{aligned} \tag{Q.8}$$

where $\{\alpha\} = \frac{1-q^\alpha}{1-q}$ is so called q-number and $\hat{\alpha} \equiv \frac{\{\alpha\}\{\alpha+1\}}{\{2\alpha+2\}}$, are generators of the q-deformed or quantum sl_{2q}-algebra. The operators (Q.8) after multiplication by some factors, become

$$\tilde{j}^0 = \frac{q^{-\alpha}}{q+1}\frac{\{2\alpha+2\}}{\{\alpha+1\}}\tilde{J}^0_\alpha ,$$

$$\tilde{j}^\pm = q^{-\alpha/2}\tilde{J}^\pm_\alpha ,$$

and span the quantum algebra sl_{2q} with the standard commutation relations [6][6],

$$\begin{aligned} \tilde{j}^0\tilde{j}^+ - q\tilde{j}^+\tilde{j}^0 &= \tilde{j}^+ , \\ q^2\tilde{j}^+\tilde{j}^- - \tilde{j}^-\tilde{j}^+ &= -(q+1)\tilde{j}^0 , \\ q\tilde{j}^0\tilde{j}^- - \tilde{j}^-\tilde{j}^0 &= -\tilde{j}^- . \end{aligned} \tag{Q.9}$$

The algebra (Q.9) is known in literature as *the second Witten quantum deformation* of sl_2 in the classification of C. Zachos [8]).

In general, for the quantum sl_{2q} algebra there are no polynomial Casimir operators (see, for example, Zachos [8]). However, in the representation (Q.8) a relationship between generators analogous to the quadratic Casimir operator appears

$$q\tilde{J}^+_\alpha\tilde{J}^-_\alpha - \tilde{J}^0_\alpha\tilde{J}^0_\alpha + (\{\alpha+1\} - 2\hat{\alpha})\tilde{J}^0_\alpha = \hat{\alpha}(\hat{\alpha} - \{\alpha+1\}) .$$

If $\alpha = n$ is a non-negative integer, then (Q.8) possesses a finite-dimensional irreducible representation in the Fock space (cf.(A.1.6))

$$\mathcal{P}_n(\tilde{b}) = \langle 1, \tilde{b}, \tilde{b}^2, \ldots, \tilde{b}^n \rangle , \tag{Q.10}$$

of the dimension $\dim \mathcal{P}_n = (n+1)$.

[6]For discussion see [7] as well

References

[1] Y. F. Smirnov and A. V. Turbiner, "Lie-algebraic discretization of differential equations", *Modern Physics Letters* **A10**, 1795-1802 (1995), ERRATUM-*ibid* **A10**, 3139 (1995);
"Hidden sl_2-algebra of finite-difference equations, Proceedings of IV Wigner Symposium, World Scientific, 1996, N.M. Atakishiyev, T.H. Seligman and K.B. Wolf (Eds.), pp. 435-440

[2] A.M. Perelomov, "Generalized coherent states and its applications", Nauka, 1987 (in Russian)

[3] A. González-Lopéz, N. Kamran and P.J. Olver, "Quasi-Exactly-Solvable Lie Algebras of the first order differential operators in Two Complex Variables", *J. Phys.* **A24** (1991) 3995–4008; "Lie algebras of differential operators in two complex variables", *American J. Math.* **114** (1992) 1163-1185

[4] L. Brink, A. Turbiner and N. Wyllard, "Hidden Algebras of the (super) Calogero and Sutherland models," J. Math. Phys. **39** (1998) 1285–1315. hep-th/9705219

[5] M.A. Shifman and A.V. Turbiner, "Quantal problems with partial algebraization of the spectrum", *Comm. Math. Phys.* **126** (1989) 347-365

[6] O. Ogievetsky and A. Turbiner, " $sl(2,\mathbf{R})_q$ and quasi-exactly-solvable problems", Preprint CERN-TH: 6212/91 (1991) (unpublished)

[7] A. V. Turbiner, "Lie algebras and linear operators with invariant subspace," in *Lie algebras, cohomologies and new findings in quantum mechanics* (N. Kamran and P. J. Olver, eds.), AMS, vol. 160, pp. 263–310, 1994;
"Lie-algebras and Quasi-exactly-solvable Differential Equations", in *CRC Handbook of Lie Group Analysis of Differential Equations*, Vol.3: New Trends in Theoretical Developments and Computational Methods, Chapter 12, CRC Press (N. Ibragimov, ed.), pp. 331-366, 1995

[8] C. Zachos, "Elementary paradigms of quantum algebras", AMS *Contemporary Mathematics,* **134**, 351-377; J. Stasheff and M. Gerstenhaber (eds.), AMS, 1991

Instituto de Ciencias Nucleares, UNAM,
Apartado Postal 70-543,
04510 Mexico D.F., Mexico
On leave of absence from the
Institute for Theoretical and Experimental Physics,
Moscow 117259, Russia
E-mail address: `turbiner@axcrnb.cern.ch`,
E-mail address: `turbiner@roxanne.nuclecu.unam.mx`